NATO ASI Series

Advanced Science Institutes Series

A series presenting the results of activities sponsored by the NATO Science Committee, which aims at the dissemination of advanced scientific and technological knowledge, with a view to strengthening links between scientific communities.

The Series is published by an international board of publishers in conjunction with the NATO Scientific Affairs Division

A Life Sciences B Physics	Plenum Publishing Corporation London and New York
C Mathematical and Physical Sciences D Behavioural and Social Sciences E Applied Sciences	Kluwer Academic Publishers Dordrecht, Boston and London
F Computer and Systems Sciences G Ecological Sciences H Cell Biology	Springer-Verlag Berlin Heidelberg New York London Paris Tokyo Hong Kong

Series H: Cell Biology Vol. 44

The ASI Series Books Published as a Result of Activities of the Special Programme on CELL TO CELL SIGNALS IN PLANTS AND ANIMALS

This book contains the proceedings of a NATO Advanched Research Workshop held within the activities of the NATO Special Programme on Cell to Cell Signals in Plants and Animals, running from 1984 to 1989 under the auspices of the NATO Science Committee.

The books published as a result of the activities of the Special Programme are:

Vol. 1: **Biology and Molecular Biology of Plant-Pathogen Interactions.** Edited by J.A. Bailey. 1986.
Vol. 2: **Glial-Neuronal Communication in Development and Regeneration.** Edited by H.H. Althaus and W. Seifert. 1987.
Vol. 3: **Nicotinic Acetylcholine Receptor: Structure and Function.** Edited by A. Maelicke. 1986.
Vol. 4: **Recognition in Microbe-Plant Symbiotic and Pathogenic Interactions.** Edited by B. Lugtenberg. 1986.
Vol. 5: **Mesenchymal-Epithelial Interactions in Neural Development.** Edited by J.R. Wolff, J. Sievers, and M. Berry. 1987.
Vol. 6: **Molecular Mechanisms of Desensitization to Signal Molecules.** Edited by T.M. Konjin, P.J.M. Van Haastert, H. Van der Starre, H. Van der Wel, and M.D. Houslay. 1987.
Vol. 7: **Gangliosides and Modulation of Neuronal Functions.** Edited by H. Rahmann. 1987.
Vol. 9: **Modification of Cell to Cell Signals During Normal and Pathological Aging.** Edited By S. Govoni and F. Battaini. 1987.
Vol. 10: **Plant Hormone Receptors.** Edited by D. Klämbt. 1987.
Vol. 11: **Host-Parasite Cellular and Molecular Interactions in Protozoal Infections.** Edited by K.-P. Chang and D. Snary. 1987.
Vol. 12: **The Cell Surface in Signal Transduction.** Edited by E. Wagner, H. Greppin, and B. Millet. 1987.
Vol. 19: **Modulation of Synaptic Transmission and Plasticity in Nervous Systems.** Edited by G. Hertting and H.-C. Spatz. 1988.
Vol. 20: **Amino Acid Availability and Brain Function in Health and Disease.** Edited by G. Huether. 1988.
Vol. 21: **Cellular and Molecular Basis of Synaptic Transmission.** Edited by H. Zimmermann. 1988.
Vol. 23: **The Semiotics of Cellular Communication in the Immune System.** Edited by E.E. Sercarz, F. Celada, N.A. Mitchison, and T. Tada. 1988.
Vol. 24: **Bacteria, Complement and the Phagocytic Cell.** Edited by F.C. Cabello and C. Pruzzo. 1988.
Vol. 25: **Nicotinic Acetylcholine Receptors in the Nervous System.** Edited by F. Celementi, C. Gotti, and E. Sher. 1988.
Vol. 26: **Cell to Cell Signals in Mammalian Development.** Edited by S.W. de Laat, J.G. Bluemink, and C.L. Mummery. 1989.
Vol. 27: **Phytotoxins and Plant Pathogenesis.** Edited by A. Graniti, R.D. Durbin, and A. Ballio. 1989.
Vol. 31: **Neurobiology of the Inner Retina.** Edited by R. Weiler and N.N. Osborne. 1989.
Vol. 32: **Molecular Biology of Neuroreceptors and Ion Channels.** Edited by A. Maelicke. 1989.
Vol. 33: **Regulatory Mechanisms of Neuron to Vessel Communication in the Brain.** Edited by F. Battaini, S. Govoni, M.S. Magnoni, and M. Trabucchi. 1989.
Vol. 35: **Cell Separation in Plants: Physiology, Biochemistry and Molecular Biology.** Edited by D.J. Osborne and M.B. Jackson. 1989.
Vol. 36: **Signal Molecules in Plants and Plant-Microbe Interactions.** Edited by B.J.J. Lugtenberg. 1989.
Vol. 39: **Chemosensory Information Processing.** Edited by D. Schild. 1990.
Vol. 41: **Recognition and Response in Plant-Virus Interactions.** Edited by R.S.S. Fraser. 1990.
Vol. 43: **Cellular and Molecular Biology of Myelination.** Edited by G. Jeserich, H. H. Althaus, and T. V. Waehneldt. 1990.
Vol. 44: **Activation and Desensitization of Transducing Pathways.** Edited by T. M. Konijn, M. D. Houslay, and P. J. M. Van Haastert. 1990.

Activation and Desensitization of Transducing Pathways

Edited by

T. M. Konijn

Zoölogisch Laboratorium, Kaiserstraat 63,
POB 9516, 2300 RA Leiden, The Netherlands

M. D. Houslay

Institute of Biochemistry, The University
of Glasgow, Glasgow G12 8QQ, Scotland

P. J. M. Van Haastert

Biochemisch Laboratorium, Nijenborgh 16,
POB 9747 AG Groningen, The Netherlands

Springer-Verlag Berlin Heidelberg New York
London Paris Tokyo Hong Kong
Published in cooperation with NATO Scientific Affairs Division

Proceedings of the NATO Advanced Research Workshop on Activation and Desensitization of Transducing Pathways held at Noordwijkerhout, The Netherlands, May 25–27, 1989

ISBN-13:978-3-642-83620-6 e-ISBN-13:978-3-642-83618-3
DOI: 10.1007/978-3-642-83618-3

Library of Congress Cataloging-in-Publication Data
NATO Advanced Research Workshop on Activation and Desensitization of Transducing Pathways (1989: Noordwijkershout, Netherlands) Activation and desensitization of transducing pathways / edited by T.M. Konijn, M.D. Houslay, P.J.M. van Haastert. p. cm.—(NATO ASI series. Series H, Cell biology; vol. 44) "Published in cooperation with NATO Scientific Affairs Division." "Proceedings of the NATO Advanced Research Workshop on Activation and Desensitization of Transducing Pathways held at Noordwijkershout, the Netherlands, May 25–27, 1989"—T.p. verso.
Includes bibliographical references.
ISBN-13:978-3-642-83620-6 (U.S.: alk. paper)
1. Cellular signal transduction—Congresses. 2. Second messengers (Biochemistry)—Congresses. I. Konijn, T.M. II. Houslay, Miles D. III. North Atlantic Treaty Organization. Scientific Affairs Division. IV. Haastert, P.J.M. van. V. Title. VI. Series. QP517.C45N37 1989 674.87'6—dc20 90-9800

Softcover reprint of the hardcover 1st edition 1990

2131/3140-543210 – Printed on acid-free-paper

Contents

AUXIN-COUPLED SYSTEMS

ION CONDUCTANCE

CELLULAR SYSTEMS

Participants

NATO Workshop on Activation and Desensitization of Transducing Pathways, held at Noordwijkerhout, The Netherlands, 24th-27th May, 1989.

H. BARBIER-BRYGOO, Centre National de la Recherche Scientifique, Physiologie cellulaire végétale, Bâtiment 15, Avenue de la Terrasse, F-91190 Gif-Sur-Yvette, France.

J.L. BOYER, Dept. of Pharmacology, Univ. of North Carolina, School of Medicine, Chapel Hill, NC 27514, USA.

M.G. CARON, Dept. of Physiology, Duke Univ. Med. Center, Durham, NC 27710, USA.

M. CHABRE, CNRS - Institut de Pharmacologie, Route des Lucioles, Sophia Antipolis, F-06560 Valbonne, France.

R.B. CLARK, Graduate School of Biomedical Sciences, Univ. of Texas Health Science Center, POB 20334, Houston, TX 77225, USA.

F. CLEMENTI, Dipartimento di Farmacologia, Università di Milano, Via Vanvitelli 32, I-20129 Milano, Italy.

B. COOKE, Dept. of Biochemistry, Royal Free Hospital School of Medicine, Rowland Hill Street, London NW3 2PF, England.

J.D. CORBIN, Howard Hughes Medical Institute, Dept. of Molecular Physiology and Biophysics, Vanderbilt School of Medicine, Nashville, TN 37232, USA.

P.N. DEVREOTES, The Johns Hopkins University, School of Medicine, 725 N. Wolfe Street, Baltimore, Maryland 21205, USA.

C.P. DOWNES, Smith, Kline & French Research Ltd, The Frythe, Welwyn, Hertfordshire, England AL6 9AR.

J.E. DUMONT, Faculté de Médicine, Campus Hôpital Erasme, Route de Lennik 808, B-1070 Bruxelles, Belgium.

G. EPHRITIKHINE, Centre National de la Recherche Scientifique, Physiologie cellulaire végétale, Bâtiment 15, Avenue de la Terrasse, F-91190 Gif-Sur-Yvette, France

M.D. HOUSLAY, Inst. of Biochemistry, The Univ. of Glasgow, Glasgow G12 8QQ, Scotland.

R. IYENGAR, Dept. of Pharmacology, Mount Sinai School of Medicine, One Gustave Levy Place, New York, NY 10029, USA.

K.H. JAKOBS, Pharmakologisches Inst. der Univ. Heidelberg, Im Neuenheimer Feld 366, D-6900 Heidelberg, FRG.

D. KLÄMBT, Botanisches Institut der Univ. Bonn, Meckenheimer Allee 170, D-5300 Bonn 1, FRG.

T.M. KONIJN, Zoölogisch Laboratorium, Kaiserstraat 63, POB 9516, 2300 RA Leiden, The Netherlands.

K.R. LIBBENGA, Botanisch Laboratorium, Nonnenstreeg 3, POB 9516, 2300 RA Leiden, The Netherlands.

J. MATO, Metabolismo Nutrición y Hormonas, Fundación Jiménes Diaz, Avda de los Reyes Católicos 2, 28040 Madrid 3, Spain.

C. MULLE, Molecular Neurobiology, Institut Pasteur, 28 Rue du Dr. Roux, 75724 Paris Cedex 15, France.

R. PREMONT, Dept. of Pharmacology, Mount Sinai School of Medicine, One Gustave Levy Place, New York, NY 10029, USA.

P. SCHAAP, Zoölogisch Laboratorium, Kaiserstraat 63, POB 9516, 2300 RA Leiden, The Netherlands.

M.I. SIMON, Calif. Inst. of Technology, Division of Biology 147-75, Pasadena, CA 91125, USA

B.E. SNAAR-JAGALSKA, Zoölogisch Laboratorium, Kaiserstraat 63, POB 9516, 2300 RA Leiden, The Netherlands.

P.J.M. VAN HAASTERT, Biochemisch Laboratorium, Nijenborgh 16, 9747 AG Groningen, The Netherlands.

B. ZBELL, Botanisches Institut der Universität, Im Neuenheimer Feld 360, D-6900 Heidelberg, FRG.

Hostess: Eva Ludérus

Projection: Cor Schoen

Recording discussions: Bert Van Duijn, Fanja Kesbeke and Ton Bominaar

Secretary: Mieke Rozenboom

Preface

This book reports the proceedings of a small NATO-supported workshop held at Noordwijkerhout, in the Netherlands, during 25-27th. May 1989. The aim of this workshop was for a panel of experts to present data and analyse future developments in the rapidly expanding field of signal transduction processes in biological systems. This field will have considerable importance for the burgeoning biotechnology and biomedical/ health-care areas, well into the 21st century. Its pervasive nature was explored in the various representatives of the workshop, who were involved in investigating a diverse array of creatures from micro-organisms and slime moulds through to plants and animal cells. In all such creatures we find that cellular signalling systems are of crucial importance in allowing such creatures to survive, adapt, grow and differentiate. Intriguingly, the fundamental components of the signalling systems employed show a high degree of conservation, although their usage is clearly individually tailored to particular creatures and situations.

One clear observation, has been the major impact that molecular biology has made to this field over the past few years. This has provided the nascent beginnings of relating the fine structure of components of signal transduction systems to their particular functions. The proteins involved in signal transduction processes are normally found in vanishingly small quantities, which makes for distinct difficulties in their analysis. However, molecular biological techniques coupled with new developments in dealing with membrane proteins should, in the future, allow us to look forward to producing considerable quantities of these proteins for X-ray crystallographic analysis.

We have known for some time that a diverse array of proteins was involved in the various signal transduction processes. However, molecular biological analyses have indicated that the true diversity of such systems is far greater than we ever envisaged. The functional significance of these observations and the mechanisms whereby signal transduction systems interact together, to form networked arrays, will provide a considerable challenge for the future.

The Editors May 1989

Chemotaxis and chemosensing

ACTIVATION AND DENSENSITIZATION IN BACTERIAL CHEMOTAXIS

Melvin I. Simon
Biology Division, 147-75
California Institute of Technology
Pasadena, California 91125 USA

Abstract

Bacterial chemotaxis allows cells to respond to over 35 different chemicals in their environment. There are transmembrane cell surface receptors that bind specific ligands. The bound state of the receptor modifies the activity of an intracellular protein kinase that can phosphorylate two chemotaxis system components, the CheY and the CheB proteins. When CheY is phosphorylated it modulates bacterial flagellar rotation and thus effects the cells' movement. When CheB is phosphorylated it increases its ability to modify the cell surface receptor and change its sensitivity. A model for the integration of these steps into a scheme that can explain the dynamic range, rapid response and sensitivity of the bacterial sensory system is described. This represents one example of a much larger group of "two component" sensory systems in bacteria.

Keywords

Signal transduction/chemotaxis/protein kinase/transphosphorylation

Activation and desensitization in bacterial chemotaxis

All cells respond to changes in the chemical and physical parameters of their environment. Bacteria provide us with an interesting model in which to examine the mechanisms that cells use to detect changes in concentration of specific chemicals as a function of time and to convert these signals into metabolic changes that result in the organism's response to its external environment (Adler 1975, Stewart and Dahlquist 1987).

Bacteria are exquisitely sensitive to approximately 30 to 40 different chemical substances. They can regulate their motility so that

NATO ASI Series, Vol. H 44
Activation and Desensitization of Transducing Pathways
Edited by T. M. Konijn, M. D. Houslay, P. J. M. Van Haastert

the cells accumulate in regions where there are higher concentrations of specific attractants and away from regions where there are high concentrations of specific compounds, which have been termed "repellents", that are generally deleterious to the bacteria. We have been studying bacterial chemotaxis for a number of years and we believe that this system will provide both insight into the specific signal transduction process in bacteria and also into a variety of other sensing mechanisms that are used by bacteria and other organisms. In addition to the chemotaxis response, bacteria respond to specific nutrient and environmental changes by changing their pattern of gene expression. For example, changes in osmolarity lead to changes in the synthesis of specific outer membrane protein molecules. There are more than twenty regulatory systems known that respond to changes in the environment by altering gene expression. Recent work has suggested that all of these systems are related by similar regulatory mechanisms (Bourret et al. 1989). This implies that the cell's sensory apparatus can use a variety of interacting proteins in many different ways to transduce signals from cell surface receptors into specific intracellular biochemical responses. Our work over the past 10 years has concentrated on the components of the bacterial chemotaxis system in an effort to understand how those components are "wired" to yield the circuits that regulate bacterial motility.

Bacteria are too small to sense concentration gradients by measuring differences in concentration across the length of the cell body and, therefore, the organism measure changes in concentration in the medium as a function of time. Thus, the cell presumably can make a measurement at one point and then after swimming multiple body lengths measure again and compare the two measurements. Thus, if the concentration of a specific attractant is increasing as a function of time the cells will continue to rotate their flagella in the same direction. However, if the concentration of specific attractants decreases drastically or if the concentration of a specific repellent increases as a function of time the cell can reverse flagella rotation and this reversal leads to a tumbling motion and a change in the direction of bacterial swimming. The cell can then again test to determine whether conditions have changed sufficiently so that it will continue to swim in the new direction. Thus, by biasing a random walk, via the mechanism of controlling the frequency of flagellar-filament rotation-reversals, the cell makes progress up a gradient of attractants or away from higher concentrations of repellents (Macnab 1987).

The bacterial system allows the extensive use of genetics to analyze the nature of the components involved in chemotaxis. We can isolate a variety of mutants that are capable of swimming but are incapable of chemotaxis and identify the affected gene product. Over a number of years many of the components that are involved in bacterial chemotaxis have been identified, cloned and sequenced (Steward and Dahlquist 1987, Macnab 1987).

Components of the chemotaxis system

A number of genes that are required for the complete chemotaxis response have been characterized. These include a series of cell surface receptors. Four receptors form an homologous family that has been characterized and shown to be responsible for interaction with specific ligands required for the observed chemotaxis responses. The four receptors are the TAR, TSR, TRG, and TAP proteins (Boyd et al. 1983, Krikos et al. 1983, Russo and Koshland 1983, Bollinger et al. 1984). Each of these is specific for a different subset of chemicals. Thus, for example, the TAR receptor can bind aspartate or glutamate and the TSR receptor interacts with serine. The TRG receptor interacts with periplasmic proteins that specifically bind galactose or ribose and the TAP receptor interacts with dipeptides. All of these receptor proteins have homologous structures and similar dispositions in the plasma membrane.

The receptor can be subdivided into a number of distinct functional domains. The amino acid sequences of the different receptor proteins are most divergent at their N-terminal ends. There is a great deal of evidence that suggests that this portion of the receptor is found in the periplasmic space on the outside of the cell membrane; e.g., mutations in this region of the molecule can interfere with the binding of specific ligands, and chimeric genes that include the periplasmic portion of the specific receptor show the ligand binding characteristics of that receptor (Krikos et al. 1985, Park and Hazelbauer 1986). Thus, the N-terminal half of the receptor is most probably responsible for ligand binding. There are also two transmembrane regions; these have the classical characteristics of membrane-spanning polypeptides. They are composed of a stretch of hydrophobic amino acids bounded on either end by charged amino acids. Approximately 50% of the amino acids of transmembrane receptor proteins are in the cytoplasmic portion of the cell. This cytoplasmic fraction of the receptor is made up of three functionally distinct subregions. The component which is closest to the membrane spanning region has been called the "linker region." This portion of the molecule seems to be involved in information transmission. Mutations in this region have been shown to compensate for mutations in the transmembrane region that affect transmembrane transmission of information (Oosawa and Simon 1986). We have no idea how the ligand binding event which occurs on the outside of the cell is transduced or transmitted to the cytoplasmic portion of the cell. The two most obvious hypotheses are either that ligand binding induces a change in the multimeric state of the receptor or that ligand binding induces a specific conformational change which is transmitted through the membrane.

The rest of the cytoplasmic portion of the receptor can be divided into two functional regions. One is composed of two stretches of apparently helical peptide containing glutamine and glutamate residues that can be modified by chemotaxis specific proteins (Kehry et al. 1983, Terwilliger and Koshland 1984, Boyd et al. 1983). Two chemotaxis specific proteins are responsible for controlling methylation of this region of the receptor. CheR is a methyltransferase (Springer and Koshland, 1977) and CheB is a methylesterase and a deamidase (Stock and Koshland, 1978, Kehry et al. 1983). Methylation of this region of the receptor is important for the ability of the organism to "adapt" to relatively high concentrations of attractant (Springer et al. 1979). Thus, it is found that if the concentration of the specific chemical that binds to one of the receptors is increased, the cell immediately responds to this increase in concentration by suppressing the reversal of the flagellar filament. However, after a period of time the organism "adapts" to the attractant and it resumes flagella rotation reversals. Adaptation appears to correspond to the process of receptor desensitization that is found in many different sensory systems. It has the effect of restoring a pattern of swimming and tumbling and it allows the sensory system to respond to changes over a large "dynamic range" of ligand concentrations. Furthermore, an individual receptor can be desensitized; e.g., Tar may adapt to aspartate concentrations while other receptors are unaffected and remain fully responsive. The kinetics of this adaptation process correlates very well with the kinetics of methylation of a specific receptor by the CheR gene product. Furthermore, mutations which remove CheR or CheB also lead to the loss of the ability of the cell to adapt to high concentrations of attractants or repellents. Thus, when the concentration of attractant is changed the cell responds by suppressing tumbles and continues to suppress tumbles until the attractant concentration is lowered or repellents are added.

The process of adaptation is relatively slow and extends over ten seconds to minutes; in contrast, when the cells are exposed to a change in attractant or repellent concentration they respond immediately, i.e., within a fraction of a second they change their frequency of flagella rotation. This immediate rapid response has been termed the excitation response and it appears to be regulated by four different gene products, CheA, CheW, CheY, and CheZ. It is clear that these gene products control the responsiveness of the chemotaxis system and generate the excitation signal. However, the mechanism by which they integrate responses from a number of receptors and the mechanisms involved in generating the excitation signal and feedback to the receptors through the CheR and CheB gene products remains obscure (Stewart and Dahlquist 1987).

Mechanism of excitation

A great deal of evidence from physiological studies of bacterial responsiveness indicates that the four genes *cheA*, *cheW*, *cheY* and *cheZ* contribute directly to the generation of the excitation signal. This signal has a very short latency time, i.e., the time from the application of ligand to the surface of the cell until the first detection of an excitation response is on the order of 200 milliseconds (Segall et al. 1982). A great deal of experimental work involving allele-specific suppression of mutations in the CheY gene showed that the compensatory mutations occurred in genes for proteins that formed the region of the flagellar motor responsible for controlling the direction of rotation of the flagellar apparatus (Parkinson et al. 1983, Yamaguchi et al. 1986). These studies initially suggested that it was the CheY gene product that directly interacted with the flagellar switch. Similar kinds of studies indicated that the CheZ gene product might also be directly involved in regulating the interaction of CheY with the switch region. Other experiments also corroborated the notion that the CheY gene product was responsible for generating the increase in the frequency of flagellar filament rotation reversal (Kuo and Koshland 1987, Wolfe et al. 1987, Ravid et al. 1986). On the other hand, experiments with the CheA and CheW gene products suggested that they might be involved in integrating signals from cell surface receptors and that they might be required for transmission of the excitation signal from the transmembrane receptor to the flagellar apparatus (Conley et al. 1989).

In order to directly look for the nature of the interactions in this pathway we have purified all of the cytoplasmic proteins involved in signal processing. When these proteins were tested for their ability to be modified in the presence of ATP, we found that the CheA gene product was autophosphorylated (Hess et al. 1987). When incubated with [γ-^{32}P]ATP, 1 mole of phosphate was transferred to 1 mole of the CheA polypeptide. Phosphorylated CheA was found to be relatively stable and it could be separated from unincorporated ATP by gel filtration chromatography and stored in the phosphorylated form. When the phosphorylated protein was added to the other gene products involved in chemotaxis it was found that phosphorylated CheA could transfer phosphate to either CheY or CheB. The transfer of phosphate was very rapid. CheY-phosphate was found to be unstable; it rapidly autodephosphorylated with a half life of approximately five to fifteen seconds (Hess et al. 1988b, 1989). The CheB protein was also rapidly phosphorylated and rapidly dephosphorylated. The rate of CheB dephosphorylation was even faster than that of CheY. Phosphorylation of CheB apparently increases its activity as a methylesterase (Stock et al. 1989). Thus, phosphorylation may regulate receptor modification and this could constitute part of the feedback from the signaling process to the receptor. CheY, on the other hand, must interact directly with the motor and our working hypothesis is that the CheY protein can exist in two conformational states. In one state it could <u>not</u> react with the "switch" region of the motor and in the other, it can react with the "switch". Phosphorylation of CheY would drive the protein into the activated state and thus increase its ability to interact with the "switch". CheY-phosphate interaction with the switch would then generate the tumble reaction. This scheme has one difficulty, i.e, the observed half-life of CheY-phosphate is much too long to account for the short half life of the excitation signal. In pursuing this question, we found that if CheY was incubated in the presence of CheZ, the rate of CheY dephosphorylation increased very rapidly (Hess et al. 1988b, 1989). CheZ may act directly as a phosphatase to remove phosphoryl groups from CheY, or its function may simply be to activate CheY and thus enhance its rate of autodephosphorylation.

These observations taken together all suggest that CheA phosphorylation can lead to CheY phosphorylation, and CheY-phosphate, if it exceeds a certain threshold, can then presumably interact with the switch at the motor and generate flagella reversals. The CheZ protein acts essentially as a filter to decrease the ambient level of CheY-phosphate and thus make sure that the signal exceeds a certain threshold before the motor can "switch". The model also allows us to understand how feedback is generated. It is clear that phosphorylation of CheB by CheA could account for feedback to specific receptors, particularly if both CheB and CheR recognize a specific conformation of the receptor. Thus, if ligand dissociated from the aspartate receptor and receptor changed conformation leading to an excitation signal, i.e., the activation of CheA phosphorylation, that activated form of the receptor could be recognized by the CheB protein and immediately be demethylated to reduce its sensitivity to ligand. Activation of the CheB protein would increase the rate of demethylation. At the same time the CheR protein would reverse the process by methylating all of the receptors. If, as has been shown, the CheR protein recognizes the conformation of the receptor, then it could remethylate all of the unliganded receptors but not affect the ligand-bound receptor, leading to a net demethylation of the liganded receptor and leaving the rest of the system in a sensitized state. There is in fact a great deal of physiological evidence that suggests that upon interaction with ligand there is an immediate demethylation followed by remethylation of the receptors not involved in signalling. Thus, all of the physiological and genetic observations fit very well with the model that is sketched in Figure 1.

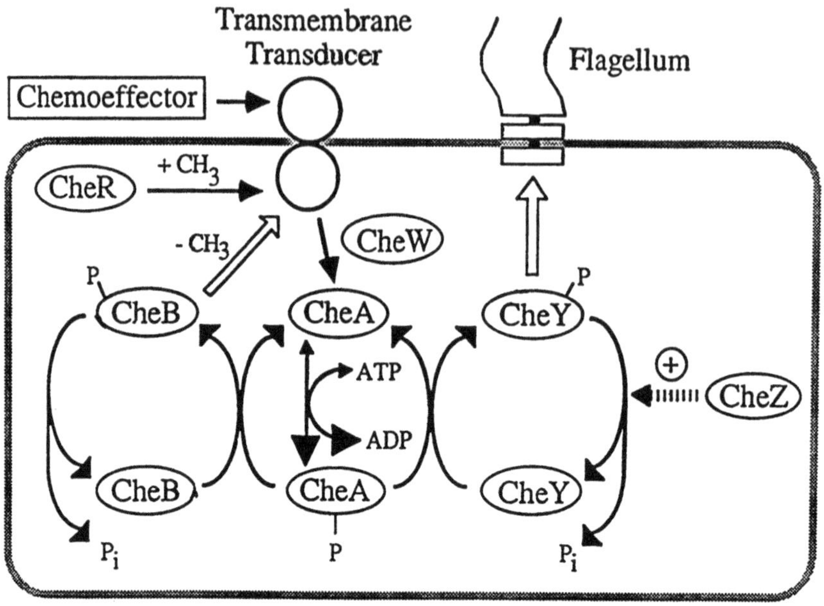

Figure 1. Model for excitation and adaptation responses

The role of the receptor

The model presented in Figure 1 postulates that the receptor somehow interacts with the CheA protein; the nature of this interaction is not specified. In order to test experimentally for an interaction between the cell surface receptor and the phosphate transfer system, we made preparations of _Escherichia coli_ cell membranes derived from bacteria that had all of their receptor genes deleted (Borkovich et al. 1989). Membrane preparations were also made from bacteria that had only the Tar receptor gene and from another strain that had the Tar receptor gene with a specific mutation that was dominant and caused the cells to tumble all the time. Finally, a fourth preparation of membranes was obtained from cells that carried the Tar receptor gene with a dominant mutation that caused the cells to swim smoothly all the time. The receptor mutants appear to "lock" the receptor in one signaling form or the other. When washed membrane preparations obtained from these cells were used to reconstitute the phosphorylation system we found that the rate of CheA phosphorylation and transfer to CheY was totally dependent on the presence of the CheW protein and the state of the cell surface receptor. Thus, in the absence of _Tar_ there was no stimulation of the endogenous CheA autophosphorylation rate. Nor was there any dependence of the rate of transfer of phosphate to CheY on the presence of the CheW protein. On the other hand, when membrane preparations containing wild-type Tar receptor or preparations of membranes that carried the "tumble" Tar receptor were reconstituted with the CheA, CheW, and CheY protein, there was an enormous increase in the rate of phosphorylation of CheY; the rate of transfer of phosphate through CheA to CheY increased 300-fold. This reaction required the presence of the CheW protein and an activated state of the receptor. In this case, the activated state of the receptor is that of the mutant receptor in the "tumble" form or the presence of the wild-type receptor in the absence of the aspartate ligand. Thus, the reconstituted preparation shows the properties that we would expect on an activated receptor. In the absence of bound ligand, receptor is able to markedly increase phosphorylation. In fact, if the receptor is titrated with aspartate, micromolar concentrations of aspartate inhibit the ability of the receptor to activate the CheA protein. In all of these reactions, the CheW protein is required. It presumably acts as a coupling factor between the receptor and CheA. These results are summarized in Table 1.

Table 1.
Phosphorylation of CheA and/or CheY in the presence of the other components

Purified Proteins	Membrane Source	Counts per minute incorporated	
		CheA	CheY
CheA	Minus receptor	2061	0
CheA	Wild type	1803	0
CheA	Tumble	1388	0
CheA	Smooth	1596	0
CheA + CheW	Minus receptor	2399	0
CheA + CheW	Wild type	2041	0
CheA + CheW	Tumble	2285	0
CheA + CheW	Smooth	1888	0
CheA + CheY	Minus receptor	0	982
CheA + CheY	Wild type	0	2380
CheA + CheY	Tumble	24	3192
CheA + CheY	Smooth	0	727
CheA + CheW + CheY	Minus receptor	11	1571

CheA + CheW + CheY	Wild type	726	10885
CheA + CheW + CheY	Tumble	331	7923
CheA + CheW + CheY	Smooth	37	1902
CheA + CheY + CheZ	Minus receptor	4	28
CheA + CheY + CheZ	Wild type	63	72
CheA + CheY + CheZ	Tumble	32	161
CheA + CheY + CheZ	Smooth	20	31
CheA + CheW + CheY + CheZ	Minus receptor	0	13
CheA + CheW + CheY + CheZ	Wild type	0	760
CheA + CheW + CheY + CheZ	Tumble	0	394
CheA + CheW + CheY + CheZ	Smooth	0	136

Reactions were performed under the standard conditions with 100 μM ADP, 4 μl membranes, 40 pmoles CheA, 40 pmoles CheW, 50 pmoles CheY, and 10 pmoles CheZ where indicated. Reaction time was 2.5 minutes.

The results of these reconstitution studies allow us to make a relatively complete model for the flow of information from the receptor to the flagellar apparatus. While there are still large gaps in our understanding of this system we can now begin to try to relate the chemical activities of each of the components of the system to chemotaxis function. We have very little idea of how ligand binding generates the appropriate conformational change that puts the receptor in an activated form nor do we know if the ligand bound form of the receptor has a direct activity, e.g., the ability to inhibit CheA phosphorylation. It appears clear that receptors in the unliganded form or in the form that interacts with repellent can, in the presence of CheW, activate CheA phosphorylation and phosphate transfer to CheY. Exactly what is the role of the signalling portion of the receptor and how does it interact with CheA and CheW to generate the active form of the CheA phosphorylating protein? How does CheY interact with the switch region at the base of the flagellum? What is the role of CheZ; does it interact with CheY both on the switch and off the switch? There are a great many mechanistic questions that we can now directly address.

In chemical experiments, the precise nature of the phosphorylation reactions involved in the chemotaxis system are beginning to become clear. CheA autophosphorylation results in the formation of a phosphoramidate bond at a specific histidine residue in the N-terminus of the CheA protein. A fragment of CheA containing this phosphohistidine residue can transfer phosphate to CheY (Hess et al. 1988a). Evidence is beginning to emerge from a number of systems that the residue in the CheY system that is phosphorylated is an aspartyl residue, which forms an acyl phosphate. The structure of the CheY protein is essential for the rapid autodephosphorylation of aspartyl phosphate; when CheY is denatured the acyl phosphate is stabilized. Thus, the excitation signal is a result of a chain of phosphorylations that starts with a conformational change at the receptor and leads eventually to an altered state of the CheY protein that can interact with the switch. This general principle, i.e., that signal transduction is embodied in the transient stabilization of different conformations of a series of proteins, is a generalization that holds for many kinds of signaling systems, both in bacteria and eukaryotic cells.

Generalized role of phosphorylation and signal transduction in prokaryotes

One interesting connection that has emerged from our understanding of the chemotaxis system is that a similar system of phosphate transfer appears to be widely used in bacteria for regulating not only transient information transduction but also signal transduction that is required for gene activation. Table 2 shows a list of systems that all show similar regions of amino acid sequence in proteins that are apparently involved in phosphate transfer required for regulating gene expression in response to various external changes in the environment of the cell.

Table 2

Two-component regulatory systems of bacteria

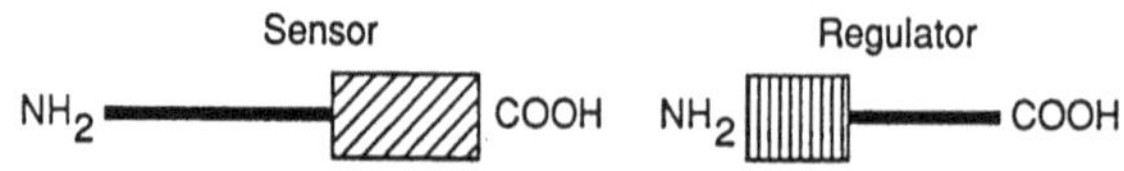

Species	Sensor Protein	Sensor Location	Regulator Protein	Regulator Transcription	Function controlled
Escherichia coli, *Salmonella typhimurium*	CheA	Cytoplasm	CheB, CheY	No No	Chemotaxis
Bradyrhizobium sp. RP501, *E. coli*, *Klebsiella pneumoniae*, *Rhizobium meliloti*	NtrB	Cytoplasm	NtrC	Yes	Nitrogen assimilation
E. coli	EnvZ	Membrane	OmpR	Yes	Porin expression
Agrobacterium tumefaciens	VirA	Membrane	VirG	Yes	Transformation of plant host
Rhizobium leguminosarum	DctB	Membrane?	DctD	Yes	C_4 dicarboxylic acid transport
E. coli	PhoM, PhoR	Membrane? Membrane?	PhoB	Yes	Phosphate assimilation
E. coli	CpxA	Membrane	SfrA	?	Aerobiosis? Conjugation?
E. coli	UhpB	Membrane	UhpA	Yes	Sugar phosphate transport
R. meliloti	FixL	Membrane?	FixJ	Yes	N_2 fixation
Bacillus subtilis	DegS	Cytoplasm?	DegU	Yes	Degradative enzymes
B. subtilis	?		SpoOA SpoOF	Yes Yes	Sporulation

This family has been called the two component regulatory system; it involves a sensor and a regulator and is relatively common. One of the best described of these systems (Magasanik 1988) involves the regulation of the enzyme glutamine synthetase. Protein synthesis in this system is regulated in a number of ways, however, one mechanism for regulating protein synthesis involves the phosphorylation of a protein called NTRC. The NTRC protein has many characteristics in common, including homologous polypeptide sequences, with the CheY protein and other regulator proteins. The other component of this system, NTRB, is an autophosphorylating kinase which responds to a receptor that is sensitive to the nitrogen supply in the cell. NTRB can be activated to autophosphorylate and to transfer phosphate to the NTRC protein. NTRB has homologous sequences that it shares with the CheA protein. It is interesting that signaling systems that involve relatively long-lived messages such as those required for regulating gene expression, and systems that regulate short-term processes such as those involved in chemotaxis all operate with a common mechanism.

The half life of the transient second message is presumably controlled by the protein structure of the regulator component. One of the other two component systems that has been studied extensively is the system that regulates the cells' response to osmolarity. In that case, the phosphorylated regulator protein has a half life on the order of 1.5 hours (Silhavy, personal communication). On the other hand, CheY protein, even in the absence of the CheZ protein, dephosphorylates in 5-15 seconds. The kinetic characteristics of the regulator proteins and the stability of the phosphorylated state are tuned to their particular biological function. The polypeptide environment surrounding the acyl phosphates in these regulator molecules must have evolved to give the system the optimum half life that is necessary for its signaling function.

It may well be that analogous systems operate in eukaryotic cells. It is clear that phosphorylation is very important in regulating processes in eukaryotic cells. However, the most common sites of phosphorylation in those cells are serine, threonine, and tyrosine. Some examples of acyl phosphate formation in eukaryotic systems that are involved in energy transduction have been found. However, a role for acyl phosphorylation of proteins in transduction in eukaryotic cells has not been established. It may be that processes in eukaryotic cells are generally much slower and therefore they do not use unstable acyl or phosphoramidate intermediates. On the other hand, we may not have seen these kinds of phosphorylation systems because they haven't been looked for carefully. In any event, whether the eukaryotic systems are homologous and involve mechanisms similar to those found in prokaryotes, or if the identity of the particular phosphorylated amino acids is different, there will certainly be analogies between the mechanisms and the strategies used both in the prokaryotic and eukaryotic systems. We can continue to learn in prokaryotic systems about the logic involved in designing signal transducing circuits and presumably some of that understanding will apply to eukaryotic cells as well.

REFERENCES

Adler, J (1975) Bacterial chemotaxis. Ann Rev Biochem 44: 341-356

Bollinger, J, Park, C, Haryama, S and Hazelbauer, G (1984) Structure of the Trg protein: homologies with and differences from other sensory transducers of Escherichia coli. Proc Natl Acad Sci USA 81: 3287-3291

Borkovich, KA, Kaplan, N, Hess, JF and Simon, MI (1989) Transmembrane signal transduction in bacterial chemotaxis involves ligand-dependent activation of phosphate group transfer. Proc Natl Acad Sci USA 86: 1208-1212

Bourret, RB, Hess, JF, Borkovich, KA, Pakula, AA and Simon MI (1989) Protein phosphorylation in chemotaxis and two-component regulatory systems of bacteria. J Biol Chem 264: 7085-7088

Boyd, A, Kendall, K and Simon MI (1983) Structure of the serine chemoreceptor in Escherichia coli. Nature 301: 623-626

Conley, MP, Wolfe, AJ, Blair, DF and Berg, HC (1989) Both CheA and CheW are required for reconstitution of chemotactic signalling in Escherichia coli. J Bacteriol 171: 5190-5193

Hess, JF, Oosawa, K, Matsumura, P and Simon MI (1987) Protein phosphorylation is involved in bacterial chemotaxis. Proc Natl Acad Sci USA 84: 7609-7613

Hess, JF, Bourret, RB and Simon MI (1988a) Histidine phosphorylation and phosphoryl group transfer in bacterial chemotaxis. Nature 336: 139-143

Hess, JF, Oosawa, K, Kaplan, N and Simon, MI (1988b) Phosphorylation of three proteins in the signalling pathway of bacterial chemotaxis. Cell 53: 79-87

Hess, JF, Bourret, RB, Oosawa, K, Matsumura, P. and Simon, MI (1989) Protein phosphorylation and bacterial chemotaxis. Cold Spring Harbor Symp Quant Biol 53: 41-48

Kehry, MR, Bond, MW, Hunkapiller, MW and Dahlquist, FW (1983) Enzymatic deamidation of methyl-accepting chemotaxis proteins in Escherichia coli catalyzed by the cheB gene product. Proc Natl Acad Sci USA 80: 3599-3603

Krikos, A, Mutoh, N, Boyd, A and Simon, MI (1983) Sensory transducers of E. coli are composed of discrete structural and functional domains.

Cell 33: 615-622

Krikos, A, Conley, PM, Boyd, A, Berg, HC and Simon, MI (1985) Chimeric chemosensory transducers of Escherichia coli. Proc Natl Acad Sci USA 82: 1326-1330

Kuo, SC and Koshland, DE Jr (1987) Roles of cheY and cheZ gene products in controlling flagellar rotation in bacterial chemotaxis of Escherichia coli. J Bacteriol 169: 1307-1314

Macnab, RM (1987) Motility and chemotaxis. In: Neidhart, FC (ed) Escherichia coli and Salmonella typhimurium: Cellular and molecular biology. American Society for Microbiology, Washington, DC, p 732

Magasanik, B (1988) Reversible phosphorylation of an enhancer binding protein regulates the transcription of bacterial nitrogen utilizing genes. Trends Biochem Sci 13: 475-479

Oosawa, K and Simon, MI (1986) Analysis of mutations in the transmembrane region of the aspartate chemoreceptor in Escherichia coli. Proc Natl Acad Sci USA 83: 6930-6934

Park, C and Hazelbauer, G (1986) Mutations specifically affecting ligand interaction of the Trg chemosensory transducer. J Bacteriol 167: 101-109

Parkinson, JS, Parker SR, Talbert, PB and Houts, SE (1983) Interactions between chemotaxis genes and flagellar genes in Escherichia coli. J Bacteriol 155: 265-274

Ravid, S, Matsumura, P and Eisenbach, M (1986) Restoration of flagellar clockwise rotation in bacterial envelopes by insertion of the chemotaxis protein CheY. Proc Natl Acad Sci USA 83: 7157-7161

Russo, AF and Koshland, DE Jr (1983) Separation of signal transduction and adaptation factors of the aspartate receptor in bacterial sensing. Science 220: 1016-1020

Segall, JE, Manson, MD and Berg, HC (1982) Signal processing times in bacterial chemotaxis. Nature 296: 855-857

Springer, MS, Goy, MF and Adler, J (1979) Protein methylation in behavioral control mechanisms and in signal transduction. Nature 280: 279-284

Springer, WR and Koshland, DE Jr (1977) Identification of a protein methyltransferase as the cheR gene product in the bacterial sensing system. Proc Natl Acad Sci USA 74: 533-537

Stewart, RC and Dahlquist, FW (1987) Molecular components of bacterial chemotaxis. Chem Rev 87: 997-1025

Stock, AM, Wylie, DC, Mottenen, JM, Lupas, AN, Ninfa, EG, Ninfa, AJ, Schutt, CE and Stock, JM (1989) Phosphoproteins involved in bacterial signal transduction. Cold Spring Harbor Symp Quant Biol 53: 49-57.

Stack, JB and Koshland DE Jr (1978) A protein methylesterase involved in bacterial sensing. Proc Natl Acad Sci USA 75: 3659-3663

Terwilliger, TC and Koshland, DE Jr (1984) Sites of methyl esterification and deamination on the aspartate receptor involved in chemotaxis. J Biol Chem 259: 7719-7725

Wolfe, AJ, Conley, P, Kramer, TJ and Berg, HC (1987) Reconstitution of signaling in bacterial chemotaxis. J Bacteriol 169: 1878-1885

Yamaguchi, S, Aizawa, S-I, Kihara, M, Isomura, M, Jones, CJ and Macnab, RM (1986) Genetic evidence for a switching and energy-transducing complex in the flagellar motor of Salmonella typhimurium. J Bacteriol 168: 1172-1179

DISCUSSION

Houslay: I presume that there are mutants in CheY and CheB, where you actually lose the phosphate acceptors, because in CheY you blocked thumbling and in CheB you blocked desensitization.

Simon: Yes, however the situation is not completely finished. Work on NTRC indicated that aspartate residues were phosphorylated. So we have been mutagenizing all of the aspartate residues in CheY. We have done four of the most conserved ones so far and the protein still gets phosphorylated nonetheless it cannot support chemotaxis so it is clear that we do not completely understand the role of phosphorylation. But the work is going on and has not been completed yet. It is possible that there are multiple sites of phosphorylation. The role of phosphorylation may be to stabilize a specific form of the protein and there may be alternative ways in which they can occur. But the answer to the question is, that a simple corre-lation that explains the behavior of all the mutants has not emerged.

Devreotes: Just to follow up on that, why not just isolate the phosphopeptide?

Simon: It turns out that the aspartyl phosphates are very unstable and therefore are relatively difficult to isolate. The nice thing about histidine phosphate was that we could get 20 or 30 % of the original phosphate on the peptide that we isolated. But with aspartate phosphate, we only found 1% or 2 % and so we do not believe this result and we would rather do it by mutagenesis. Boris Magasanik used borohydride to reduce the phosphorylated residue and then demonstrated that this was aspartyl phosphate, but we would rather do it by mutagenesis.

Corbin: I was struck by 2 pictures that you described in bacteria, that I am not familiar with in mammals. I'd like to discuss these. One is the transfer of phosphate in the ping pong kind of mechanism. In the systems that I work in, we do not see these kind of things. The other is the methylation as a way of regulating, and I have not seen that in mammals. Do you think they are present in mammals and if not why not?

Simon: Well let me say 2 things: first about the methylation; a number of examples of methylation that might be involved in controlling things are just beginning to emerge. In a recent paper about the assembly of laminin the process seems to correlate with carboxymethylation; Another example that should be familiar to a lot of people in this room is RAS modification, and again it is not clear that it is involved in regulation, but there is some evidence that 3 amino acids are clipped from the end of RAS or perhaps even from the τ-components of the G-protein. The C-terminal amino acid (often a cystein residu) then gets carboxymethylated, so this is another example of carboxymethylation. So far as I know it is not absolutely clear that it is involved in regulation. Now the other part of the question that is why wouldn't you see the kind of phosphate transfer found in bacteria in mammalian systems. Two things, first I don't know how carefully it has been looked for. It turns out that tyrosine phosphates and serine phosphates are relatively stable, so there is less incentive to go looking for unstable phosphates. The other thing is the time scales, for example when we are talking about in terms of a response in gene expression i.e., from the time that bacteria sense a change the environment to the time it turns on it's genes, involves a period of a couple of minutes. On the other hand in eukaryotes we are talking about time scales of 45 minutes or something like that. The message that remains is a stable message, the life cycle of the cell is 24 hours. So I think that one of the possibilities is that, (this is speculation) the difference is that the bacteria just live for a much shorter period of time and have to do all of these things very rapidly and therefore use this formation of unstable phosphates and the transfer mechanism, while in eukaryotes the signals last for longer time and even the adaptation in tyrosine kinase signals are signals that go on, that one would like to last for a longer period, however I don't really know the answer.

Chabre: You said that the receptor only has 2 α-helices transmembrane and 2 large domains. Is there any idea that the receptor is a monomer or do they aggregate like the EGF-receptor.

Simon: Good question, I do not think that the question is answered. There are some people who think that the question is answered. The evidence now, particularly from Koshland's laboratory is that the receptor exists as a dimer in the membrane whether that dimer associates to form tetromers or dissociates to form monomers. I do not think that the case is closed. A lot of people think that what happens is that a conformational change is transmitted through the membrane, but again there is no conclusive evidence as far as I know. We thought we had good evidence, but it was all very indirect genetic evidence. A number of labs are doing cross linking studies, but everybody knows how dangerous those experiments are particularly if chimerization is short lived.

Chabre: You know the problem is, it is hard to believe, to understand how a signal could be transmitted by a tiny helix, so one thinks what happens is, the ligand binds and so the external domain changes conformation and aggregates and this induces the internal domain.

Simon: I like that very much. An interesting experiment was recently done in Inoyues' laboratory; we started to do this together with M. Inoyue. He took the N-terminal of the aspartate receptor and put it on to the C-terminal end of the Env-receptor. The Env-receptor turns on porin genes. Now using the chimeric receptor, he can add aspartate to turn on the porin genes. So presumably the aspartate C-terminal ligand binding domain is sending a signal to the autophosphorylating kinase in the EnvZ-domain. It is very difficult for me to see how a structural change would work. Why would the chimera respond to the same structural change as the original protein. On the other hand if aggregation alone was enough to generate the signal, it would make sense. So I like to think more and more about the aggregation idea, but there is no conclusive evidence. I think it interesting to hear in the rest of the meeting what the people, who work on 7-pass membrane receptors say, because I cannot see how those things could work by aggregation, they must work on conformational change.

Klämbt: I was impressed, even I am not very familiar with the bacterial system, that you found membrane receptors accepting small molecule ligands, and at the same time, as I saw, ligand bound to the binding proteins. Is there something known about the binding proteins for the carbohydrates and maltose?

Simon: There is a great deal known about them. The tar transmembrane receptor and the Trg receptor bind the maltose binding protein and the very well characterized ribose binding the galactose binding protein. I think there are crystal structures known for both of these proteins. Mutations in those binding proteins and in the corresponding receptor, show allele specific suppression, indicating that there is specific interaction. You can show by mutation that if you knock out the maltose binding proteins, or the galactose binding protein you loose sensitivity for the one sugar but not for the other. So I think there is lots of good evidence that the binding proteins can work to bind directly to ligand and subsequently to the transmembrane receptor.

Klämbt: There are some homologies between these binding proteins?

Simon: There are 2 cases that I know. One case is the maltose binding protein and aspartate both bind to the same receptors. In that case people have been able to make specific mutations in the receptor that knock out the aspartate interaction and other mutations that knock out maltose binding protein interactions suggesting that there are 2 separate sites on the receptor. There is also some interaction between those sites. So the ligand binding region of the receptor can bind multiple ligands. In the case of the galactose binding protein and ribose binding protein, I am not sure whether they compete for the same site or they bind to separate sites, I just don't remember, but that is known.

Devreotes: Just one other question about the model. In the model, does the occupied receptor influence the role of phosphorylation?

Simon: That is a point I did not talk about. I talked about the signal, the 'tumble signal', that one gets from the unliganded receptor. We think that there is also a 'smooth signal' that results from ligand binding the evidence is based on mutants that signal smooth swimming and are dominant. Since they are dominant, they must be doing something i.e. signalling smooth swimming. There is some evidence that this involves the CheW product but we do not know how it works. It may be that the receptor has two forms, in one form it just sequesters the CheW protein and keeps it from activating CheA.

Devreotes: But in the reconstitution experiment, where you can see an unoccupied receptor is there any activity?

Simon: We have not done the right experiments; we must demonstrate the dominant effect of receptor mutants in vitro. Right now it is a technical problem; we cannot do the appropriate experiment with membranes; we are trying to purify the receptor to get rid of the membrane, so that we can use just clean receptors to do the reconstitution and then hopefully we will see this effect.

ADAPTATION OF CHEMOATTRACTANT ELICITED RESPONSES IN DICTYOSTELIUM DISCOIDEUM

Roxanne Vaughan, Ronald Johnson, Michael Caterina, and Peter Devreotes
Department of Biological Chemistry
The Johns Hopkins University School of Medicine
Baltimore, MD 21205 U.S.A.

INTRODUCTION

During the developmental phase of the Dictyostelium life cycle, up to 10^6 unicellular amoebae undergo aggregation to produce a multicellular structure. Aggregation is initiated by pulses of cAMP secreted by cells in the center of a territory. Surrounding cells respond by migrating up the concentration gradient toward the aggregation center and concomitantly synthesizing and secreting additional cAMP. Maximal synthesis of cAMP occurs within a few minutes of stimulation, then the rate rapidly declines due to desensitization. Extracellular phosphodiesterase degrades the secreted cAMP, and the cells subsequently resensitize, or regain the ability to be stimulated. In this way, waves of cAMP are propagated throughout the population, resulting in orderly chemotactic aggregation (Devreotes, 1982). The natural waves or oscillations are also essential for the expression of many developmentally regulated genes (Devreotes, 1982). Adaptation and subsequent resensitization of cAMP-induced processes are thus essential for multiple facets of development in this organism.

Several of the components involved in the transmembrane signal transduction pathway of Dictyostelium have recently been identified. The system displays remarkable similarity to signal transduction pathways of higher eukaryotes.

Cell Surface Receptors

Binding of cAMP to cells occurs via a cell surface

NATO ASI Series, Vol. H 44
Activation and Desensitization of Transducing Pathways
Edited by T. M. Konijn, M. D. Houslay, P. J. M. Van Haastert

integral membrane protein. The receptor has recently been cloned and the deduced amino acid sequence indicates that it is a member of the seven transmembrane domain receptor family which includes rhodopsin and the beta-adrenergic receptors (Klein et al., 1988). Similar to these receptors, the C-terminal region of the cAMP receptor is hydrophilic and contains multiple serine and threonine residues which are potential sites for ligand-induced phosphorylation. Agonist-induced receptor phosphorylation and its relationship to adaptation is discussed in detail below. Another common feature of these receptors is that they exert their actions through G-proteins.

G-Proteins

The involvement of G-proteins in cAMP signal transduction pathways has been shown by guanine nucleotide inhibition of cAMP binding (Van Haastert, 1987), *in vitro* regulation of receptor stimulated adenylate cyclase by guanyl nucleotides (Theibert and Devreotes, 1986), and by analysis of mutants.

Two groups of aggregation-deficient mutants have provided important insights into the roles of G-proteins in the cAMP-induced responses in *Dictyostelium*. One of the mutants, termed *frigid A*, does not undergo chemotaxis, and exogenously applied cAMP does not induce expression of developmental genes (Kesbeke et al., 1988). Although the cells express surface cAMP receptors, guanyl nucleotides do not inhibit cAMP binding in membranes. Receptor-stimulated adenylate and guanylate cyclase activities are absent *in vivo*, but guanyl nucleotides can activate adenylate cyclase in broken cells (Van Haastert et al., 1987). These findings indicate that *frigid A* cells have a defect between receptors and G-proteins that prevents activation of chemotactic and gene expression pathways, and that a still unknown component produced in this pathway is required for *in vivo* adenylate cyclase activation.

In contrast, the second group of aggregation deficient mutants, termed *synag*, does exhibit chemotaxis to exogenously applied cAMP, and cAMP can induce the expression of some developmental genes. However, synag mutants cannot produce cAMP in vivo in response to extracellular cAMP or in vitro in response to guanyl nucleotides (Theibert and Devreotes, 1986; Pupillo et al., 1989). Since cAMP receptors and unregulated adenylate cyclase activities are normal in these cells, the defect presumably occurs between the adenylate cyclase catalytic unit and the G-protein which regulates it. This also indicates that different G-proteins regulate adenylate cyclase and the chemotaxis/gene expression pathways.

These results have been combined into the working model of signal transduction pathways shown in Figure 1. Current work

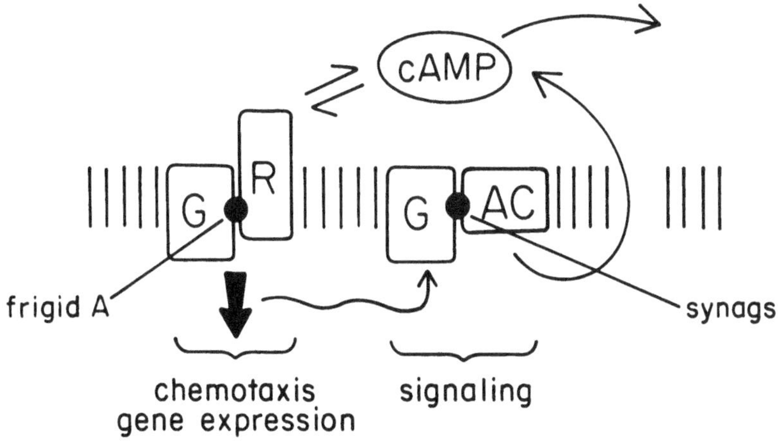

Figure 1. Proposed model for cAMP transmembrane signaling. (R) cAMP surface receptor; (G) G-proteins; (AC) adenylate cyclase; black circles indicate presumed sites of interruption in the signal transduction pathway in the frigid A and synag mutants.

in this laboratory is aimed at refining and expanding our knowledge of these processes. The recent cloning in this laboratory of two G-protein alpha subunits and one G-protein beta subunit (Pupillo et al., 1989) will increase the ability to analyze signal transduction events in normal and mutant cells.

Receptor Phosphorylation

Adaptation, or the tendency of a response to subside in the presence of constant stimulus, is a phenomenon common to many cell types. In Dictyostelium, several cAMP-induced processes undergo adaptation, including adenylate cyclase activation, myosin phosphorylation, and cell shape changes (Berlot et al., 1985; Mato et al., 1977; Theibert and Devreotes, 1986). These processes reach maximum activity 1-2 minutes after cAMP stimulation and then adapt with half times of about 2 minutes. A growing body of evidence from a variety of eucaryotes has implicated phosphorylation of stimulated receptors as an important element of the adaptation response (Sibley et al., 1987). This also appears to be the case in Dictyostelium (Vaughan and Devreotes, 1988).

A. Kinetics and Dose Response

The cAMP chemotaxis receptor appears as a doublet on SDS-PAGE when identified by cAMP photoaffinity labeling or by immunoblotting. The lower mobility form (R form; M_r=40000) is isolated from resting cells while a higher mobility form (D form; M_r=43000) appears in stimulated cells. This change in electrophoretic mobility is reversible and occurs spontaneously as cells undergo endogenous cAMP oscillations (Devreotes and Sherring, 1985; Vaughan and Devreotes, 1988). Metabolic labeling studies have shown that the electrophoretic shift is associated with a 5- to 20-fold increase in the amount of phosphate incorporated into the protein (Klein et al., 1985; Vaughan and Devreotes, 1988).

Figure 2 shows the kinetics of receptor phosphorylation and dephosphorylation upon addition and removal, respectively, of cAMP. Stimulus induced phosphorylation was detectable within 5 seconds, had a half-time of 45 seconds and was

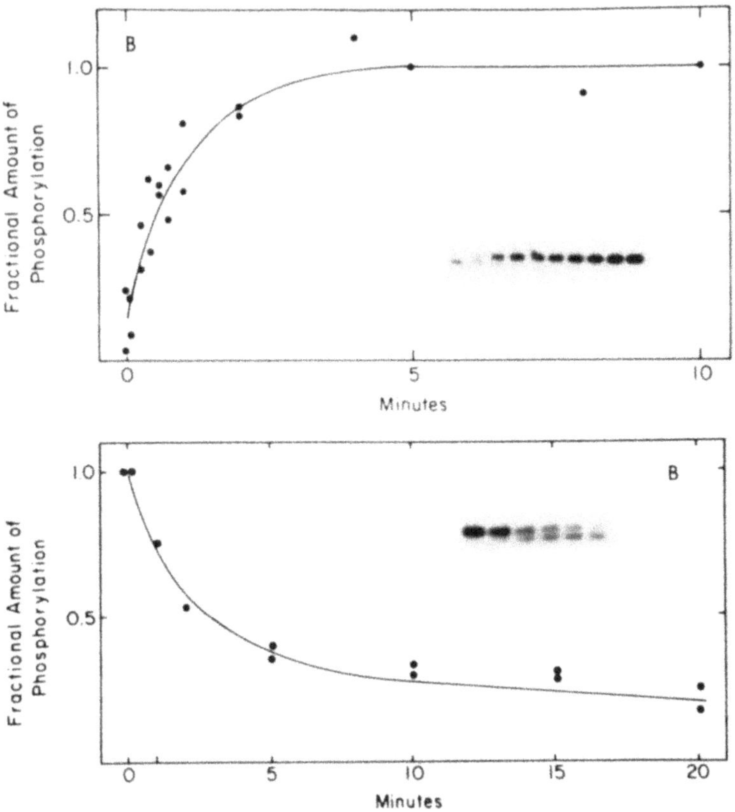

Figure 2. Upper panel. Kinetics of receptor phosphorylation. Developed cells were incubated with caffeine to block cAMP signaling and labeled with $^{32}P_i$. Aliquots of cells were removed before and at intervals after stimulation with 10^{-7} M cAMP. Samples were immunoprecipitated and analyzed by SDS-PAGE, autoradiography, and densitometric scanning. Insets show the 35-48 kD region of autoradiographs in which samples were taken at 0, 5, 15, 25, 35, 45, 60, 120, 300, and 600 sec after addition of cAMP. Graph shows the quantitative representation of the data.

Lower panel. Receptor dephosphorylation. Cells were labeled with $^{32}P_i$ and stimulated with cAMP to produce the D forms of the receptor. The cells were then washed twice with 0° buffer to remove cAMP. Dephosphorylation was initiated by diluting the cells into a 4-fold volume of 22^oC buffer. Samples were removed at 0, 2, 5, 10, 15, and 20 minutes after dilution. Inset shows autoradiograph of immunoprecipitated receptor, graph shows representation of densitometric scan of autoradiographs.

completed by 4-5 minutes. No change in the overall level of phosphorylation then occurred as long as the cAMP levels remained constant. All of the increases in phosphorylation occurred on the D form; no change in the phosphorylation state of the R form was observed. When receptors were fully phosphorylated and the cAMP was then removed by washing, dephosphorylation was detected within 30 seconds, had a half-time of 2 minutes and plateaued by about 10 minutes. These kinetics are compatible with the kinetics of adaptation of the processes mentioned above, which adapt with half-times of 1-2 minutes, and with their deadaptation rates which occur with half-times of 2-4 minutes.

The dose response of phosphorylation to cAMP is shown in Figure 3. Half-maximal phosphorylation occurs at about 5 nM cAMP and saturation occurs at 100 nM cAMP. A characteristic feature of adaptation is that a subsaturating stimulus induces

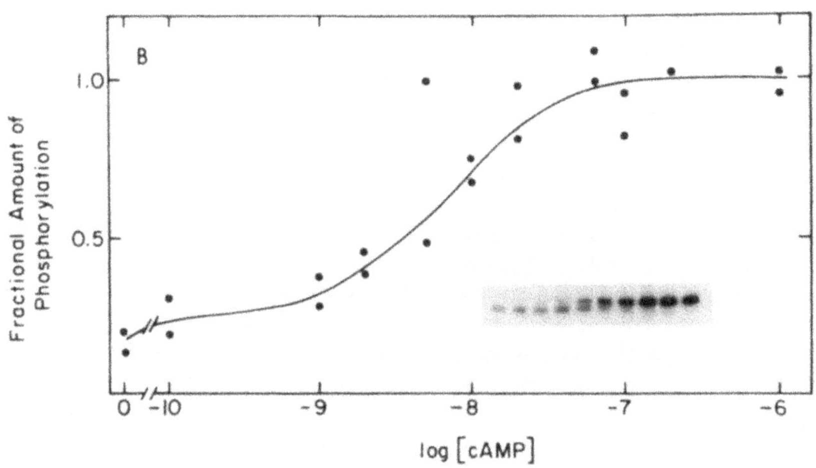

Figure 3. Dose response of receptor phosphorylation to cAMP. Cells were labeled with $^{32}P_i$, then stimulated for 10 minutes with various concentrations of cAMP. The inset shows immunoprecipitated receptors from cells that had been stimulated with 0, 0.1, 1, 2, 5, 10, 20, 50, 100, and 1000 nM cAMP.

a sub-maximal response which then subsides, and a further dose of higher magnitude can induce a further response (Devreotes, 1982). Therefore phosphorylation and electrophoretic shift of

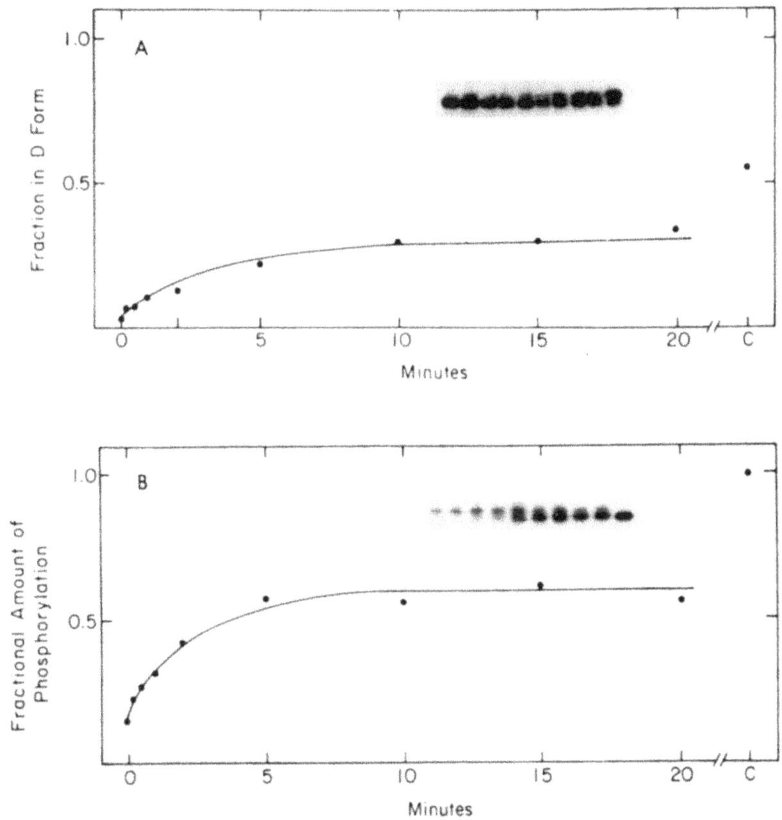

Figure 4. Receptor modification at a subsaturating dose. Parallel sets of cells were prepared with or without $^{32}P_i$ labeling for 45 minutes. Each set was then stimulated with 5 nM cAMP, and aliquots were removed at the indicated times. After the last time point was taken the remaining cells were stimulated with 10^{-7} M cAMP for 3 additional minutes. The graphs show the averages of two independent experiments. Upper panel. Kinetics of receptor shift determined by immunoblotting. The inset shows one of the immunoblots used to generate the graph. The lanes contain samples taken at times 0, 0.17, 0.5, 1, 2, 5, 10, 15, and 20 minutes after stimulation. The final lane corresponding to the point labeled C on the graph is the sample prepared from cells stimulated with the saturating dose. Lower panel. Kinetics of receptor phosphorylation. The inset shows the autoradiograph of immunoprecipitated $^{32}P_i$-labeled receptors taken at the same time points shown in panel A. The final lane is the sample prepared from cells stimulated with the saturating dose.

the receptor was examined at the half-maximal dose of 5 nM cAMP followed by a saturating dose (Figure 4). The upper panel shows the receptor protein analyzed by immunoblotting. The fraction of receptors in the D form reaches a steady-state level of about 0.30 within 5-10 minutes which is maintained until the cAMP concentration is raised to 100 nM cAMP. Then a further increase in the fraction of receptors in the D form occurs (point C). Phosphorylation levels (lower panel) follow a similar pattern, reaching an intermediate level which is maintained until the stimulus concentration is raised (point C) resulting in a further increase. The immunoblot data demonstrate that during persistent stimulation with a subsaturating concentration of cAMP, only a fraction of the receptors become modified, (i.e., the receptors in the D form); however, this leads to complete attenuation of the sub-threshold response. At the subsaturating dose, a fraction of the receptors remain in the R form and are not phosphorylated. These receptors are capable of triggering a further response. This response then also adapts as these receptors become phosphorylated.

A mathematical model termed the "Adapting Box", was developed to explain how receptor modification might control adaptation (Knox et al., 1986). The receptor forms R and D bind cAMP with equal affinity. The distribution between the possible states of the receptor is represented by the scheme illustrated below.

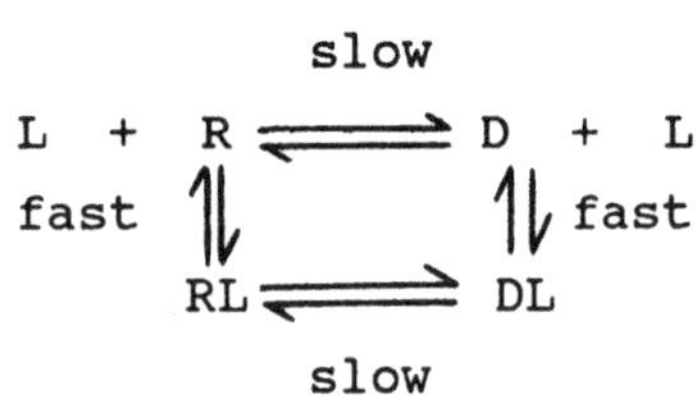

It is assumed that each of these four states of the receptor contributes to its overall "activity". When the condition of exact adaptation is imposed, state RL is found to be most active; states R and DL are intermediately active; and state D is least active. Based on the observed kinetics of receptor modification and demodification, the model accounts quantitatively for the observed properties of adaptation and deadaptation. The "Adapting Box" has recently been extended to account for cAMP oscillations and wave propagation.

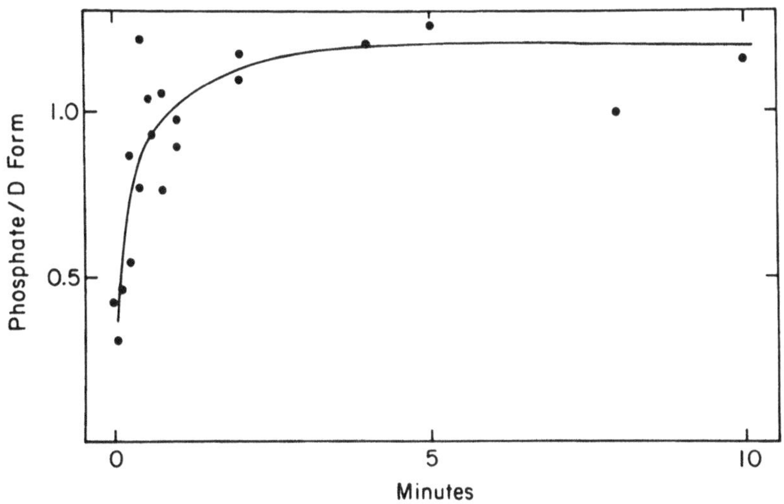

Figure 5. Specific activity of D form after cAMP stimulation. The data shown in Figure 2 were used to calculate the specific activity of $^{32}P_i$-labeled D form. The amount of $^{32}P_i$ in the D form at each time point was determined by densitometic scanning. This amount (as a percent) was divided by the fraction of receptor in D form at that time point (also a percent) as determined by immunoblotting.

The specific activity of the receptor D form as a function of time after stimulation was determined by dividing the relative amount of phosphate/D form by the relative amount of protein/D form using the data in Figure 2. This analysis (Figure 5) shows that about 40% of the maximal amount of phosphate is present on the D form at the earliest assayable times after stimulation. Once the receptor is in the D form, phosphate continues to be incorporated into the protein until the maximal level is attained at 2-3 minutes. We have estimated that stimulated receptors contain 4-5 moles of phosphate/mole protein (Klein et al., 1985). The specific activity data thus suggest that phosphorylation on one or two sites is associated with the mobility shift and the remainder of the sites become phosphorylated after the shift is complete.

B. Receptor Kinase(s)

Very little is known about the protein kinase(s) responsible for receptor phosphorylation in Dictyostelium. Treatment of cells with caffeine, which prevents synthesis of intracellular cAMP does not block receptor phosphorylation, and treatment of cells with phorbol esters and DAG analogs does not affect receptor phosphorylation, suggesting that kinases A and C are not involved. This is in contrast to the muscarinic cholinergic, alpha-adrenergic, and beta-adrenergic receptors in which some of the phosphorylation appears to be mediated by these protein kinases (Sibley et al., 1987). Responses induced by folic acid (another chemoattractant) (Van Haastert, 1983), adapt completely independently of those induced by cAMP; moreover, folic acid does not induce cAMP receptor phosphorylation. These observations may indicate that phosphorylation of the cAMP receptor is caused by a kinase which recognizes only the agonist-occupied form, as suggested for rhodopsin and beta-adrenergic receptors (Sibley et al., 1987), but more work remains to be done to test this idea.

C. Phosphopeptide Mapping

All of the G-protein-linked, 7-transmembrane receptors that have been cloned have a C-terminal hydrophilic section rich in serine and threonine which is presumed to be the phosphorylated region. The Dictyostelium receptor also fits this pattern with a large C-terminal region containing 18 serines and 9 threonines (Klein et al., 1987). Two serines and 3 threonines are also found in the putative third intracellular loop. Many of the serines have nearby acidic residues which have been implicated in other systems as important receptor kinase recognition sites (Thomson and Findlay, 1984).

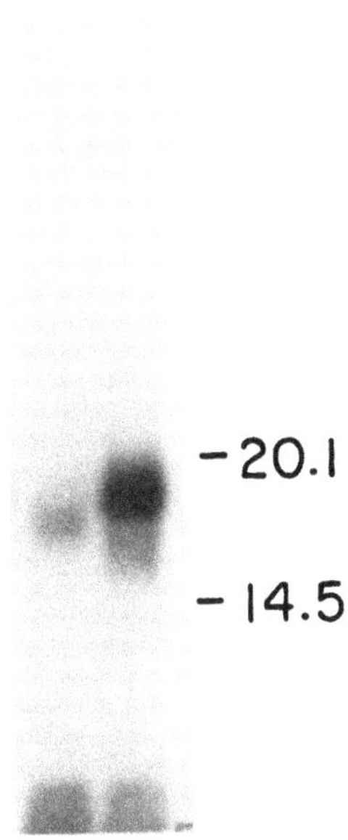

Figure 6. Localization of phosphorylation sites to the C-terminal domain. Cells were labeled with $^{32}P_i$, with or without cAMP stimulation. Membranes were prepared and resuspended in lysis buffer plus 100 µg/ml trypsin. After 30 minutes 150 µg/ml soy bean trypsin inhibitor was added and samples were re-centrifuged. The resulting supernatants were immunoprecipitated using antiserum specific for the C-terminal 18 amino acids of the receptor. Lanes A and B, post-trypsin supernatants generated from basal and stimulated membrane samples.

The basal and stimulated phosphorylation sites of the cAMP receptor have been partially mapped using controlled proteolysis. Light trypsinization of $^{32}P_i$-labeled membranes prepared from cells before and after cAMP stimulation results in the release of a soluble M_r=19000 phosphopeptide which is immunoprecipitable with receptor antiserum and contains both basal and stimulated phosphorylation sites (Figure 6).

Furthermore, this fragment can be immunoprecipitated by an antiserum specific for the C-terminal 18 amino acids of the receptor, demonstrating that the large C-terminal region is intracellular and that most of the phosphorylated residues are found there. Purification of this fragment and further localization of the sites are currently underway.

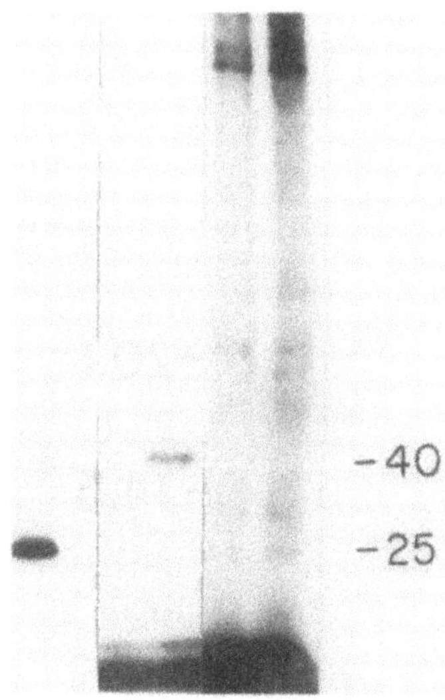

Figure 7. Deletion mutant of cAMP receptor. A 3′ deletion of the cAMP receptor cDNA was cloned into a plasmid under the control of the actin 6 promotor. Cells transformed with this vector were grown clonally and analyzed for cAMP photoaffinity labeling and receptor phosphorylation. Lane 1, membrane preparation of cells treated with 8-N_3-^{32}P-cAMP; Lane 2, same as Lane 1 with 1 mM cAMP included during binding reaction; Lanes 3 and 4, immunoprecipitation of membrane preparation from cells labeled with $^{32}P_i$, before and after cAMP stimulation, respectively; Lanes 5 and 6, total membrane preparation of $^{32}P_i$-labeled cells before and after cAMP stimulation, respectively.

Additional evidence that the tail region is the major site of phosphorylation comes from the analysis of cells transformed with a C-terminal deleted receptor. The receptor cDNA was truncated at amino acid 315, which eliminates the 77 C-terminal amino acids, including 13 serines and 4 threonines. The remaining portion of the cDNA encodes the putative transmembrane domains and 54 amino acids of the cytoplasmic tail, including 5 serines and 5 threonines. The truncated cDNA was cloned into a plasmid vector containing a constitutive promotor and used to transform wild-type Dictyostelium. The resulting transformants were capable of binding ^{3}HcAMP at 2-3 fold higher levels than wild type (not shown). cAMP photoaffinity labeling of these cells demonstrated the presence of the endogenous receptors at 40 kD and a more intensely labeled band at 25 kD (Figure 7). Receptor phosphorylation in these cells was also monitored using our standard protocol. Because the receptor antiserum used to immunoprecipitate the samples may not recognize the truncated receptor, whole membrane preparations from the transformed cells were also analyzed. As shown in Figure 7 no detectable basal or stimulated phosphorylation of the truncated receptor was observed in either immunoprecipitates or membrane preparations. Since a small amount of phosphorylation on the truncated receptors might not be readily visible in the whole membrane preparations, we are currently developing N-terminal specific antisera capable of immunoprecipitating the truncated receptors.

CONCLUSIONS

The unique advantages of Dictyostelium as a model system are allowing rapid progress in determining with more precision the molecular events involved in signal transduction. Current projects underway in this laboratory include attempts to eliminate the receptor and G-protein subunits by antisense transformation or gene disruption, the overproduction of these proteins, and expression and analysis of deletion mutants and chimeric proteins. These techniques will be useful tools for determining the precise mechanism by which receptor phosphorylation mediates adaptation. The importance of specific phosphorylated sites can be determined individually and in combinations by using site-directed mutagenesis, and the roles of G-proteins or other accessory proteins can be analyzed to ascertain the specific molecular interactions regulated by phosphorylation.

In view of the many similarities in signal transduction pathways already known to exist between Dictyostelium and higher cells, it is likely that adaptation mechanisms will also be similar. The experimentally manipulatable nature of Dictyostelium may provide an excellent system for determining with more precision the molecular events underlying this important regulatory mechanism.

REFERENCES

Berlot C, Spudich J, and Devreotes P (1985) Cell 43:307-314

Devreotes P (1982) In: Loomis WF (ed) The Development of Dictyostelium discoideum. Academic Press, Orlando Florida, pp 117-168

Devreotes P and Sherring J (1985) J Biol Chem 260:6378-6384

Kesbeke F, Snaar-Jagalska BE, and Van Haastert PJM (1988) J Cell Biol 2:521-528

Klein P, Sun T, Saxe C, Kimmel A, Johnson R, and Devreotes P (1988) Science 241:1467-1472

Klein P, Vaughan R, Borleis J, and Devreotes P (1987) J Biol Chem 262:358-364

Knox BE, Devreotes PN, Goldbeter A, and Segel LE (1986) Proc Natl Acad Sci USA 83:2345-2349

Mato J, Kreno F, Van Haastert P, and Konijn T (1977) Proc Natl Acad Sci USA 74:2348-2351

Pupillo M, Klein P, Vaughan R, Pitt G, Lilly P, Sun T, Devreotes P, Kumagai A, and Firtel R (1988) Cold Spring Harbor Symp on Quant Biol, Vol. LIII, pp 657-665

Sibley DR, Benovic JL, Caron MG, and Lefkowitz RS (1987) Cell 48:913-922

Theibert A and Devreotes P (1986) J Biol Chem 261:15121-15125

Thompson P and Findlay JBC (1984) Biochem J 220:773-780

Van Haastert P (1987) J Biol Chem 262:7700-7704

Van Haastert PJM (1983) Biochem Biophys Res Comm 115:130-136

Van Haastert PJM, Snaar-Jagalska BE, and Janssens PMW (1987) Eur J Biochem 162:251-258

Vaughan R and Devreotes P (1988) J Biol Chem 263:14538-14543

DISCUSSION

Simon: I wondered if you were looking at different receptors to do Northerns, to look at the expression of different genes at different stages of development.

Devreotes: We have one set of Northerns, using full length probes, at high stringency and what we see is transient expression of each of the receptors as a function of time and development. The first one is similar to what I've showed for cAR 1 and then cAR 3 appears, and then cAR 2 appears later in development. But I have to view this with a little bit of caution; we used full length probes and we really didn't evaluate the extent of hybridization between the different ones. But now we make very specific oligonucleotides, which can't cross-hybridize, with which we want to repeat the experiment.

Simon: But it couldn't get any better. One possibility is that the antisense is just blocking the first stage in the transition and therefore blocking the rest of it.

Devreotes: That's the other possibility. It's a pleiotropic effect; you need cAR 1 to get cAR 2 and 3, in effect the antisense may be specific and even if you do knock out cAR 1 you may knock out 2 and 3; so the thing to do is try to knock out 2 and 3, the later ones, and see if you get a differential effect.

Clark: I wondered if you have looked at in vitro the phosphorylation at all, and what kinases might be involved in these phosphorylations; G protein and receptor.

Devreotes: The G protein is pretty new, so we haven't looked at that. On the receptor, we can carry out phosphorylation, but so far we haven't been able to separate the receptor and the kinase, so if we make cytoskeletons it contains both receptor and the kinase activity, and we can get stimulation if we prepare these cytoskeletons from cells which have been stimulated. We don't have a kinase assay yet in which we can separate the two. I expect it to be similar to the rhodopsin or B-ark system, but that's as far as we have gone so far. We are trying to make a synthetic substrate now so we can look at it a little better.

Simon: I just wondered in general about the use of antisense. Did you try to do transient antisense as well as just recombine the gene back into the cell line?

Devreotes: We didn't try transient.

Simon: Do you know about the half life of the receptor or can you guess?

Devreotes: That's really never been done carefully with pulse labeling. If you treat the cells with cycloheximide, the receptor disappears with a time course of about 3 or 4 hours or something like that, that's a very crude estimate but these are stable cell lines. I mean one thing we know about it is we thought at first that since the antisense is on an actin promotor, there is a lot of antisense RNA already present in the cells (which we can show) before you induce the cells and the receptor message. But that doesn't seem to be necessary because we can do it now with the receptor promotor itself. If we take the receptor gene just turn the coding sequence around backwards. So even if the antisense RNA appears at the same time as the induction of sense RNA, it works as efficiently as the actin promotor.

Simon: I think that people have been trying to do the antisense experiments with G proteins, but the big problem in animal cells has been the half life of the G protein since it seems to stay around for a long time. This looks tremendously efficient.

Devreotes: I don't know why that is, this is not the only case, even abundant proteins like myosin have been effectively antisensed in *Dictyostelium*.

Clementi: When you get the oscillation of the two types of receptor, the 40 and the 43, do you have also change in affinity for cAMP of this receptor?

Devreotes: It's hard to measure, there is a number of different assays for the receptor. The total number of receptors certainly has not changed. Can you comment on that Peter? (Van Haastert).

Van Haastert: Well in fact you did the experiments.

Devreotes: Well that was ammonium sulfate.

Van Haastert: Well, Peter mentioned that there are several ways to measure binding of cAMP to *Dictyostelium* cells, one of them is in ammonium sulfate, in fact in saturated ammonium sulfate; this is the best assay to measure cAMP binding in *Dictyostelium*, because the rate of dissociation is decreased from seconds to several minutes which results in an affinity of something in the picomolar range. Peter measured the affinity of the receptor with that assay after the receptor was phosphorylated, and observed that the affinity of the receptor is reduced. Of course it is not completely physiological, at least the assay is not. We have measured the affinity of the receptor with another type of assay in just phosphate buffer, and find that the affinity of the receptor is decreased, probably faster than phosphorylation of the receptor. We have interpreted that, as an indication that the receptor couples with a G protein. Thus originally the receptor is in a high affinity state and during the binding on intact cells the affinity of the receptor is reduced to the low affinity state. Whether this lower affinity state represents the phosphorylated state or the activated receptor-G-protein complex is presently unknown.

Corbin: Correct me if I'm wrong, but don't you add ammonium sulfate at the end of the incubation, so that you test your changing concentration before adding ammonium sulfate.

Van Haastert: That is what we did originally. We first incubated the cells with radioactive cAMP and after the incubation was terminated the ammonium sulfate was added. Then you stabilize the binding reaction, that's what we said, but Peter found out that you can do it much more simple and put just your cells, the ammonium sulfate and the radioactive cAMP together.

Devreotes: It's really strange, at first we thought it was trapping of the ligand receptor complex in a stable form, but in fact if you put the cells first in ammonium sulfate, than add the ligand, it will bind, but now with much higher affinity. The on-rate is slowed, the off-rate is dramatically slowed, but you have an identical specifity for analogues.

Clementi: Is this receptor internalized?

Devreotes: We have some photographs, using receptor fluorescent antibodies, worked on by Wang Mei and Pauline Schaap. You see receptors rearrange themselves in a way that looks like being internalized, so if you look at the fluorescent micrographs initially receptors are in a ring round the cells and then after 10 or 15 minutes stimulation this staining is decreased and there is a lot of what looks like internalized staining. But since it is fluorescence I can't be sure that's not just clustered on the cell surface. Concurrent with this is, this is done with fairly high concentrations of ligand, there is a large decrease in number of surface binding sites. So you can cause down-regulation of about 70 or 80 %.

Iyengar: Is there any similarity between the cAMP binding region of the regulatory subunit and the receptor?

Devreotes: Not on amino acids. We don't have the crystals.

Iyengar: Do you know where cAMP binds to the receptor. Is it in the hydrophobic core?

Devreotes: We don't know that yet.

Simon: I guess that the third loop is involved in G protein interaction, is that a very highly conserved region? I noticed the N-terminal and end C-terminal were very diverse in the different receptor sequences. What about the G protein coupling loop?

Devreotes: Among the three receptors it is highly conserved, but not more than all the other membrane domains and other loops.

Clark: Is the idea that one of those receptors stimulates guanylate cyclase?

Devreotes: There is a guanylate cyclase activation by cAMP that lasts for several seconds before it turns off. There is about an 10-fold increase in cGMP which peaks about 10 seconds after stimulation and is brought down by PDE by 30 seconds. If you look at the analogue specificity of that stimulation, it's identical to what you measure for cAMP binding to the receptors and it's gone in the frigid A mutant; you no longer see cGMP increase, so we presume this couples to guanylate cyclase, although now, since there are three receptors it will be interesting to see whether the same one couples to adenylate cyclase and guanylate cyclase.

Clark: As you purified have you ever tried to follow the GC activity?

Devreotes: When we purified the receptor we did it by purifying that phosphorylated band, it was just solubilized in SDS.

Schaap: You have shown that receptor phosphorylation correlates with adenylate cyclase adaptation, so it's most likely what you are looking at there is the A-sites, but then Gα 2 in current hypothesis, is linked to the receptor B-sites, so considering that it is not so surprising that the kinetics of Gα2 phosphorylation are different from the kinetics of the receptor phosphorylation, because actually you are looking at two different signal transduction systems. What do you think about it?

Devreotes: Well I do not know if the kinetics are that different in time other than that one is transient and the other one is persistent. I do not find that either one is extremely rapid as you might expect where it involved activation in guanylate cyclase. This Gα2 phosphorylation is transient but, if you look at cyclic GMP at the same time scale it is not rapid.

Schaap: So your feeling is not that the receptor phosphorylation may be the mode of adaptation for the adenylate cyclase response and that G protein phosphorylation may be the mode of adaptation for, for instance, the guanylate cyclase response or the IP3 response.

Devreotes: No I mean I do not think that G protein phosphorylation could be the mode of adaptation for anything because it is transient. So if you look at either guanylate cyclase adaptation or at adenylate cyclase adaptation it persists as long as the stimulus is present. And you have to remove the stimulus, let the cells rest, and reapply it. Peter showed for cyclic GMP increase in cells as well as cAMP. I do not see how Gα 2 which loses its phosphorylation, could be involved in adaptation.

Zbell: Is anything known about the molecular mechanism of the cAMP translocator and its modulation.

Devreotes: It is temperature sensitive. That is about it.

THE INOSITOLCYCLE OF DICTYOSTELIUM DISCOIDEUM

Peter J.M. Van Haastert*, Anthony A. Bominaar*, Jeroen Van der Kaay*, Richard Draijer˙, Louis C. Penning˙, Edwin Roovers˙, Martinus M. De Vries˙, Ard A. Vink˙, Fanja Kesbeke˙, and B. Ewa Snaar-Jagalska˙.

*, Department of Biochemistry, University of Groningen, Nijenborgh 16, 9747 AG Groningen, the Netherlands.
˙, Cell Biology and Genetics, Zoological Laboratory, University of Leiden, Kaiserstraat 63, 2311 GP Leiden, the Netherlands.

Transmembrane signal transduction is characterized largely by the interaction between its components: ligand, receptor on the surface of cells, G-protein subunits at the inner face of the plasma membrane, and effector enzymes. The effector enzymes may vary widely depending on the organism and the ligand, and include adenylate cyclase, guanylate cyclase, phospholipase C, and ion channels. The consequence of these interactions is the production of intracellular second messengers, such as cAMP, cGMP, inositol 1,4,5-trisphosphate [Ins(1,4,5)P_3], diacylglycerol, Ca^{2+}, and K^+. Besides the interaction between these proteins that generate second messengers, there also excists an extensive interaction between the second messenger systems such that one system modulates or rules another system. The main problem for the elucidation of transmembrane signal transduction is probably to understand how the flow of information proceeds through this complicated network of interacting molecules. Mutants have been shown to be very useful to unravel complex biochemical pathways. Microorganisms are most advantageous in this respect, since they are easy to grow, have short generation times, and have a relatively small genome which is expressed in a haploid stage. Transmembrane signal transduction has been studied extensively in the eukaryotic microorganism Dictyostelium discoideum, and appears to be very similar to signal transduction in higher eukaryotes (see 1). Cyclic AMP is the extracellular signal in Dictyostelium, to be compared with the hormone in mammalian cells. Cyclic AMP is detected by surface receptors that have the classical seven putative transmembrane spanning domains of

NATO ASI Series, Vol. H 44
Activation and Desensitization of Transducing Pathways
Edited by T. M. Konijn, M. D. Houslay, P. J. M. Van Haastert

receptors that interact with G-proteins (2). The effector enzymes are adenylate cyclase, guanylate cyclase, and phospholipase C; the second messengers interact with target enzymes, such as protein kinases, calcium channels, and cytoskeletal components. The two main cellular functions of extracellular cAMP in Dictyostelium are chemotaxis to bring the amoeboid cells in a multicellular structure, and cell type specific gene expression to induce cell differentiation in this structure.

In this paper we describe our recent work on the characterization of the inositol cycle in Dictyostelium. The metabolism of $Ins(1,4,5)P_3$ was investigated in vivo and in vitro, and the stimulation of $Ins(1,4,5)P_3$ production by receptor and G-protein agonists was demonstrated. The function of the inositol cycle was established mainly by using a mutant which appears to lack a functional G_α-subunit that activates phospholipase C. Finally, we demonstrate that the elevated levels of $[^3H]Ins(1,4,5)P_3$ in RAS-THR^{12}-transformed cells have an unexpected source, i.e. the enhanced formation of PtdInsP.

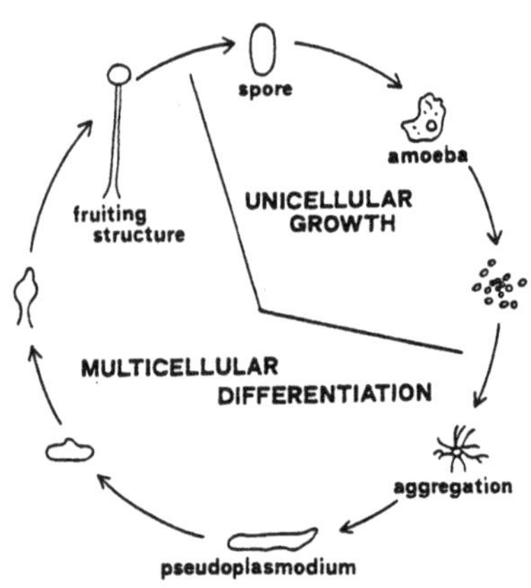

Fig. 1. The life cycle. Dictyostelium cells live in the soil as single amoebae and feed on bacteria. Exhaustion of the food source induces a developmental program: cells aggregate by means of a chemotactic reaction to a compound which is secreted by the starving cells. The aggragate may contain upto 100,000 cells, which are organized in a spatial pattern; about one third of the cell mass at the front differentiate to prestalk cells, while about two third differentiate to prespore cells. By means of morphogenetic movement and final differentiation of the cells a fruiting body is formed which is composed of spores that are embedded in a slime droplet on top of a cylinder of vacuolized dead stalk cells.

Metabolism of Ins(1,4,5)P_3 *in vitro*

In mammalian cells Ins(1,4,5)P_3 is degraded by a 5-phosphatase yielding Ins(1,4)P_2, which is further dephosphorylated via Ins4P to inositol. The major part of Ins(1,4,5)P_3, however, is phosphorylated to Ins(1,3,4,5)P4 which is then dephosphorylated by the same 5-phosphatase to Ins(1,3,4)P3; this InsP_3 isomer is then metabolized by a complex pattern of phosphorylations and dephosphorylations (see Downes et al. in this volume). Pilot experiments in *Dictyostelium* revealed that the metabolism of Ins(1,4,5)P_3 could be different than in mammalian cells.
The dephosphorylation of Ins(1,4,5)P_3 was elucidated in homogenates of *Dictyostelium*, using a mixture of [2-^{3}H]Ins(1,4,5)P_3 and [4,5-^{32}P]Ins(1,4,5)P_3 followed by chromatography of the products on Dowex columns (3). The rationale of this experiment is that the ^{3}H-radioactivity is associated with the inositol structure, whereas the ^{32}P-radioactivity is present predominantly at the 5-position (85%); therefore, detection of ^{3}H-radioactivity describes the extend of dephosphorylation, whereas the detection of ^{32}P-radioactivity describes the specificity of the dephosphorylation. It was readily observed that the major part of Ins(1,4,5)P_3 dephosphorylation did not occur at the 5-position as in mammalian cells, because essentially all ^{32}P-radioactivity was retained in the InsP_2 product. We concluded that the product was either Ins(1,5)P_2 or Ins(4,5)P2. The InsP_2 product was purified, further dephosphorylated by a *Dictyostelium* lysate, and the InsP product was analysed and identified by HPLC as Ins4P. Thus the major route of Ins(1,4,5)P_3 dephosphorylation in *Dictyostelium* was identified as Ins(1,4,5)P_3 --> Ins(4,5)P_2 --> Ins4P --> Ins.

This route of Ins(1,4,5)P_3 dephosphorylation is present exclusively in the cytosol and dephosphorylates about 80% of the Ins(1,4,5)P_3. The other 20% of Ins(1,4,5)P_3 dephosphorylation is mediated by the mammalian route: Ins(1,4,5)P_3 --> Ins(1,4)P_2 --> Ins4P --> Ins. Between 30-50 % of the Ins(1,4,5)P_3 5-phosphatase is present in the membrane fraction; All the other phosphatases are present in the cytosol fraction (3).
Since Ins(1,4,5)P_3 and Ins(1,4)P_2 are both dephosphorylated at the 1-position, we investigated the substrate specificity of this enzyme reaction; the same was done for the dephosphorylation of

$Ins(1,4,5)P_3$ and $Ins(4,5)P_2$ at the 5-position (Bominaar et al., unpubl. res.). The cytosol was partially purified by DEAE-cellulose chromatography, and the dephosphorylation of $Ins(1,4,5)P_3$, Ins(1,4)P2, $Ins(4,5)P_2$, Ins1P, and Ins4P was investigated. To get a complet picture of the specificity profile, the dephosphorylation of $Ins(1,3,4,5)P_4$ and $Ins(1,3,4)P_3$ was also determined, but the products of the reaction were not identified. The results are summarized in table I. There exist probably at least six enzyme activities that are involved in the deposphorylation of $Ins(1,4,5)P_3$ to Ins. Enzyme 1 is an $Ins(1,4,5)P_3$ 5-phosphatase that also dephosphorylates $Ins(1,3,4,5)P_4$, but not $Ins(4,5)P_2$; the enzyme may be very similar to the mammalian $Ins(1,4,5)P_3$ 5-phosphatase. Enzyme 2 dephosphorylates exclusively $Ins(1,4,5)P_3$ at the 1-position. Enzyme 3 is a minor activity and not very stabile; preliminary results suggest that it dephosphorylates $Ins(1,4,5)P_3$ first at the 4-position and then at the 5-position; other inositol-phosphates are not degraded by this column fraction, suggesting that the enzyme is very specific.

Table I. Enzymology of $Ins(1,4,5)P_3$ dephosphorylation.

Enzyme #	name	Elution	Reaction	Co-substrate	No Substrate
1	145-5Pase	0.1 M	145 -> 14	1345	45
2	145-1Pase	wash	145 -> 45	-	1345, 134, 14
3	*	0.3 M	145 -> 1	-	1345, 14, 134
4	14-1Pase	0.1 M	14 -> 4	134	1345, 145
5	45-5Pase	0.1 M	45 -> 4	-	1345, 145
6	monoPase	0.1 M	InsxP->Ins	$x \neq 2$	x=2

* $Ins(1,4,5)P_3$ 4->5 bisphosphatase;
The high speed supernatant of a <u>Dictyostelium</u> lysate was chromatographed over a DEAE cellulose column that was eluted with a gradient of NaCl. The eluting fractions were assayed for the dephosphorylation of $Ins(1,3,4,5)P_4$, $Ins(1,4,5)P_3$, $Ins(1,3,4)P_3$, $Ins(4,5)P_2$, $Ins(1,4)P_2$, Ins1P, Ins3P, and Ins4P; the product of all reactions were identified [with the exception of the reactions with $Ins(1,3,4,5)P_4$ and $Ins(1,3,4)P_3$].

Enzyme 4 dephosphorylates Ins(1,4)P_2 at the 1-position; Ins(1,3,4)P_3 is also a substrate, but Ins(1,3,4,5)P_4 and Ins(1,4,5)P_3 are not. The enzyme may be very similar to the mammalian inositolpoly-phosphate 1-phosphatase. The 5th enzyme dephosphorylates Ins(4,5)P_2 at the 5-position. This enzyme is probably specific and does not dephos-phorylate Ins(1,4,5)P_3 or Ins(1,3,4,5)P4; the separation from the Ins(1,4,5)P_3 5-phosphatase was not complete however. Finally, Ins1P and Ins4P are dephosphorylated by an inositolmonophosphate phosphatase; this enzyme is very similar to the mammalian mono-phosphatase. In summary, Dictyostelium has probably at least six phosphatases that participate in the dephosphorylation of Ins(1,4,5)P_3. Three enzymes may be similar to the mammalian counterparts: Ins(1,4,5)P_3 5-phosphatase, Ins(1,4)P_2 1-phosphatase and InsP phosphatase. Three enzymes may be unique for Dictyostelium: the Ins(1,4,5)P_3 1-phosphatase, the Ins(1,4,5)P_3 4->5- bis-phosphatase, and the Ins(4,5)P_2 5-phosphatase. It is likely that this complex dephosphorylation pattern is more than just degradation. InsP_2 isomers may have signal transducing functions. Furthermore, the trifurcation in the dephosphorylation of Ins(1,4,5)P_3 opens ways for fine-regulation, and the system could be compartmentized or under developmental control.

In contrast to the complex dephosphorylation of Ins(1,4,5)P_3, the phosphorylation of Ins(1,4,5)P_3 is extremely simple: All experimental evidence indicate that the appropriate kinases are absent. The Ins(1,4,5)P_3 3-kinase could not be detected under conditions that the rat brain enzyme was very active, and Ins(1,3,4,5)P4 and Ins(1,3,4)P_3 were not detectable in extracts from [^{3}H]inositol-labeled cells (4).

Metabolism of Ins(1,4,5)P_3 in permeabilized cells

In collaboration with Cor Schoen, University of Amsterdam, we measured the degradation of [^{3}H]Ins(1,4,5)P_3 in electropermeabilized cells. The protocol was optimized for the generation of holes that are sufficiently large for charged small molecule such as cAMP and ATP to pass, but are impermeable to enzymes (5). Intact cells do not dephosphorylate Ins(1,4,5)P_3 at a detectable rate. Permeable cells dephosphorylate Ins(1,4,5)P_3 to Ins(4,5)P_2 and Ins4P (Fig 2), confirming the results that Ins(1,4,5)P_3 is dephosphorylated in vitro mainly by the 1-phosphatase.

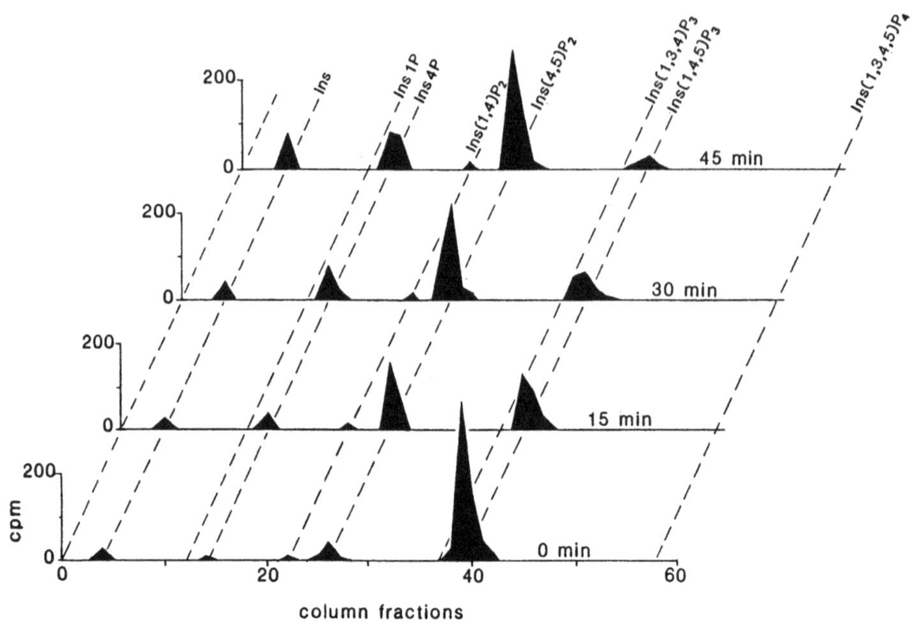

Figure 2. Dephosphorylation of [^{3}H]Ins(1,4,5)P_3 in electro-permeabilyzed Dictyostelium cells. Permeable cells were incubated with [^{3}H]Ins(1,4,5)P_3 and at the times indicated samples were withdrawn and analyzed by HPLC on a Partisil SAX column using gradient elution. The elution of authentic standers is shown.

Metabolism of [^{3}H]Inositol *in vivo*

[^{3}H]Inositol was introduced into Dictyostelium cells by electroporation; the protocol was optimized to generate very small holes, that are just large enough for inositol to pass, but are impermeable for charged molecules such as Ins(1,4,5)P_3 or ATP (4). This method of labelling cells with [^{3}H]inositol is very efficient (2.5% within 10 min) if compared with metabolic labelling (0.2% in 6 hr). The metabolism of [^{3}H]inositol was then followed by the analysis of the inositolphospholipides by thin layer chromatography and the analysis of the water soluble compounds by HPLC (Van der Kaay et al., unpubl. res.). Two HPLC systems had to be used: Ins(1,4,5)P_3 levels were analysed by reversed phase ion-pair chromatography and the other inositolphosphates by ionexchange chromatography. We have learned that a very significant amount of the radioactivity is incorporated in unidentified compound(s) that elute from Dowex columns in the $InsP_3$/$InsP_4$ fraction, from reversed phase columns

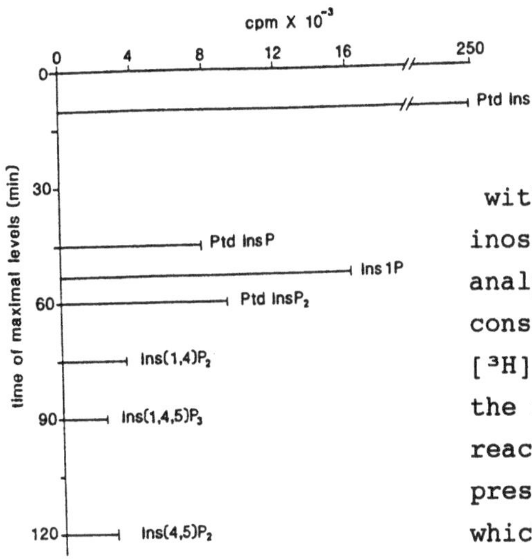

Figure 3. Dynamics of the inositol cycle. Cells were pulse labelled with [3H]Inositol by electroporation. Samples were withdrawn at 15 min interval and inositol-containing compounds were analysed. Since there is considerable secretion of [3H]Inositol containing compounds, the radioactivity in all compounds reaches a maximum. The figure presents the magnitude and time at which the maximal radioactivity in each compound was reached.

in the InsP6 fraction, and elutes only partly from the HPLC SAX column (4). It should be mentioned that we have not yet identified the inositol phospholipids in terms of polar headgroup and fatty acid composition. This is probably relevant, since Dictyostelium seems to lack arachidonic acid (6).

The results (Fig. 3) demonstrate that, after pulse-labelling with [^{3}H]inositol, the radioactivity was very rapidly incorporated into PtdIns; a maximum was obtained after 10 min followed by a decline to half peak levels at about 60 min. Radioactivity was subsequently found in the phospholipids PtdInsP and PtdInsP$_2$ with maxima at respectively 45 and 60 min after pulse labelling. This kinetics is in perfect agreement with the lipid part of the inositol cycle where PtdIns is formed from Ins and CDP-DG, and PtdInsP and PtdInsP$_2$ from the phosphorylation of PtdIns. The first watersoluble inositolphosphate co-chromatographed with Ins1P; the radioactivity was maximal at 50 min after pulselabelling, well before Ins(1,4)P$_2$ was formed. This suggests that there is a phospholipase C activity that acts on PtdIns. The second watersoluble product cochromatographed with Ins(1,4)P$_2$ and reached a maximum at 75 min. Ins(1,4)P$_2$ was formed well before Ins(1,4,5)P$_3$, suggesting that it was not formed from the dephosphorylation of Ins(1,4,5)P$_3$, but from a phospholipase C reaction acting on PtdInsP. [^{3}H]Ins(1,4,5)P$_3$ was the next

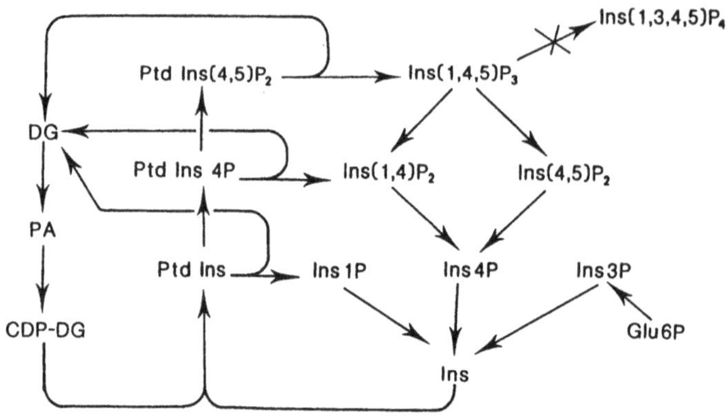

Figure 4. The inositol cycle of Dictyostelium.

watersoluble radioactive compound that could be detected with a maximum at 90 min, suggesting that it was formed from $PtdInsP_2$. Finally, the last compound was $Ins(4,5)P_2$, in accordance with its proposed $Ins(1,4,5)P_3$ source. The kinetics of Ins4P formation is presently unknown, because it was not yet possible to obtain a complete separation of the small amount of Ins4P from the bulk of Ins1P.

The results of the experiments on the metabolism of $Ins(1,4,5)P_3$ *in vitro* and inositol *in vivo* provide a nearly complete picture of the inositol cycle in *Dictyostelium*; this is summarized in Fig. 4.

Stimulation of $Ins(1,4,5)P_3$ production by cAMP and GTPγS

Cells were labelled with [^{3}H]inositol as described above, incubated for 45 min to allow for the incorporation of [^{3}H]inositol into $PtdInsP_2$, permeabilized with saponine (7), and then stimulated with the receptor agonist cAMP or the G-protein agonist GTPγS. Both agonists induce a rapid increase of [^{3}H]$Ins(1,4,5)P_3$ to about 145% of basal levels with a maximum at about 6 s after stimulation; basal levels were recovered at 20-30 s after stimulation (4). The response was also measured in unlabelled cells from which $Ins(1,4,5)P_3$ was extracted and the levels measured by isotope dilution assay (8). The results were similar (Fig. 5), but now absolute $Ins(1,4,5)P_3$ levels were known: the mean intracellular concentration increased from 3.3 uM to a maximum of 5.5 uM. The increase is relatively small, but the absolute increase is considerable, taking into account the

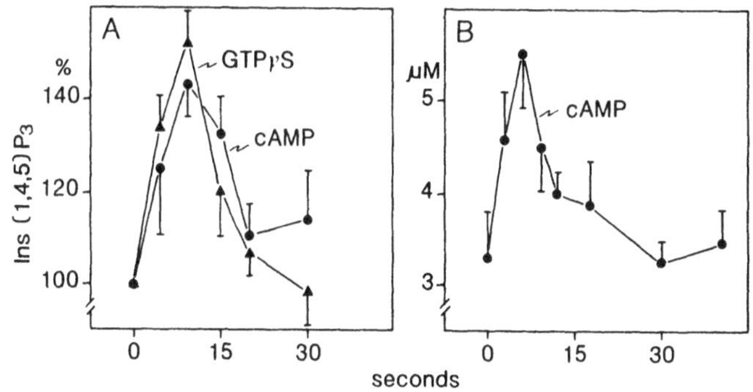

Fig 5. Ins(1,4,5)P_3 levels after stimulation with cAMP or GTPγS. A, relative levels of [^{3}H]Ins(1,4,5)P_3 after stimulation of saponin-permeabilyzed cells with cAMP or GTPγS. B, absolute Ins(1,4,5)P_3 levels of cells after stimulation with cAMP.

submicromolar concentrations of Ins(1,4,5)P_3 that induce Ca^{2+} release.

Since the pool size of Ins(1,4,5)P_3 is about ten times smaller than the pool of PtdInsP_2, it is not unexpected that we have never been able to detect a receptor stimulated decrease of PtdInsP_2. Unfortunately this implies that we have no complete evidence that the increase of Ins(1,4,5)P_3 is derived from a receptor and G-protein stimulated phospholipase C activity.

Function of inositol cycle in Dictyostelium

Several mutants of Dictyostelium were analysed for the receptor and G-protein stimulation of Ins(1,4,5)P_3 production (Van Haastert et al., unpubl. res.). Mutants of the fgd A complementation group (9) are defective in both cAMP and GTPγS-stimulation of Ins(1,4,5)P_3 levels (Table II). Biochemical characterization of this mutant has demonstrated defects in the interaction between receptor and G-protein (10, 11), due to the low expression of a gene that codes for a G-protein alpha subunit (12). These results strongly suggest that the defective G-protein is involved in the receptor mediated activation of phospholipase C. This hypothesis should be confirmed by experiments which measure a GTP-stimulation of phospholipase C activity in vitro. Asuming that the hypothesis is correct, the phenotype of the mutant provides the function of the receptor stimulated inositol cycle in Dictyostelium. The chacteristics of mutant fgdA are summarized in table III. cAMP receptors are present and functional in terms of binding, covalent modification and down-regulation. However, none of the second messenger responses is induced in the mutant, and cAMP does not induce chemotaxis or differentiation. Thus there is a complete blockade of all signal transduction in mutant fgdA. It was therefore unexpected to find that GTP-stimulation of adenylate cyclase was still normal in membranes from the mutant (10). This suggests that the defective G-protein is not directly involved in the regulation of adenylate cyclase. This observation leads to two conclusions: There must be another G-protein, and the G-protein that stimulates phospholipase C is indirectly involved in the regulation of adenylate cyclase.

Table II. cAMP and GTPγS-stimulation of Ins(1,4,5)P_3 levels in mutant fgdA.

Wild-type and mutant cells were labelled with [^{3}H]inositol, permeabilized with saponin and stimulated with cAMP or GTPγS. [^{3}H]Ins(1,4,5)P_3 levels were determined by HPLC.

Stimulus	Ins(1,4,5)P_3 levels (% of control) Wild type	fgd A
cAMP	146±11 *	101±4 NS
GTPγS	148±13 *	102±8 NS

*, significant at P<0.01, n=4; NS, not significant.

Table III. Charaterization of mutant fgdA.

- Surface cAMP receptors are present
- Normal cAMP-induced recptor phosphorylation
- Normal cAMP-induced receptor down-regulation
- No cAMP-induced cAMP formation
- No cAMP-induced cGMP formation
- No cAMP-induced chemotaxis
- No cAMP-induced differentiation
- Normal GTPγS-stimulation of adenylate cyclase
- No GTPγS-stimulation of Ins(1,4,5)P_3 production

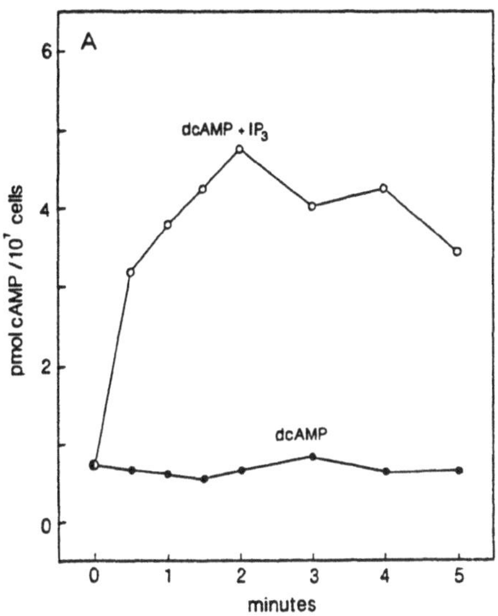

Figure 6. Rescue of adenylate cyclase stimulation in mutant <u>fgdA</u> by Ins(1,4,5)P_3. Sapponin permeabilyzed mutant cells were stimulated with the receptor agonist 2'dcAMP and Ins(1,4,5)P_3 and the accumulation of cAMP levels was measured.

This last hypothesis was tested by the experiment shown in Fig. 6. Mutant <u>fgdA</u> cells were permeabilized with saponin and stimulated with cAMP in the absence or presence of Ins(1,4,5)P_3, and the production of cellular cAMP levels was measured. cAMP or Ins(1,4,5)P_3 alone did not alter cellular cAMP levels, but a strong increase was observed when both compounds were present. Subsequent experiments suggest that cAMP acts on the surface receptors, whereas Ins(1,4,5)P_3 is probably required intracellularly.

The function of the RAS oncogene

<u>Dictyostelium</u> cells have at least two genes that are highly homologous with mammalian RAS protooncogenes (13-15). One gene is present mainly during growth, while the other is present predominantly during starvation when cells are sensitive to extracellular cAMP signals. Cells have been transformed with a vector containing the RAS gene that codes for a protein with a GLY^{12} --> THR^{12} point mutation. These cells have a subtle abberant phenotype, forming multiple tips on the developing fruiting bodies (16). Cells also show an enhanced adaptation of the cAMP-stimulated activation of guanylate cyclase (17). Recently it was shown that Ins(1,4,5)P_3 levels were increased 3-5 fold in the RAS-THR^{12} transformants (18). Using a more accurate assay for the determination of [^{3}H]Ins(1,4,5)P_3 in [^{3}H]inositol labelled cells by HPLC and of absolute Ins(1,4,5)P_3 concentrations by the isotope dilution assay, we reevaluated the

inositol cycle in RAS-THR[12] transformed cells. Indeed, we found that there was a small but significant increase of $[^3H]Ins(1,4,5)P_3$ radioactivity in the RAS-THR[12] cells relative to wild-type cells. However, the absolute $Ins(1,4,5)P_3$ concentration was not significantly different between the cell lines (data not shown). This suggests that the metabolism of $[^3H]$inositol is altered, and that the observed effects are kinetic and not concentration effects. Therefore, cells were pulse labelled with $[^3H]$inositol, and the kinetics of labelling in all inositol-containing compounds was followed (Van der Kaay et al., unpubl. res.). It appeared that the radioactivity in the various compounds was different in the RAS-THR[12] cells than in the control cell line (Table IV). Some compounds had more and other compounds had less radioactivity. The pattern is quite meaningful: more radioactivity was found in PtdInsP, $PtdInsP_2$, $Ins(1,4,5)P_3$, $Ins(1,4)P_2$, and $Ins(4,5)P_2$ whereas less radioactivity was found in PtdIns and Ins1P. These data should be compared with the inositolcycle of *Dictyostelium* (Fig. 3), which suggests that there is more $Ins(1,4,5)P_3$ because there is more

Table IV. Dynamics of the inositol cycle in RAS-THR transformed cells.
Control cells, and RAS-THR[12] transformants were labelled with $[^3H]$inositol, and inositolphospholipids and inositolphosphates were extracted and analyzed. The data shown are at about 45 min after electroporation.

Compound	Fraction in wild-type	Fraction in RAS-THR[12] (as % of wild-type)
inositol	23.0	110 ± 10 NS
Ins1P	2.51	63 ± 17 **
$Ins(1,4)P_2$	0.55	200 ± 51 **
$Ins(4,5)P_2$	0.44	173 ± 51 *
$Ins(1,4,5)P_3$	0.36	163 ± 30 **
PtdIns	39.4	79 ± 3 ***
PtdInsP	1.22	224 ± 34 **
$PtdInsP_2$	1.44	151 ± 37 *

Significance (n=4): *, $P<0.05$; **, $P<0.01$; ***, $P<0.001$

$PtdInsP_2$, and there is more $PtdInsP_2$ because there is more PtdInsP; the same is true for $Ins(1,4)P_2$ and $Ins(4,5)P_2$. It also appears that there is less Ins1P because there is less PtdIns. The only conversion that shows a completely different behavior in the $RAS\text{-}THR^{12}$ transformant relative to the wild-type cells is the conversion of PtdIns to PtdInsP, which seems to proceed about three fold faster in the $RAS\text{-}THR^{12}$ cells than in the wild-type cells. These results imply that either the PtdIns kinase shows higher activity or that the substrate PtdIns (formed in the ER) is converted more rapidly to the enzyme (present in the plasma membrane).

These results nicely show that it is not always easy to find the altered step in a complex cyclic metabolic pathway; they clearly demonstrate that measurements of $Ins(1,4,5)P_3$ levels alone are not sufficient.

Conclusions

The microorganism *Dictyostelium* has an inositol cycle that is reminicent to the inositol cycle of mammalian cells. The main differences are the metabolism of $Ins(1,4,5)P_3$: i) the phosphorylation to $Ins(1,3,4,5)P_5$ seems to be absent (and thus also the complex metabolism of this compound); ii) the dephosphorylation of $Ins(1,4,5)P_3$ is more complex and involves three additional enzyme activities that have not yet been described in mammalian cells. The inositol cycle is probably the major signal transduction pathway in *Dictyostelium* as was demonstrated with mutant *fgdA* that presumably lacks the functional G protein to activate phospholipase C.

We have not yet completely identified the inositol cycle of *Dictyostelium*. During the analysis of the watersoluble compounds we detected large quantities of radioactivity that could not be identified; these compounds may contain upto 50% of the watersoluble radioactivity. Furthermore, it is still largely unknown how $InsP_6$ is formed in *Dictyostelium*. Finally, the identification of some compounds is badly needed, such as of the fatty acid and polar headgroup of the inositolphospholipids. Whereas the identification of the inositol cycle may be on its way, the identification and characterization of the participating enzymes are still in its infancy. The enzyme phospholipase C has not been detected yet in *in vitro* assays (19). Finally, the precise function of Ins(1,4,5)P_3 in the cell is not completely known; it releases Ca^{2+} from non-mitochondrial stores (7), but other functions cannot be excluded. This is suggested by the complex dephosphorylation of Ins(1,4,5)P_3. These details of the inositol cycle may be relevant since the mutant fgdA suggests that the inositol cycle rules signal transduction, chemotaxis and differentiation in Dictyostelium.

Acknowledgements
This study was supported by a grant of the C. and C. Huygens Fund which is subsidized by the Netherlands Organization for Scientific Research.

References

1. Janssens, P.M.W., and Van Haastert, P.J.M. (1987). Molecular basis of transmembrane signal transduction in Dictyostelium discoideum. Microbiol. Rev. 51, 396-418.
2. Klein, P., Sun, T.J., Saxe, C.L., Kimmel, A.R., and Devreotes, P.N. (1989). A chemoattractant receptor controls development in Dictyostelium discoideum. Science, 241, 1467-1472.
3. Van Lookeren Campagne, M.M., Erneux, C., Van Eijk, R., and Van Haastert, P.J.M. (1988). Two dephosphorylation pathways of inositol 1,4,5-trisphosphate in homogenates of the cellular slime mould Dictyostelium discoideum. Biochem.J. 254, 343-350.
4. Van Haastert, P.J.M., De Vries, M.J., Penning, L.C., Roovers, E., Van der Kaay, J., Erneux, C., and Van Lookeren Campagne, M.M. (1989). Chemoattractant and guanosine 5'-[.-thio]triphoshate induce the accumulation of inositol 1,4,5-trisphosphate in Dictyostelium cells that are labelled with [^{3}H]inositol by electroporation. Biochem.J. 258, 577-586.
5. Schoen, C.D., Arents, J.C., Bruin, T., and Van Driel, R. (1989). Intracellular localization of secretable cAMP in relaying Dictyostelium cells. Exp. Cell Res. 181, 51-62.
6. MacDonald, J.I.S., and Weeks, G. (1985). The biosynthesis and turnover of lipid during the differentiation of Dictyotelium discoideum. Biochim.Biophys.Acta 834, 301-307.

7. Europe-Finner, G.N., and Newell, P.C. (1986). Inositol 1,4,5-trisphosphate induces calcium release from a non-mitochondrial pool in amoebae of Dictyostelium. Biochim.Biophys.Acta 887, 335-340.

8. Van Haastert, P.J.M (1989) Determination of inositol 1,4,5-trisphosphate levels in Dictyostelium by isotope dilution assay. Anal. Biochem. 177, 115-119.

9. Coukell, M.B., Lappano, S., and Cameron, A.M. (1983). Isolation and characterization of cAMP unresponsive (frigid) aggregation-deficient mutants of Dictyostelium discoideum. Dev. Genet. 3, 283-297.

10. Kesbeke, F., Snaar-Jagalska, B.E., and Van Haastert, P.J.M. (1988). Signal transduction in Dictyostelium fgdA mutants with a defective interaction between surface cAMP receptor and a GTP binding regulatory protein. J. Cell Biol. 107, 521-528.

11. Snaar-Jagalka, B.E., Kesbeke, F. Pupillo, M., and Van Haastert, P.J.M. (1988). Immunological detection of G protein α subunits in Dictyostelium discoideum. Biochem. Biophys. Res. Commun. 156, 757-761

12. Kumagai, A., Pupillo, M., Gunderson, R., Mike-Lye, R., Devreotes, P.N., and Firtel, R.A. (1989). Regulation and function of G_{α} protein subunits in Dictyostelium. Cell, 57, 265-275.

13. Reymond, C.D., Gomer, R.H., Mehdy, M.C., and Firtel, R.A. (1984). Devolopmental regulation of a Dictyostelium gene encoding a protein homologous to mammalian ras protein. Cell 39, 141-148.

14. Pawson, T., Amiel, T., Hinze, E., Auersperg, N., Neave, N., Sobolewski, A., and Weeks, G. (1985). Regulation of a ras-related protein during development od Dictyostelium discoideum. Mol. Cell. Biol. 5, 33-39.

15. Robbins, S.M., Williams, J.G., Jermyn, K.A., Spiegelman. G.B., and Weeks, G. (1989) Growing and developing Dictyostelium cells express different ras genes. Proc. Natl. Acad. Sci. USA. 86, 938-942.

16. Reymond, C.D., Gomer, R.H., Nellen, W., Theibert, A., Devreotes, P.N., and Firtel, R.A. (1986). Phenotypic changes induced by a mutated ras gene during the devolopmental of Dictyostelium transformants. Nature, Lond. 323, 340-343.

17. Van Haastert, P.J.M., Kesbeke, F., Reymond, C.D., Firtel, R.A., Luderus, E., and Van Driel, R. (1987) Aberant transmembrane signal transduction in Dictyostelium cells expressing a mutated ras gene Proc. Natl. Acad. Sci. USA. 84, 4905-4909.

18. Europe-Finner, G.N., Luderus, M.E.E., Small, N.V., Van Driel, R., Reymond, C.D., Firtel, R.A., and Newell, P.C. (1988). Mutant ras gene induces elevated levels of inositol tris- and hexakisphosphate in Dictyostelium. J. Cell Sci. 89, 13-20.

19. Irvine, R.F., Letcher, A.J., Brophy, P.J., and North, M.J. (1980). Phosphatidylinositol-degrading enzymes in the cellular slime mould Dictyostelium discoideum. J. Gen. Microbiol. 121, 495-487.

DISCUSSION

Iyengar: Do you think that the Gα-2 is working directly on the phospholipase C and on adenylate cyclase, or do you think that the IP3 feeds back to stimulate cAMP?

Van Haastert: I think that there is a GTP-binding protein, but I don't know whether I have to call that a G-protein that regulates adenylate cyclase. We know from other mutants, for instance mutant Synag 7, that there is also another cytosolic protein involved in the regulation of adenylate cyclase, which is absent in mutant Synag 7. It may be that part of the inositol cycle, - IP3 itself, one of its degradation products or effector products -, is in fact making the communication possible between the receptor, the GTP-binding protein, the cytosolic factor and the adenylate cyclase. So I think it is more complicated than for instance in the β-adrenergic system.

Houslay: You showed with the frigid A mutant that you could get cAMP sensitivity back by adding IP3. Did you do 2 other things? 1. Did you try elevating calcium and 2. did you add diacylglycerol instead of IP3?

Van Haastert: We did not add calcium because calcium inhibits the activation of the enzyme, so that makes the experiment a little bit difficult. I know that I should do it, but I just don't like the experiment. The other experiment, we didn't add diacylglycerol.

Simon: What does Gα1 do?

Van Haastert: Perhaps I can feed back that question to Peter.

Devreotes: We don't know really, but when you overexpress it about 20-fold cells are multinucleated.

Simon: So it is not cross talking at all with Gα2, or providing a stimulation of adenylyl cyclase, but it seems to be involved with the morphology of the growth control.

Devreotes: Gα1 is expressed constitutively, whereas Gα2 is expressed after starvation at the same time as the receptor and adenylate cyclase. So I cannot say that Gα1 is not involved in activation of adenylate cyclase, but it does not seem to be that way. We haven't got good antisense of Gα1 or Gα2.

Simon: But in the absence of Gα2, Gα1 cannot provide either the IP3 or cAMP response.

Mato: Is the defect in the RAS transvected cells specific for the PI-lipids or also for other phospholipids like phosphatidylcholine? What happens with the levels of diacylglycerol in those cells?

Van Haastert: These are quite recent results, we have not yet measured diacylglycerol levels. We have briefly analysed the other phospholipids in ^{32}P-labeled cells, but if you want to do that completely satisfying, you should also measure the specific activity of ^{32}P in ATP and the kinetics of labelling, and we have not done that. So, as far as the data go, there is no difference in the other phospholipids.

Cooke: If arachidonic acid is not present in the phospholipids, does that mean that the whole of the phospholipase A2 prostaglandin leukotrien system is not present in your cells?

Van Haastert: One of the problems in assaying phospholipase C activity is the very high activity of phospholipase A; I don't know whether it is A1 or A2 or both. So if you incubate a homogenate with radioactive Ptd Ins P2 then the first water soluble product is glycerolphospoinositol. This product is not found if the reaction starts with Ptd Ins P so there is first PLA reaction(s) and then a dephosphorylation.

Cooke: But arachidonic acid is missing?

Van Haastert: Yes.

Cooke: But you don't know which.

Van Haastert: No, that is one of the experiments that we want to do in the near future; to completely identify the Ptd Ins P2 in terms of fatty acids and polar head group.

Downes: Can I just turn you back to the experiment where you looked at the appearance of label in each of the different inositol compounds. What you measured was just the amount of radioactivity in each one?

Van Haastert: Yes.

Downes: Forgive me when I am wrong, but it seems to me that you cannot conclude from that, which was labeled first. You need to know the specific radioactivity of each one or some other feature which relates not just the radioactivity, but the mass of that compound; I mean if the masses are very different you have no basis to conclude in which compound the label is first.

Van Haastert: All these compounds are labeled to a maximum and then decline again. So they are all transient.

Downes: So in that case the data that are relevant are basically the whole time course of each.

Van Haastert: Yes, of course; but if I draw a whole time course of all compounds the audience would not see anything.

Downes: So what you gave, was the peak time.

Van Haastert: The bell-shaped curve for each of the compounds is similar.

Downes: That was what I was trying to get out in terms of whether or not the label was incorporated to a maximum or not. Secondly, you had Ins 1 P as the first IP-product, did you demonstrate that it was indeed Ins 1 P or could it be some other isomer?

Van Haastert: It could have been another isomer, but it is not Ins 4 P, because it elutes at the position of Ins 1 P, which is different from the elution of Ins 4 P.

Downes: But you only know that it is different from Ins 4 P.

Van Haastert: That is correct; if you have the hypothesis that it is Ins 5 P, I cannot exclude that possibility.

Downes: I don't have a hypothesis; I only want to say that if it is not Ins 1 P it clearly cannot have come from Ptd Ins.

Devreotes: Do you know anything about the timing of the rescue of frigid A. Can you add IP3, then wash and then add cAMP, or must IP3 and cAMP to be added at the same time? The order might tell you how IP3 might be working.

Van Haastert: No I have no indication about the kinetics of these responses. The only thing I know is, that the receptor agonist or IP3 alone don't work.

Iyengar: What is the nature of the cytosolic factor that is missing in synag 7; are you looking for a β-subunit?

Devreotes: There is a β-subunit actually, because the sequence looks very much like the human one, but it is on a different linkage group than the mutation of synag 7, so it cannot be the cytosolic factor.

Iyengar: So you are looking at another component after the G-protein before it reaches the cyclase; is that the take home message?

Van Haastert: Yes, it looks lake that. Ewa Snaar-Jagalska has done experiments with the mutant and its reconstitution with wild-type cytosol. The factor appears to function as an activator of GTPτS-binding and GTP-ase activity, more or less as GAP in the RAS-protein. That is how the experiments worked out. I don't want say that it has something to do with RAS or GAP.

Caron: I think I have missed the point from your comment that cAMP and IP3 have to be added from the outside and these are whole-cell experiments; does that imply that there is a IP3 receptor on the cell?

Van Haastert: No, the experiment was done with permeabilized cells.

Caron: Sorry, so that is what I missed.

Van Haastert: IP3 probably has to enter the cells but the cAMP has to bind to the cell surface receptor.

Simon: Without getting too far, but do you think it is possible that a GAP-like protein is operating on the GTP-ase activity of the G-protein or of the RAS system? I shouldn't have asked it, sorry.

Van Haastert: Of course you can ask me. If I do not like the question, I just do not answer it. It is hard to say, because, the RAS protein in *Dictyostelium* does not appear to be involved in the regulation of adenylate cyclase. This cytosolic protein, however, is clearly involved in the regulation of adenylate cyclase. So I think that this cytosolic protein is not interacting with RAS. But that is all I can say about it.

Houslay: Is this cytosolic protein related to ARF or another GTP-binding protein.

Devreotes: I think Peter has done some experiments, showing that it is not. It is not a GTP-binding protein or the site of GTP-action is not on that protein.

Van Haastert: Well you remember my experiments better than I do. What comes out of the experiment is, that the activity that stimulates adenylate cyclase in a GTP-dependent manner, is in the membranes and not in the cytosol. When you add GTPτS to mutant membranes you get no activation. When you remove all GTPτS, then add wild-type cytosol you will get activation. In another experiment, if you add GTPτS to the cytosol, incubate, rapid spin through a gel filtration column, then add that back to native membranes of the mutant you get no stimulation of cyclase. That indicates that the target of GTP stimulation of adenylate cyclase is in the membranes; it does not exclude the possibility that this cytosolic fraction is still a GTP-binding protein.

Iyengar: Let us assume that there is another protein that is required for the activation of cyclase, is it the same Gα that is involved in interacting with this protein as is involved in activating PLC; so is the bifurcation of PLC and cyclase pathway occurring at the level of Gα?

Van Haastert: We know from the mutant analysis that the frigid A mutant is completely normal in GTP stimulation of cyclase in membranes, so that path is correct. The synag 7 mutant is totally defective in GTP stimulation of adenylate cyclase, but is completely normal in the receptor and GTPτS stimulation of IP3 production. So it appears to us that this cytosolic factor does not interact with the pathway that is essential for IP3 production. I think that the bifurcation between PLC and adenylate cyclase is before the G protein. Peter and I always discuss whether it is at the beginning, so that there are different receptors for stimulating adenylate cyclase and PLC. I think there are different receptors, but the molecular genetics should give the final answer.

Cyclic nucleotide-coupled systems

Glucagon activates two distinct signal transduction systems in hepatocytes, which leads to the desensitization of G-protein-regulated adenylate cyclase, the phosphorylation and inactivation of G_i-2 and the phosphorylation and stimulation of a specific cyclic AMP phosphodiesterase

Miles D. Houslay, Mark Bushfield, Elaine Kilgour, Brian Lavan, Suzanne Griffiths, Nigel J. Pyne, Eric K.-Y. Tang & Gregory J. Murphy

Molecular Pharmacology Group, Department of Biochemistry, University of Glasgow, GLASGOW G12 8QQ, Scotland, UK

Summary

A transient increase in the intracellular concentrations of cyclic AMP occurs as a result of the challenge of hepatocytes with glucagon. This event is determined by the initial rapid activation of adenylate cyclase, which is responsible for the production of cyclic AMP within the cell. Following on from this we observe the desensitization of adenylate cyclase; the A-kinase-mediated activation of the 'dense-vesicle', high affinity cyclic AMP phosphodiesterase; the phosphorylation and functional inactivation of the inhibitory G-protein G_i-2 and the establishment of a 'selective' insulin-resistant state. These events identify 'interplay' or 'cross-talk' occurring between distinct cellular signalling systems.

These normal physiological controls are identified as being affected in the insulin resistant states that characterize the pathophysiological states conditions of diabetes and obese.

The locus of the glucagon-induced desensitization phenomenon affecting adenylate cyclase is at the the coupling interface between the glucagon receptor and the stimulatory G-protein G_s

We, and others, have shown that challenge of intact hepatocytes with glucagon leads [see 1] to a transient increase in the intracellular accumulation of cyclic AMP (fig. 1) which was not accompanied by any marked efflux of cyclic AMP from these cells. If glucagon was administered in the presence of inhibitors of cyclic AMP phosphodiesterase activity then, whilst the magnitude of accumulation of cyclic AMP was amplified, cyclic AMP production appeared to almost cease some 5 minutes later. The lesion which gives rise to such phenomena must therefore lie primarily at the site of production of cyclic AMP, namely with the control of adenylate cyclase. This is desensitization.

We wished then to determine if challenge of hepatocytes with glucagon

NATO ASI Series, Vol. H 44
Activation and Desensitization of Transducing Pathways
Edited by T. M. Konijn, M. D. Houslay, P. J. M. Van Haastert

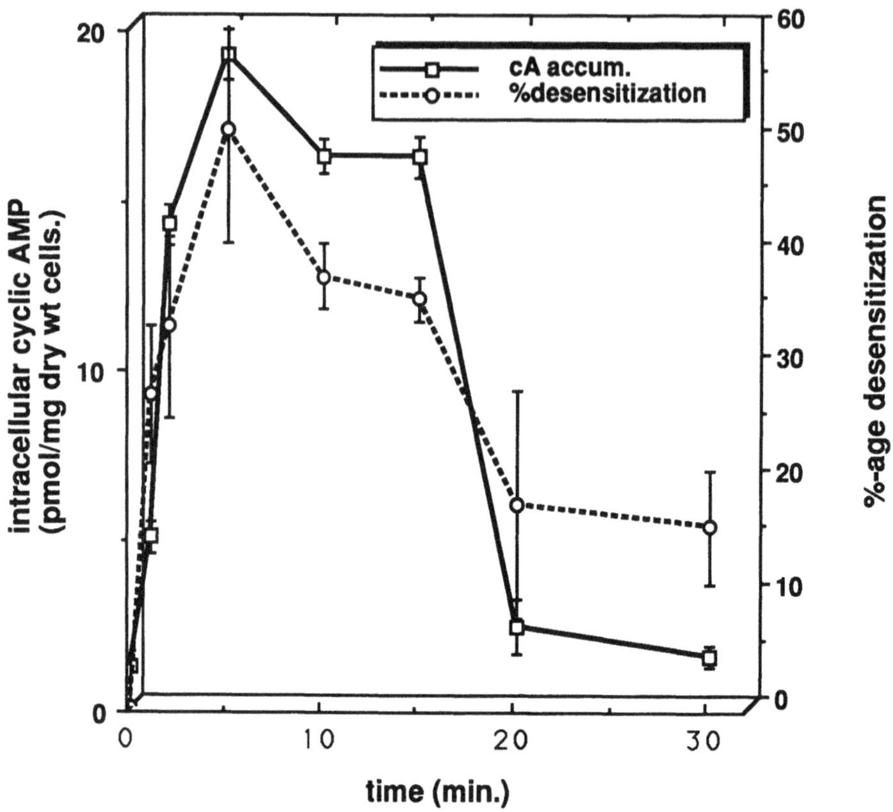

Figure 1. Demonstrates the effect of challenge of intact hepatocytes with 10nM glucagon on the intracellular accummulation of cyclic AMP and the degree of desensitization.

could lead to any stable, and thus identifiable, change in adenylate cyclase function which would allow us to characterize the desensitization process. In order to try an identify this, hepatocytes were first challenged with glucagon (10nM) for various times, harvested, disrupted and a washed membrane fraction prepared for the assay of adenylate cyclase (fig. 2). Using such washed membrane fractions, we were able to determine [1] that the ability of glucagon to stimulate the adenylate cyclase activity in membranes from glucagon-pre-treated cells was markedly reduced when compared with the activity observed in washed membranes from hepatocytes which had not been pre-treated with glucagon (control cells). The reduction in the ability of glucagon to stimulate adenylate cyclase activity in membranes from glucagon pre-treated cells was about 50% (cf. control), despite there being no change in basal (unstimulated) adenylate cyclase activity. This indicated that the desensitization process was due to a functional

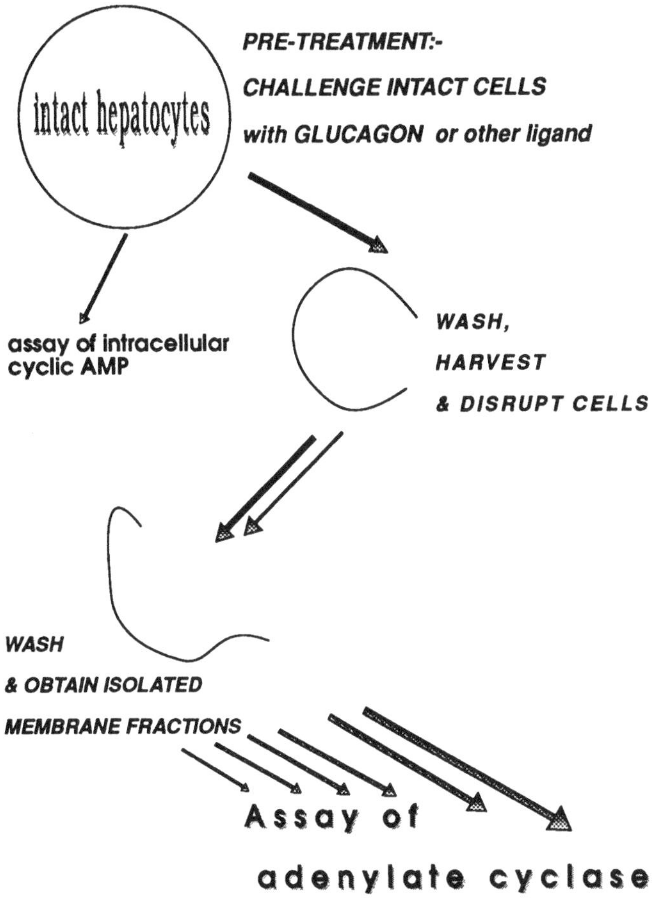

Figure 2. Shows the scheme used for pre-treating hepatocytes with hormones prior to assessing intracellular cyclic AMP concentrations and the preparation of a washed membrane fraction.

uncoupling of the glucagon receptor from adenylate cyclase stimulation. Such a process was shown by us [1] to be both time-dependent, with a $t_{0.5}$ of 1.5-2min and dose-dependent, with an EC_{50} of $4x10^{-10}$M for the action of glucagon.

The molecular lesion which gives rise to the desensitization process could occur at a number of points, with some possibilities being summarized in fig.3. We have assessed the possibility of lesions occurring at the catalytic unit of adenylate cyclase by assessing either forskolin-stimulated or basal activity of this enzyme, with no detectable changes being observed. The coupling of G_s to adenylate cyclase has been assessed by measuring Na(Al)F-stimulated activity, again with no detectable changes being noted. The lesion thus appears to be sited at the coupling of the glucagon receptor to G_s. This could result from either loss of receptor binding or receptor internalization, however, evaluation of both of these various possibilities showed quite clearly that both receptor numbers on the plasma membrane and their ability to bind glucagon are unaffected over the timescale of the desensitization process [1]. The lesion which leads to desensitization appears thus to be due to a modification of the coupling process which links the glucagon receptor to G_s. This could reflect the attenuation of the functioning of either the glucagon receptor or the stimulatory G-protein G_s, which serves to couple the glucagon receptor to adenylate cyclase. A further possibility is that the inhibitory G-protein G_i could be activated constitutively, although we show below (*vide infra*) that this is not the case.

Elevations in the intracellular concentrations of cyclic AMP do not trigger the desensitization of adenylate cyclase but can attenuate the rate of resensitization

The glucagon-mediated desensitization of adenylate cyclase cannot be

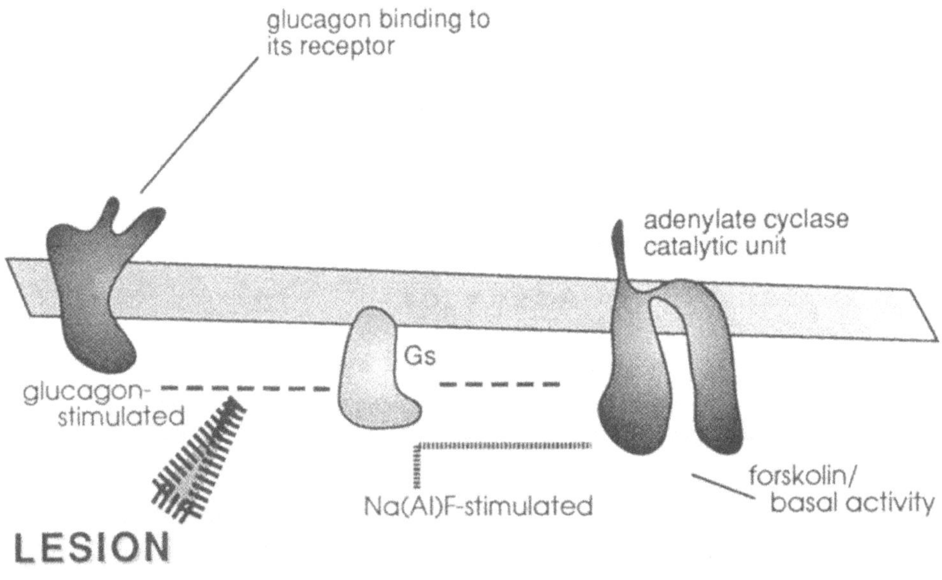

Figure 3. Schematic showing the point of the lesion occurring in desensitization at the glucagon receptor-Gs coupling interface. Also the components and means of assessing individual function.

elicited by the addition of permeant analogues of cyclic AMP, such as dibutyryl cyclic AMP and 8-bromo cyclic AMP. This is despite the fact that when such compounds are given to intact hepatocytes they can activate processes mediated through protein kinase A [1]. Furthermore, when the cyclic AMP phosphodiesterase inhibitor IBMX was added [1] to glucagon-treated hepatocytes, then despite the fact that cyclic AMP concentrations were some 5-fold greater than in cells which were not treated with this agent, the half-time for the onset of desensitization was similar ($t_{0.5}$ of 1.5-2min).

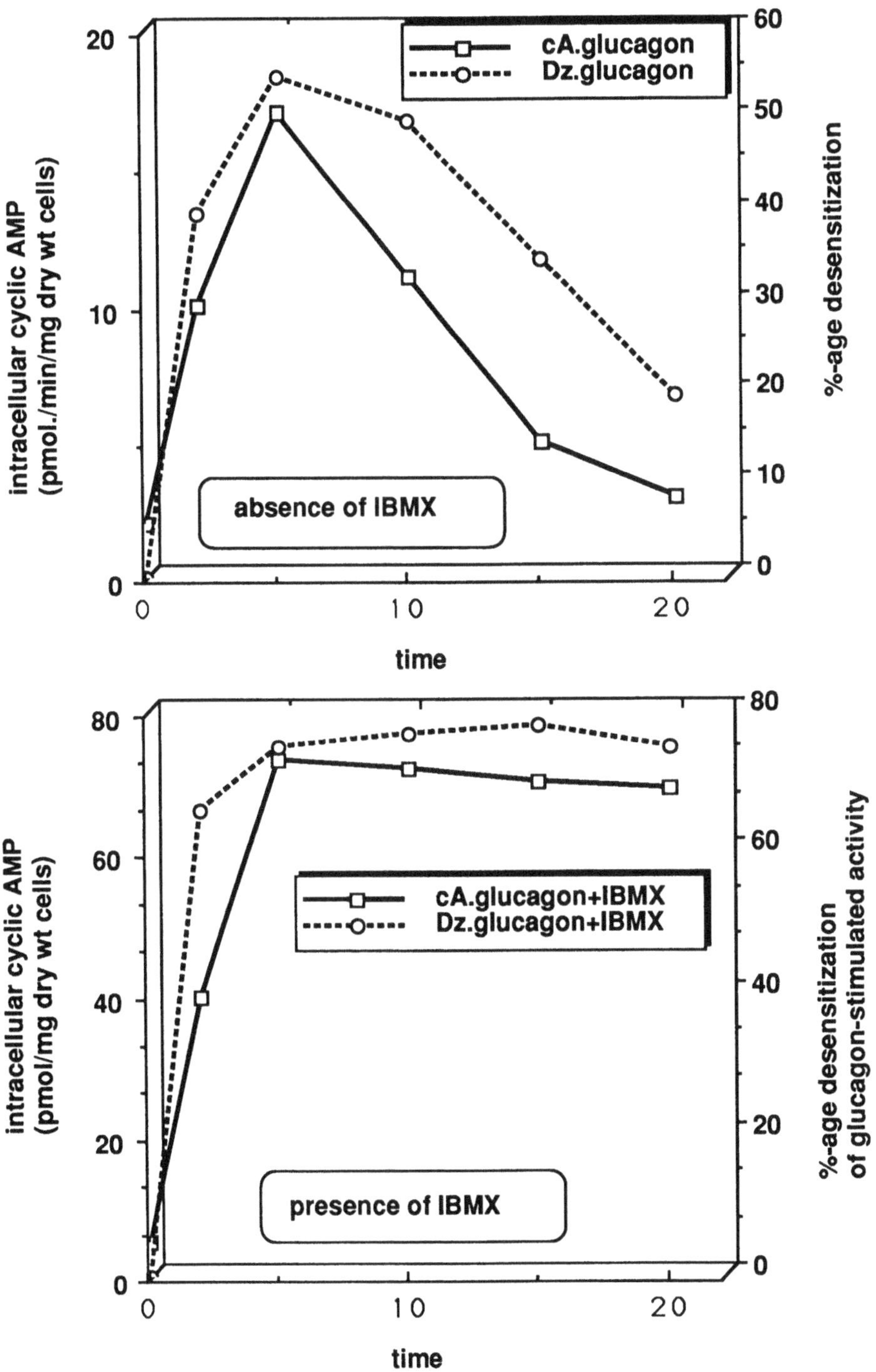

Figure 4. Shows the accummulation of cyclic AMP and the degree of desensitization seen in hepatocytes treated with 10nM glucagon. In the right hand panel 1mM-IBMX was also present.

We did note, however, that the desensitization of adenylate cyclase, elicited by challenge of hepatocytes with glucagon, appeared in itself to be transient (fig. 1), with a slow resensitization phase ensuing [1]. Such an observation did not seem to be consistent with our observation that the glucagon-mediated increase in intracellular cyclic AMP almost ceased after 5 minutes, in IBMX-treated hepatocytes. In order to investigate this we determined the degree of desensitization observed in hepatocytes treated with glucagon in the presence of IBMX. This showed quite clearly that under such conditions that the resensitization process was blocked (fig. 4). Such an action was not unique to IBMX and could be mimicked by the compound Ro-20-1724, a non-methyl xanthine inhibitor of cyclic AMP phosphodiesterase activity. By carrying out experiments at different concentrations of these inhibitors, we were able to determine that the degree of resensitization of adenylate cyclase was inhibited as a function of the intracellular concentration of cyclic AMP with EC_{50} values of 22-26µM [2]. Whilst phosphodiesterase inhibitors can elicit this blockade of the resensitization effect in the presence of glucagon, they are incapable of doing this when the ligands used to elicit desensitization (*vide infra*) are not able to activate adenylate cyclase [2]. This is because, under such conditions, intracellular cyclic AMP concentrations are not elevated abover basal levels (circa 1µM).

The glucagon-mediated desensitization of adenylate cyclase is elicited via the activation of C-kinase

That permeant cyclic AMP analogues cannot elicit desensitization, together with the fact that the EC_{50} values for the glucagon-mediated desensitization of adenylate cyclase were an order of magnitude lower than those noted for glucagon's ability to both activate adenylate cyclase and to increase the intracellular concentration of cyclic AMP, suggests very strongly that desensitization is mediated by a cyclic AMP-independent process [1].

This we have shown quite clearly [3] using [1-*N*-α-trinitro-phenylhistidine, 12-homoarginine] glucagon, which is an analogue of glucagon called TH-glucagon. This ligand is capable of eliciting desensitization in intact hepatocytes at concentrations as low as 10^{-10}M, yet is incapable of either activating adenylate cyclase or increasing the intracellular concentration of cyclic AMP even at concentrations as high as 10^{-5}M [3].

A clue as to the molecular mechanism of desensitization was provided by our observations [4] that treatment of intact hepatocytes with the tumour-promoting phorbol ester TPA appeared to uncouple adenylate cyclase activity from stimulation by glucagon and thus to mimic the desensitization process. Now TPA can activate protein kinase C, by virtue of its ability to mimic the action of the natural agonist diacyl glycerol (DAG) and indeed we can now show that exposure of hepatocytes to synthetic diacylglycerols can indeed mimick the desensitization process.

One well-established route for the production of DAG is through the stimulation of inositol phospholipid metabolism to produce IP_3 (inositol 1,4,5 trisphosphate) and DAG. We [5] and others [6,7, 7a] have been able to show that both glucagon and TH-glucagon cause a significant, albeit small, stimulation of inositol phospholipid metabolism, which leads to the production of IP_3 and IP_2 and the enhanced metabolism of the parent phospholipid PIP_2. We (unpublished) and others [8] have now indeed shown that both glucagon and TH-glucagon can cause a rise in the intracellular concentrations of DAG (stearoyl, arachidonyl). Thus we suggest that glucagon and TH-glucagon cause the production of diacyl glycerol and that this leads to the desensitization of adenylate cyclase.

Consistent with such a concept, we were able to show that treatment of intact hepatocytes with either vasopressin or angiotensin, both of which markedly stimulate inositol phospholipid metabolism in hepatocytes, can also mimic glucagon's ability to desensitize adenylate cyclase [3]. Importantly, challenge of hepatocytes with these ligands at concentrations where they elicited comparable effects on inositol phospholipid metabolism to glucagon achieved the desensitization effect [3]. Nevertheless, whilst glucagon appears to stimulate inositol phospholipid metabolism to levels which are 10-20 fold less than is seen with either vasopressin or angiotensin, under maximal stimulatory conditions, glucagon appears to increase intracellular DAG concentrations to values which are some 50% of that found after challenge with either vasopressin or angiotensin. Thus glucagon might also stimulates the breakdown of another phospholipid to elevate DAG levels, possibly phosphatidylcholine. Alternatively, the elevated cyclic AMP concentrations found with glucagon might attenuate the breakdown of DAG and allow it to accumulate to higher levels.

In addition, we have shown that treatment of hepatocytes with the Ca^{2+}-ionophore A23187 was also shown to be capable of mimicking the desensitization process[9]. Presumably this was because A23187 both stimulates inositol phospholipid metabolism[10] and can also be expected to activate C-kinase by increasing intracellular Ca^{2+} concentrations.

Our data has provided the first evidence towards a molecular explanation for various observations [7,11-13] which have implied that glucagon might exert effects upon target cells which were independent of any elevation of the intracellular concentration of cyclic AMP. It might, in part, account for the well-established observations that glucagon can increase the intracellular concentrations of Ca^{2+}. However, the molecular basis of such an action has led to considerable controversy. Nevertheless, there now seems to be good evidence [14] that glucagon mediates an increase in intracellular Ca^{2+} by both cyclic AMP-depedent and independent processes. The magnitude of the involvement of stimulation of inositol phospholipid, either directly or indirectly, and any role of G_s-coupled Ca^{2+} -channels has yet to be established.

Pertussis toxin blocks the glucagon-mediated desensitization of adenylate cyclase through a process which does not appear to involve the inhibitory G-protein G_i

Adenylate cyclase is believed to be under the control of two distinct G-proteins, G_s which mediates stimulatory phenomena and G_i which mediates inhibitory phenomena [15]. Treatment of hepatocytes with pertussis toxin causes the ADP-ribosylation of G_i coupled with its functional inactivation [16].

Exposure of intact hepatocytes to pertussis toxin also prevents the glucagon-mediated desensitization of adenylate cyclase [16,17]. One interpretation of this could be because desensitization was caused by the constitutive activation of G_i, perhaps mediated through an action of C-kinase.

We, however, consider this to be unlikely as such a mechanism would be expected to lead to the decreased functioning of both forskolin-stimulated and basal adenylate cyclase activity. This clearly does not occur [13]. Nevertheless, we have attempted to analyse this possibility rigorously using two different model systems. The first of these employed hepatocytes from diabetic animals which we have shown not to exhibit a functional G_i [18]. In these hepatocytes desensitization occurred exactly as found with hepatocytes from normal animals [17]. The second model that we employed was hepatocytes from immature animals [17]. In this instance levels of G_i in plasma membranes were only some 50% of those found in hepatocyte membranes from mature animals. This reduction in G_i expression is reflected in a comparable reduction in the ability of G_i to inhibit adenylate cyclase activity. Nevertheless, glucagon was found to be equally as effective in eliciting desensitization in hepatocytes from these animals as those from mature animals [17]. These experiments all militate against the constitutive activation of G_i as mediating the desensitization process. Indeed, as we shall discuss below (*vide infra*), treatment of intact hepatocytes with glucagon in fact leads to the functional inactivation of G_i [17,19].

Intriguingly, however, pertussis toxin was still capable of blocking the desensitization process in hepatocytes from diabetic animals which lacked functional G_i [17]. This implied that pertussis toxin was not exerting its action by modifying G_i. We assessed [17] the possibility that the binding (B-subunit) of pertussis toxin might be mediating these effects on the desensitization process by treating hepatocytes from normal animals with the isolated B-subunit. However, this did not block the desensitization process, although it did appear to attenuate the resensitization process.

The mechanism whereby pertussis toxin exerts its inhibitory effect on desensitization is thus unclear. Certainly, it is not restricted to glucagon-mediated desensitization of adenylate cyclase, as desensitization elicited by either vasopressin or angiotensin could be blocked similarly [17], indicating that the site of action is at a point in the mechanism which is common to all of these hormones. This action is not due to any inhibitory effect of pertussis toxin on the stimulation of inositol phospholipid metabolism by either vasopressin, angiotensin or glucagon. For, under conditions where this toxin caused the ADP-ribosylation of G_i in intact hepatocytes, such responses were unaffected [17].

Indeed, this experiment also indicates that not only is the stimulation of inositol phospholipid metabolism not affected by pertussis toxin in hepatocytes but that the two forms of G_i found in hepatocytes, namely G_i -2 and G_i -3, cannot serve as G-proteins which couple these various receptors to the stimulation of phospholipase C in these cells.

One possibility then is that pertussis toxin might elicit its blockade of the desensitization process by inhibiting a species of C-kinase involved in mediating desensitization.

Treatment of intact hepatocytes with TPA, glucagon and hormones which stimulate inositol phospholipid metabolism causes the phosphorylation of G_i-2 and the functional inactivation of "G_i"

Hepatocytes contain a functional" G_i" whose activity can be identified by the ability of either high concentrations of GTP or low concentrations of its non-hydrolysable analogue guanylyl 5'-imidodiphosphate (p[NH]ppG) to inhibit adenylate cyclase activity [16,19]. As basal adenylate cyclase activity is relatively small, then experiments are performed in the presence of the diterpene forskolin, which stimulates the catalytic unit of adenylate cyclase directly. In intact hepatocytes, P_2-purinergic receptors appear to be functionally coupled to G_i [20].

There are three known forms of pertussis toxin sensitive G_i, these are called G_i-1, G_i-2 and G_i-3. Using specific anti-peptide antibodies and Northern blotting techniques with specific 34-mer oligonucleotide probes, we can show that both G_i-2 and G_i-3, but not G_i-1, are expressed in hepatocytes.

Treatment of intact hepatocytes with pertussis toxin causes the ADP-ribosylation of these two forms of G_i and the inactivation of functional G_i as assessed by the three criteria stated above [16]. To date, however, it is not known unequivocally whether one or both of these forms of G_i serves to mediate inhibition of adenylate cyclase. However, as stated below, we have some evidence which suggests that this action may be mediated by or at least involve G_i-2. Indeed, the molecular mechanisms responsible for the expression of so-called "G_i" activity are unresolved. Certainly, there is very good evidence to show that the release of β–γ subunits from G_i can inhibit the functioning of G_s by attenuating its ability to be dissociated. However, this does not appear to explain how inhibition of adenylate cyclase occurs in systems which lack G_i, such as cyc- cells, or inhibition of forskolin-stimulated adenylate cyclase at (low) concentrations of p[NH]ppG which are insufficient to activate G_s [15].

When intact hepatocytes are treated with the phorbol ester TPA, this leads

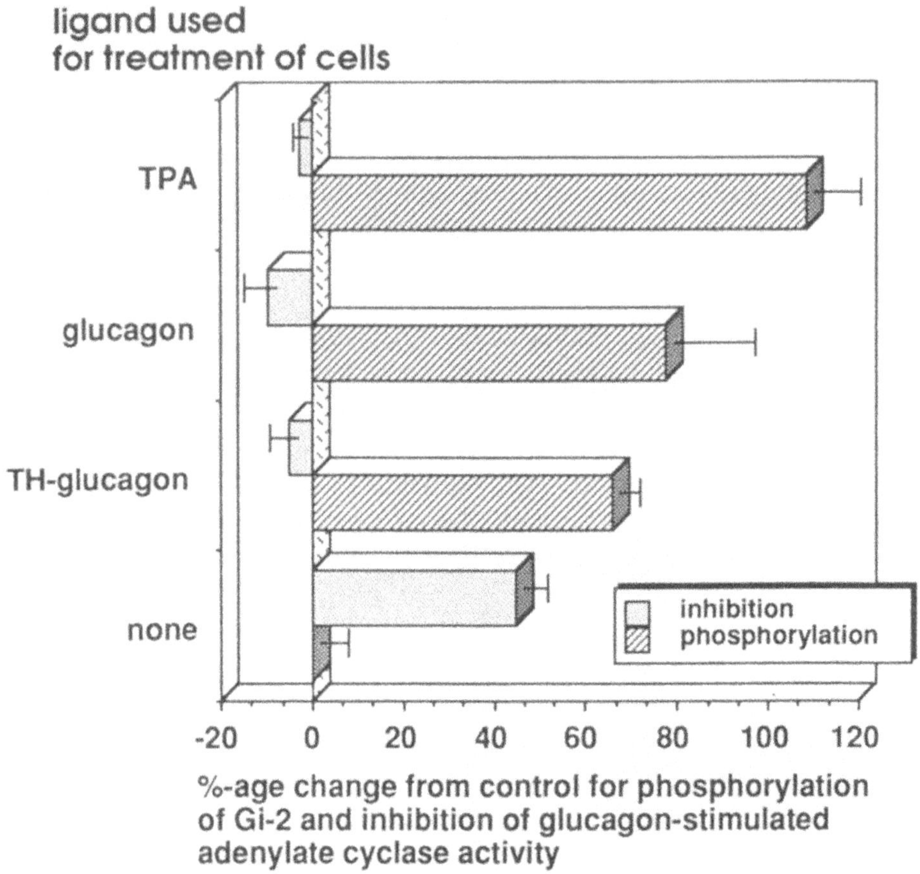

Figure 5. Treatment of intact hepatocytes with either glucagon (10nM), TH-glucagon (10nM) or the phorbol ester TPA (10ng/ml) leads to the functional inactivation of "Gi" and to the phosphorylation of the alpha subunit of Gi-2 which can be detected by immunoprecipitation from hepatocytes labelled with ^{32}P.

to the functional inactivation (fig. 5) of "G_i" [19]. Such an occurrence is accompanied by the phosphorylation of the α-subunit of what we know can show is G_i-2 [19], which we can immunoprecipitate with the antiserum AS7. We determined which form of G_i was involved because the antiserum AS7 is able to immunoprecipitate G_i-1 and G_i-2 but not G_i-3. Thus, because we can show that hepatocytes do not express G_i-1, then the phosphorylated species immunoprecipitated by AS7 must be the α-subunit of G_i-2. We have subsequently performed experiments with an anti-peptide antiserum which is specific for G_i-2 to confirm this. We have also done experiments with an anti-peptide antiserum which is specific for G_i-3 and failed to show any phosphorylation of the α-subunit of this form of G_i.

If indeed, 'G_i-like' activity, i.e. inhibition of adenylate cyclase is indeed mediated by a pertussis toxin sensitive G-protein, then these experiments suggest that the major form of G_i which exerts inhibitory effects upon adenylate cyclase in hepatocytes is in fact G_i-2 (fig. 6).

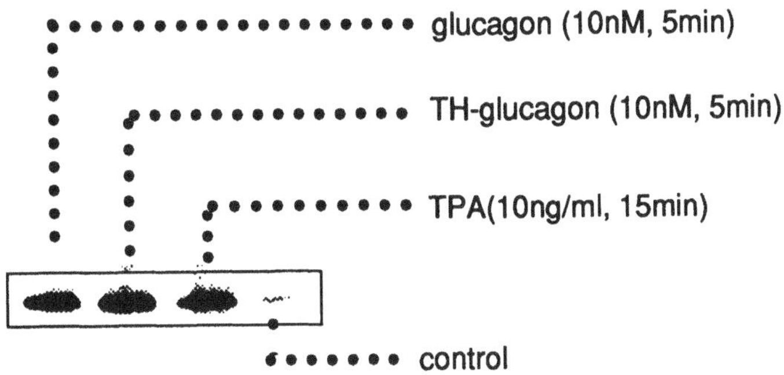

Figure 6. Shows the phosphorylation of G_i-2 upon treatment of intact hepatocytes with either glucagon, TH-glucagon or TPA. This G-protein was specifically immunoprecepitated from ^{32}P-labelled hepatocytes using an antipeptide antibody.

Challenge of intact hepatocytes with either vasopressin, angiotensin, glucagon or TH-glucagon also elicited the inactivation and phosphorylation of G_i-2. We would like to suggest that these actions are mediated as a consequence of the activation of protein kinase C.

Nevertheless, we have observed that treatment of intact hepatocytes with 8-bromo cyclic AMP also causes an increase in the phosphorylation state of G_i-2. This may be because protein kinase A can also phosphorylate the α-subunit of G_i-2. However, there does not appear to be any obvious site on this subunit for phosphorylation by protein kinase A, as determined by analysis of the protein sequence. This does not mean, however, that A-kinase cannot phosphorylate this species, as indeed using purified proteins the converse appears to be true. Thus, one possibility is that dibutyryl cyclic AMP may not be exerting this action directly through A-kinase and a second possibility is that the α-subunit of G_i-2 has two distinct phosphorylation sites, one for protein kinase A and the other for protein kinase C.

Hepatocyte membranes from animals expressing insulin-resistance show a loss of functional G_i activity

We [21] suggested in 1983 that the occupied insulin receptor might interact with members of the G-protein family and, indeed, there is now a body of evidence which has come from a number of groups which provides some support for this concept [22]. One recent aspect of this approach has focussed on the functioning of G_i in hepatocytes from insulin-resistant animals, such as are seen in diabetes and obesity as well as due to high fat diets [18,23].

As discussed above (*vide supra*), rats made diabetic with either streptozotocin or alloxan show a loss in G_i function [18]. We have attributed this to a reduction in the expression of α-G_i protein in the hepatocyte plasma membranes from such animals [18]. However, our recent studies have indicated that the quantification of α-subunit of this G_i was underestimated in membranes from hepatocytes of diabetic animals due to a change in the boyant density of the purified plasma membrane fraction used for immunoblotting. The total loss of functional G_i was not compromised in that study [18] as a crude (total) membrane fraction was used for the functional assays. We can now show by either immunoprecipitation from hepatocyte extracts or by Western blot analysis of partially purified membranes that there are similar amounts of α-G_i in plasma membranes of the normal and diabetic animals. This is consistent with there being little change in the concentrations of mRNA for either G_i-2 or G_i-3 in hepatocytes from diabetic animals. Thus, as in the obese zucker rats [23], G_i is apparently inactive in hepatocyte plasma membranes from the diabetic animals.

Interestingly, treatment of intact hepatocytes, from either diabetic animals or fatty Zucker rats, with either TPA or other ligands which activate C-kinase failed to elicit the phosphorylation of α-G_i. This, we believe, is because the inactivation of G_i in these conditions is due to its prior phosphorylation. We suggest, in the case of the diabetic animals, that this may result from the elevated plasma levels of vasopressin [24] which occur in these animals and which would be expected to cause the stimulation of inositol phospholipid metabolism in liver. Certainly this would be in accord with the elevated tissue levels of diacylglycerol which have been found in diabetic animals.

We do not construe from such studies that insulin might exert actions through G_i, but rather that activation of C-kinase may be an underlying feature of insulin resistance which happens to modify both G_i and the insulin receptor in an inhibitory fashion.

Nevertheless, glucagon desensitization was seen to occur in hepatocytes from diabetic animals [17]. This suggests that one may be able to dissociate the pathways which control G_i phosphorylation and inactivation from those which control desensitization and resensitization. One possibility is that distinct forms (isoenzymes) of protein kinase C are involved. However, we believe that the most likely point which dissociates these two systems is at the level of action of protein phosphatases which control the reversal of these inhibitory effects, i.e. the rate of resensitization and the rate of re-expression of functional G_i.

These observations are a further demonstration of the extent of the cross-talk processes which interlink cellular signalling systems.

Protein kinase A causes the phosphorylation and activation of the 'dense-vesicle' high affinity, cyclic AMP-specific phosphodiesterase

All of the various cyclic nucleotide phosphodiesterases in hepatocytes have been evaluated by us for their potential for rapid regulation through the action of glucagon and insulin. We have shown that plasma membranes possess at least two high affinity enzymes [25-27], one of which is activated by insulin through a mechanism which involves its phosphorylation on tyrosine residues [28]. Mitochondria express two species which appear to be functionally of little consequence in determining total intracellular cyclic AMP concentrations [29], smooth endoplasmic reticulum expresses one species [30], rough endoplasmic reticulum expresses three species and lysosomes and nuclei have none. We have also identified five distinct cyclic nucleotide phosphodiesterase forms in the cytosol fraction [31]. We have evidence [32] to suggest that one of these species has an activity which is rapidly stimulated by insulin in a fashion that is lost upon cell breakage and dilution. This may be the so-called cyclic GMP-activated cyclic nucleotide phosphodiesterase [33].

There is, however, another species which is associated with an as yet undefined intracellular vesicle fraction and this we have called the 'dense vesicle' phosphodiesterase [26]. It is this phosphodiesterase which can be activated by both insulin and by glucagon, albeit through distinct routes [32]. Now, glucagon can elicit the rapid activation of this 'dense vesicle' phosphodiesterase by a process which is triggered through the increase in hepatocyte intracellular cyclic AMP, achieved by glucagon [26]. Thus activation of this enzyme can be mimicked

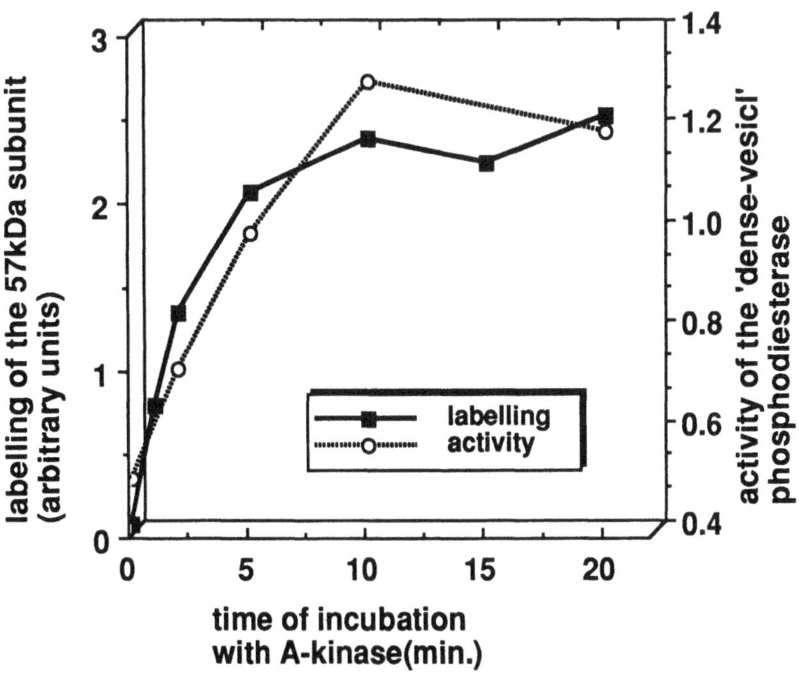

Figure 7. Shows the activation and phosphorylation of the 'dense-vesicle' cyclic AMP phosphodiesterase from rat liver membranes. In order for labelling and activation of the enzyme to be elicited by A-kinase, a pre-incubation with Mg^{2+} was essential.

by treating intact hepatocytes with any one of forskolin, cholera toxin, 8-bromo cyclic AMP and dibutyryl cyclic AMP [26]. Thus we have suggested [26] that the 'dense vesicle' phosphodiesterase may be a substrate for phosphorylation and activation by protein kinase A.

This enzyme has been purified to apparent homogeneity by us [34] and antisera prepared to it. These do not appear to cross-react with any other phosphodiesterase species [34,35]. Using these we can show, in both hepatocytes [36] and adipocytes [37], that protein kinase A treatment of a membrane fraction containing the 'dense vesicle' phosphodiesterase leads to its activation and stoichiometric phosphorylation (fig. 7).

Cyclic AMP metabolism can be simulated knowing the known kinetic properties of all the phosphodiesterases involved. This shows that activation of the 'dense vesicle' cyclic AMP phosphodiesterase has a major influence on the accumulation of cyclic AMP in hepatocytes. Essentially, the nature of its action appears to be in damping the magnitude of the rise in cyclic AMP attained. Thus, in its own way, activation of the dense vesicle' phosphodiesterase makes an important contribution to the nature of the transient increase in cyclic AMP seen as a consequence of the challenge of hepatocytes with glucagon.

Comment

Challenge of hepatocytes with glucagon appears to yield a 'simple' response, namely a transient elevation in the intracellular concentrations of cyclic AMP. This belies the complexity of the numerous enzymes involved in controlling the concentrations this key second messenger and the fact that this hormone activates another signal transduction system whose action has a profound effect on that coupled to adenylate cyclase. Indeed, it may be that such a pathway, which could be contyrolled by a distinct set of glucagon receptors, may play a key role in exerting certain of glucagon's actions, especially at low hormone concentrations where adenylate cyclase is not stimulated.

Desensitization thus appears to be mediated through the action of protein kinase C. The site of the lesion which leads to desensitization is at the coupling interface between the glucagon receptor and G_s. To date we have been unable to show that G_s becomes phosphorylated under conditions where G_i is clearly phosphorylated. Thus we would like to suggest that it is the glucagon receptor itself which provides the target for C-kinase mediated desensitization of adenylate cyclase.

Thus glucagon's action on hepatocytes involves the complex co-ordinated interplay between the various signalling systems present there. Our studies clearly show that the signalling systems for glucagon (major action on cyclic AMP), vasopressin/angiotensin (inositol phospholipid) and insulin (tyrosyl kinase) do not function in isolation. Instead, we see that the 'quality' and strength of signals passing through these systems is modulated by the function-

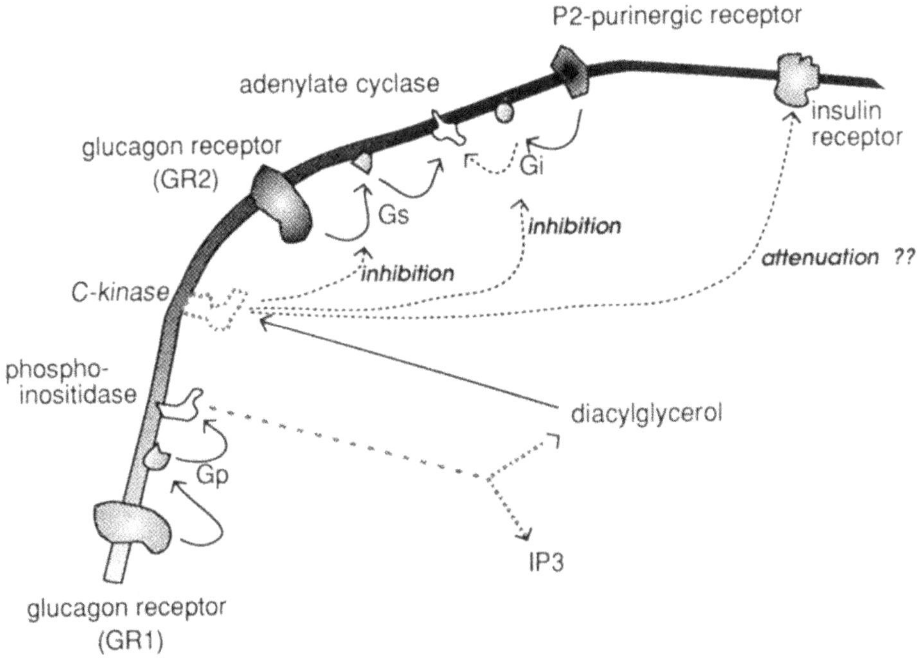

Figure 8. Scheme showing the identified areas of 'cross-talk' between distinct signalling systems in intact hepatocytes.

ing of the others (fig. 8). These may all integrate to form a 'network' capable of allowing the hepatocyte to adapt and learn from distinct physiological stresses.

Acknowledgements

We thank the MRC, AFRC, British Diabetic Association, Lipha Pharmaceuticals and the CMRF for support.

References

1. Heyworth, C.M. & Houslay, M.D. (1983) Biochem. J. 214, 93-98.
2. Murphy, G.J. & Houslay, M.D. (1988) Biochem. J. 249, 543-547.
3. Murphy, G.J., Hruby, V.J., Trivedi, D., Wakelam, M.J.O. & Houslay, M.D. (1987) Biochem. J. 243, 39-46.
4. Heyworth, C.M., Whetton, A.D., Kinsella, A.R. & Houslay, M.D. (1984) FEBS Lett. 170, 38-42.
5. Wakelam, M.J.O., Murphy, G.J., Hruby, V.J. & Houslay, M.D. (1986) Nature (Lond.) 323, 68-71.
6. Whipps, D.E., Armstron, A.E., Pryor, H.J. & Halestrap, A.P. (1987) Biochem. J. 241, 835-845.
7. Blackmore, P.F. & Exton, J.H. (1986) J. Biol. Chem. 261, 11056-11063.

7a. Charest, R., Prpic, V., Exton, J.H. & Blackmore, P.F. (1985) Biochem. J. 227, 8769-8773.

8. Bocckino, S.B., Blackmore, P.F. & Exton, J.H. (1985) J. Biol. Chem. 260, 14201-14207.

9. Irvine, F.J. & Houslay, M.D. (1988) Biochem. Pharmacol. 37, 2773-2779.

10. Takenawa, T., Homma, Y. & Nagai, Y. (1982) Biochem. Pharmacol. 31, 2663-2667.

11. Okajima, F. & Ui, M. (1976) Arch. Biochem. Biophys. 175, 549-557.

12. Birnbaum, M.J. & Fain, J.N. (1977) J. Biol. Chem. 252, 528-535.

13. Cardenas-Tanus, R. & Garcia-Sainz, J.A. (1982) FEBS Lett. 143, 1-4.

14. Mine, T., Kojima, I. & Ogata, E. (1988) 970, 166-171.

15. Houslay, M.D. (19843) Nature (lond.) 303, 133.

16. Heyworth, C.M., Hanski, E. & Houslay, M.D. (1983) Biochem. J. 222, 189-194.

17. Murphy, G.J., Gawler, D., Milligan, G., Wakelam, M.J.O., Pyne, N.J. & Houslay, M.D. (1989) Biochem. J. 259, 191-197.

18. Gawler, D., Milligan, G., Spiegel, A.M., Unson, C.G. & Houslay, M.D. (1987) Nature (Lond.) 327, 229-232

19. Pyne, N.J., Murphy, G.J., Milligan, G. & Houslay, M.D. (1989) FEBS Lett. 243, 77-

20. Okajima, F., Tokumitsu, Y., Kondo, Y. & Ui, M. (1987) J. Biol. Chem. 262, 13483-13490.

21. Houslay, M.D. & Heyworth, C.M. (1983) Trends Biochem. Sci.8, 449-457.

22. Houslay, M.D. & Siddle, K. (1989) Brit. Med. Bull. 45, 264-284.

23 .Houslay, M.D., Gawler, D., Milligan, G. & Wilson, A. (1988) Cellular Signalling 1, 9-22.

24. Brooks, D.P., Nutting, D.F., Crofton, J.T. & Share, L. (1989) Diabetes 38, 54-57.

25. Pyne, N.J., Cooper, M.E. & Houslay, M.D. (1986) Biochem.J. 234, 325-334.

26. Heyworth, C.M., Wallace, A.V. & Houslay, M.D. (1983) Biochem. J. 214,99–110

27. Houslay, M.D. (1986) Biochem. Soc. Trans 14, 183-193.

28. Pyne, N.J., Cushley, W., Nimmo, H.G. & Houslay, M.D. (1989) Biochem. J. (in press)

29 Cercek, B. & Houslay, M.D. (1982) Biochem. J. 207, 123-132.

30. Cercek, B., Wilson, S.R. & Houslay, M.D. (1983) Biochem. J. 213, 89-97.

31. Lavan, B.E., Lakey, T. & Houslay, M.D. (1989) Biochem. Pharmacol. (in press)

32. Heyworth, C.M., Wallace, A.V., Wilson, S.R. & Houslay, M.D. (1984) Biochem. J. 222, 183 187

33. Pyne, N.J. & Houslay, M.D. (1988) Biochem. Biophys. Res. Commun. 156, 290-296.

34. Pyne, N.J., Cooper, M.E. & Houslay, M.D. (1987) Biochem.J. 242, 33-42.

35. Pyne, N.J., Anderson, N., Lavan, B.E., Milligan, G., Nimmo, H.G. &Houslay, M.D. (1987) Biochem. J. 248, 897-901.

36. Kilgour, E., Anderson, N.G. & Houslay, M.D. (1989) Biochem. J. 260, 27-36.

37. Anderson, N.G., Kilgour, E. & Houslay, M.D. (1989) -in press.

DISCUSSION

Downes: I am struck by the similarity in the magnitude to TH-glucagon and glucagon by comparison with vasopressin responses; it implies to me you ought to be able to separate the dose response curves with an analogue of vasopressin. Can you separate the dose response curves for diacylglycerol (DAG) accumulation and the inositolphosphate formation?

Houslay: The DAG studies were preliminary data and seem to be really complicated because they give biphasic time courses. Our central con- clusion, however, is that glucagon will increase DAG and the TH-glucagon will also increase DAG. The time-courses are much more complicated and you definitively see differences in the form of the plots of the time-courses for vasopressin and glucagon.

Dumont: I am really astonished by the low levels of stimulation of DAG production. I mean, when you stimulate in parotid, thyroid and many organs the PIP2 cascade is stimulated by a factor of 5 or 10. So in a way I am surprised because the same thing happened with vasopressin, from the calculations I could make from your data DAG is multiplied by a factor of 1.5, glucagon 1.2 and I would say TH-glucagon 1.2 and these are really very small increments. And so what I am wondering is if you decrease the vasopressin level so as to achieve the same level of stimulation of IP3, as you do glucagon, do you still get desensitization.

Houslay: Firstly, our experiments with DAG are preliminary; whilst we believe that it goes up in the presence of glucagon, and I really don't want to say anything about the magnitude, although it agrees with Exton's results. However IP3 is something that we can measure by both the binding assay and radiochemically, if we actually do take down the vasopressin, angiotensin concentration to get comparable IP3 amounts to glucagon then we can still see desensitization.

Clark: Just a clarification, because you did not show the data, when you treat the cells with vasopressin does that give the heterologous desensitization of glucagon stimulation.

Houslay: It certainly causes the uncoupling of the glucagon receptor from Gs.

Clark: So you get the same effect treating with vasopressin as you do with glucagon.

Houslay: Yes.

Clark: I think that answers my major question. If you look at C-kinase, just to see what is going on, is it your assumption that you are activating a C-kinase with these low concentrations of vasopressin?

Houslay: Yes, that is our assumption. What we are trying to do is look for C-kinase translocation with specific antibodies and also activation.

Corbin: Are you implying that the glucagon acts through DAG and IP3 in some of its effects, in other words I was not only thinking about desensiti- zation but also about phosphorylase activation, gluconeogenesis etc.

Houslay: I am beginning to think that glucagon can perturb a number of pathways. For example, there is a considerable wealth of strong data showing calcium increases with glucagon, which appear to be partly attri-butable to elevated cAMP and partly not. If you actually look at these various results it appears as though the glucagon receptor may interact with more than one signalling pathway mediated via Gs, Gp and perhaps even Gi. So I suppose my answer is yes you are creating a pleiotropic effect of which the major one may be cAMP. As I said, though, there is an enormous literature where effects of glucagon are seen at inmeasurably low concen-trations of cAMP. Similarly, the TH glucagon data that has been done shows that adenylate cyclase is not activated, yet certain actions of glucagon can be mimicked.

Corbin: We had a thought, about 5 years ago, and I do not know if we published it, that glucagon might be acting through an unknown cAMP-independent path-way, so we did the experiment of comparing glucagon with a cAMP analogue and its effect on phosphorylase activation and we measured in the same experiment cAMP dependent protein kinase activation, so we could get some idea of whether or not the same degree of cAMP dependent PK activation would cause the same physiological response and it did. So in other words the cAMP analogue reproduced exactly the glucagon effect on those. This suggested to us that all the metabolic effects of glucagon were due to cAMP and that there was no difference. I mean if there would have been an extra IP3 or some other thing you might think you would see it, you see what I mean, so we did not see any difference, so that was our conclusion then. Correct me if I am wrong but that is what we got.

Houslay: It may be the with glucagon the major action is through cAMP and a trivial metabolic support is given through the other pathways, with desen-sitization being the important issue. However, perhaps in the absence of any A-kinase activation, as is seen for TH-glucagon, then the stimulation of cAMP independent pathways is sufficient to stimulate metabolic events as has been shown by Garcia-Sainz, Fain & Hruby.

Iyengar: You know while we have not done any desensitization studies on hepatocytes I did binding on rat liver membranes for over ten years and I never ever saw two sites, so do you still think there are two receptors or are you thinking of one receptor talking to multiple pathways.

Houslay: There is quite a large literature implying that there is more than one site for glucagon in both isolated membranes and cells from rat liver.

Iyengar: When I used pure iodinetyrosine-10-glucagon I did not see two sites, at least on rat liver membranes. You can go down into the subnanomolar range, all of the early literature that said there are multiple sites, that is because tyrosine-10-glucagon and tyrosine-13-glucagon have different affinities and till the HPLC techniques became available there were two analogues, not two receptors.

Houslay: There are a number of points, one is G-proteins could be promiscuous, I mean the β-receptor has been shown to interact with Gi, and one often sees the pertussis toxin-enhancement of stimulatory receptors. This can be accounted for by taking out a Gi input into the system. However, what is stimulating Gi, unless it is the stimulatory receptor itself. The other point to make is where is the DAG coming from: evidence from Exton's lab suggests it may be PC as well as PI. There is also a well-established calcium flux, which lots of people have showed. Do you think that could be related to Gs, maybe a calcium channel or is it due to an IP3 action. If the intracellular calcium is increased, and we can mimic desensitization for instance using a calcium ionophore, then calcium would cause a breakdown in inositolphospholipids, I mean one possibility could be that the calcium released through a channel coupled to Gs that actually causes the breakdown of PIP2.

Downes: In hepatocytes ionophore unequivocably does not break down inositol-phospholipids.

Houslay: Well there is a Biochemical Pharmacology paper saying quite strongly that it does.

Downes: Well our original data came out in Biochemical Transactions which showed very clearly that the ionophore does not stimulate at least at the level of immediate loss of lipid, PIP2.

Houslay: All I think that tells me is that you did not referee our paper which was sent back to me saying that calcium ionophore cannot be exerting its effect by increasing intracellular Ca^{2+} and this activating protein kinase C because it is well known to cause PIP2 breakdown in hepatocytes.

Downes: On the record I can say that I definitely did not referee that paper.

Iyengar: I have no problems with C-kinase phosphorylating the glucagon receptor, because in all the studies we did, it behaved very much like the β adrenergic receptor and, although we never had a phosphorylated band, I always suspected that receptor phosphorylation was driving the homologous desensitization, but what I am worried about is, are there multiple receptors, because somewhere we should have stumbled on that. In Houston we spent from `79 to '83 in making analogues resolving the various iodinated glucagons and a lot of effort and they were always so monotonous as to be boring. So I really question the notion of multiple receptors in rat liver.

Houslay: Our original, most economical, argument suggested that. However, by analogy with the vasopressin V1 and V2 receptor it could be argued that two receptors occur.

Iyengar: Am I right, isn't TH-glucagon supposed to antagonize the effects of glucagon on cAMP?

Houslay: It will do that yes.

Iyengar: It is not a full antagonist is that right?

Houslay: Well that depends, if you get into concentrations which are over micromolar then you see very small increases in cyclic AMP. However, bio-active effects occur at 0.1 nM with TH-glucagon.

Dumont: First I would like to come back to what Jackie (Corbin) said about phosphorylation and cAMP and so on. There was work in our laboratory by De Wulf and l'Ami and others where they studied hepatocytes phospho- rylation in response to various agents. And there was a clear pattern of phosphorylation induced by vasopressin angiotensin, the PIP2 cascade agonist on the one hand and an other pattern by glucagon, forskolin, cholera toxin on the other hand. And there was some overlap in the patterns. But what ever glucagon was doing was reproduced by cAMP and what ever vasopressin was doing was reproduced if I remember well by the ionophore and by TPA. There was nothing that glucagon was doing on the studied cells, that cAMP would not do, so there was no indication at least by these criteria, which is in intact cells the pattern of phosphorylation, and which shows very nicely, in dog thyroid cells, the overlap between the different pathways. There was no indication of an overlap in that case.

Houslay: I think there are two things really. We are not saying glucagon is working through DAG or some other pathway and not cAMP and we are not saying that cAMP is not the dominant pathway as I said about. However, I am sticking to my conclusion, that desensitization is clearly not mediated by cAMP, and that we can mimic it by anything that activates C-kinase. I think taking your data, if glucagon is really hitting the cAMP path hard, their effect on the other patways will be small. And then a) you are not going to see it and b) the control for that is adding cAMP with vaso-pressin, in the right ratio.

Dumont: I am puzzled and I do not like to remain with fuzzy ideas in my mind and last year in Roscoff I saw Exton telling us that glucagon does not activate the PIP2 cascade and you were not there to contradict him. And I asked him the question and he said I was right, that is all and now you are here and he is not. So now I really like to know what are your differences and why.

Houslay: We see a very small effect but it is an observable effect; Williams sees a small effect and John Exton does also - he's published two pages saying so. However, IP3 breakdown is small. Thus activation of this pathway seemed to us, at the time, the most reasonable explanation of our results. In the light of the initial results that I have shown you with DAG it seems as if those are much bigger effects as what we see with the IP3, because after all IP3 can not mediate desensitization; the real thing we are interested in, lets face it, is DAG and C-kinase activition. So to be honest I am not really overly worried about IP3, what I am worried about is DAG because that is the thing working on the kinase presumably.

Mato: The DAG might come from other lipids; glucagon can activate the transmethylation pathway which in the liver is about 50 % of the lipid, so you get more phosphatidylcholine. I do not know if TH-glucagon activates this pathway but glucagon does.

Houslay: I and Exton have shown that PC breakdown is stimulated by glucagon. It may be that the IP3 is not very relevant. What I think as relevant is the DAG, which might come from lipids other than inositol lipids.

Iyengar: Can you do these physiologically relevant experiments, like showing some C-kinase activity, I mean you treat with glucagon and show that something is happening to your C-kinase activity. Is there any measure of measuring C-kinases of cells. Like the Jim Garrison experiments and the others which convinced everybody else is that you have got proportional activation of A-kinase and everybody said O.K. glucagon works through A-kinase, so now if you come back and say some of it is going through a C-kinase, what about looking at C-kinase activity.

Houslay: I think that is going to take a while but we are trying to do this, looking at isoenzymes.

Iyengar: What difference does it make which enzyme it is,

presumably DAG is going to work on all of them.

Houslay: Yes as long as you have got the right membrane fractions and soluble fractions I mean it is not a trivial experiment, but it is one that ought to be done I agree.

Cooke: I would like to draw the parallel with what I was saying the other day about Leydig cells. I mean there is no evidence that IP3 is increased by LH but the data we presented are almost identical to yours implying that PKC is involved in desensitization of LH stimulation of cAMP production. And there you have the same situation that dibutyryl cAMP will mimic everything LH will do except the desensitization. That is not to say that there is not an other pathway involved in the desensitization of adenylate cyclase, it is an analogous situation, it is very similar.

Iyengar: The problem is not that there is desensitization, the problem is whether IP3 or DAG need to be invoked for what glucagon does, that is the point of the argument. I told you I agree that C-kinase will inactivate glucagon-receptors, no problem there.

Houslay: The definitive experiment has not been done to show that glucagon will stimulate protein kinase C activation. And I think that is going to be a very difficult problem to be solved.

Caron: I would like to return to the comment of whether one receptor can couple to more than one G-protein, and I agree with you that for the β-receptor it has been shown that it can couple to Gs and to Gi, but it couples to Gi with much reduced efficiently. It is about 15 or maximally 20 % of the response which you get with Gs. That being said I would like to mention some other data which are now coming from subtypes of receptors, say the two α2 receptors that we have expressed in cells and we are looking at different pathways and they do indeed inhibit adenylate cyclase so they presumably act through Gi and they do it with a reasonable dose-response that is high affinity and they also work on PI turnover, they also stimulate PI turnover but they do it with a much, much reduced response again, which is about 30 to 40 % of what say an α1 which naturally couples to PI turnover would do. When they do it they do it with a reduced affinity, the dose-response is shifted by sometimes as much as a hundred fold. So to say that a given receptor can talk to different pathways, but sometimes there are some differences, it is kind of amazing that you would see what you should consider the secundary pathway because it is such a small response, being more potent than the normal response. I mean that is what I find a little interesting.

Corbin: I have a question about the latter part of your talk, about the phosphodiesterase (PDE). I was at a meeting last year and talked with Vincent Margianello, who purified the adipose tissue PDE and he told me that he could not get phosphorylation of it, do you think that he purified a proteolytic breakdown product?

Houslay: No I think that the enzyme he purified is in vanishingly small amounts in the cell in both activity terms as well as in protein. It appears to be a different enzyme from ours in terms of size and properties. However, I believe that his more recent data shows he can phosphorylate it.

MULTIPLE PATHWAYS FOR GLUCAGON-INDUCED HETEROLOGOUS DESENSITIZATION OF LIVER ADENYLYL CYCLASE

Richard T. Premont and Ravi Iyengar
Department of Pharmacology
Mount Sinai School of Medicine
New York, NY 10029

Multicomponent signal transduction systems play an important role in intercellular communication. These systems, in addition to transducing external signals, are also capable of adapting their transducing capabilities such that they lower their responsiveness when they are challenged with repeated or prolonged signals. This development of tolerance is called desensitization.

The best studied of the multicomponent signal transduction systems is the hormone-stimulated adenylyl cyclase. This system consists of three components: the hormone receptor, the GTP-binding protein G_s, and the effector enzyme adenylyl cyclase. Hormone binding to the receptor results in activation of G_s. Activation of G_s involves the release of its GTP-liganded α-subunit, which in turn causes adenylyl cyclase to increase its velocity of cAMP production (Premont and Iyengar, 1989).

The Desensitization Process

Desensitization of the pathway which results in stimulation of adenylyl cyclase is a complex phenomenon. It involves several distinct reactions. These various reactions are shown below.

NATO ASI Series, Vol. H 44
Activation and Desensitization of Transducing Pathways
Edited by T. M. Konijn, M. D. Houslay, P. J. M. Van Haastert

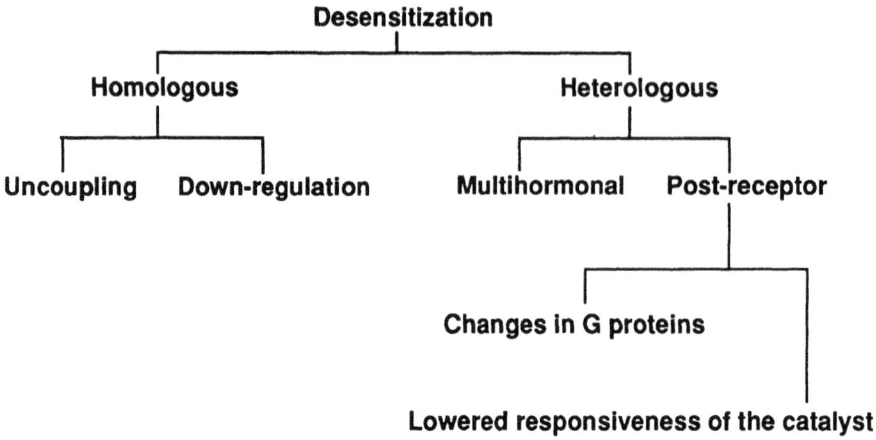

Figure 1. Schematic summary of the processes involved in desensitization of adenylyl cyclase.

Homologous desensitization results in the reduction of the receptor response which initially stimulated the system. Other modes of stimulation are not affected. Homologous desensitization initially results from uncoupling of the hormone-occupied receptor from G_s (Su *et al*, 1980). This has been shown to occur in numerous systems including the glucagon-stimulated adenylyl cyclase in liver (Premont and Iyengar, 1988). The uncoupling reaction is thought to involve a novel receptor-specific protein kinase which only phosphorylates the agonist-occupied receptor (Lefkowitz and Caron, 1988). Temporally, uncoupling is followed by loss of the receptor from the cell surface by internalization and down-regulation. In spite of this temporal relationship, down-regulation and uncoupling do not appear to be causally related since G_s is not required for the uncoupling reaction (Green and Clark, 1981) but is required for the down-regulation (Shear *et al*, 1976; Rich and Iyengar, 1989).

A further form of desensitization involves decreases in stimulation of adenylyl cyclase by numerous agents other than the one to which the system was first exposed is called heterologous desensitization. As is the case with homologous desensitization, heterologous desensitization also results from several distinct reactions. Clark *et al* (1988) have shown that the cAMP-dependent protein kinase (protein kinase A) is required for a multihormonal heterologous desensitization in which the kinase phosphorylates numerous unoccupied receptors and uncouples them from adenylyl cyclase. Lefkowitz and coworkers have directly demonstrated such protein kinase A-mediated receptor phosphorylation (Lefkowitz and Caron, 1988). We have previously observed such multihormonal cAMP-mediated desensitization in the MDCK cell line (Rich *et al*, 1984). In addition to the multihormonal aspect described by Clark and his colleagues, two types of post-receptor events can also be involved in heterologous desensitization. This chapter focuses on these latter, less well-studied pathways: modulation at the level of the G proteins, and at the level of the effector, adenylyl cyclase.

G Proteins and Heterologous Desensitization

G proteins play a central role in transducing hormonal signals to effector systems, and are an attractive target for alteration during desensitization. Modulation of coupling by phosphorylation of G proteins, while an appealing hypothesis, has not proven to be an observable phenomenon in any physiologically relevant situation. Hence, other mechanisms must be considered. Three alternatives that could result in decreased hormonal stimulation of adenylyl cyclase due to modulation at the level of G proteins are 1) decrease in the cell surface G_s, 2) increase in the cell surface $\beta\gamma$-subunits of G proteins, or 3) increase in the cell surface G_i. Since the $\beta\gamma$-subunits dampen the activation of G protein α-subunits, increased levels of $\beta\gamma$-subunits would result in decreased stimulation of G_s. Similarly, increased levels of G_i, the G protein which mediates hormonal inhibition of adenylyl cyclase, would result in decreased overall stimulation.

We have tested for these alternatives in the chick hepatocyte system. Exposure of chick hepatocytes to glucagon results in the rapid onset of heterologous desensitization as measured by decreased NaF-stimulated adenylyl cyclase activity. An interesting feature of this desensitization process(es) is its rapid reversibility under conditions where homologous desensitization is maintained (Premont and Iyengar, 1988). This rapid reversibility places temporal constraints on the mechanisms which can account for this desensitization. Degradation and *de novo* synthesis of G protein subunits would be too slow to be important in this process.

When G_s activity was measured by reconstitution with *cyc*$^-$ S49 cell membranes, it was found that there was a 25% reduction in G_s activity following glucagon treatment (Premont and Iyengar, 1989). The levels of the various G protein subunits in control and desensitized membranes were estimated by quantitative immunoblotting using sequence-specific antisera directed against defined G protein subunits. It was found that $\beta\gamma$-subunit levels were unaltered in desensitized membranes, as were the levels of α_i-subunits. Surprisingly, we were also unable to detect any significant change in the levels of α_s-subunits. However, the sensitivity in the linear range of the immunoblotting assay is such that one needs at least a 50-100% change to obtain a reliably distinct signal. Hence, if the decrease in G_s levels is proportional to the decrease in G_s activity, then our immunoblotting assay may not have adequate sensitivity to detect a 25% loss in α_s protein in desensitized hepatocyte membranes. Since the $\beta\gamma$-subunit and α_i-subunit levels appear unchanged and since the alterations required in the levels of these subunits to elicit the observed functional alteration are much larger (ie., greater than 2-fold), it seems that the most reasonable explanation for the decrease in G_s activity is the decrease in the amount of G_s in the plasma membrane available to receptors and adenylyl cyclase. Further, since the change in G_s activity is also reversed upon recovery from heterologous desensitization, sequestration of a fraction of the cell surface G_s into an intracellular pool where it is unable to participate in receptor coupling to adenylyl cyclase seems a likely mechanism for this phase of heterologous desensitization. This model is shown schematically in Figure 2.

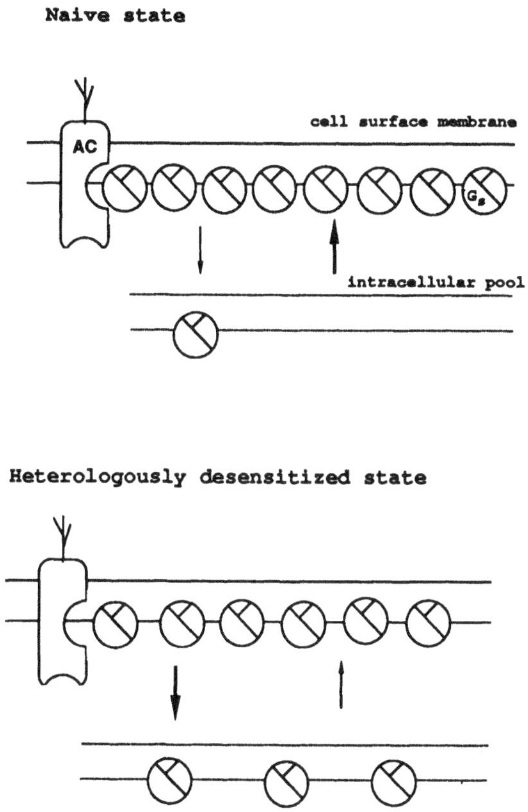

Figure 2. Translocation of G_s from the cell surface membrane to an intracellular pool where it is unable to couple receptors to adenylyl cyclase.

In this model we propose that there are intracellular stores of G_s. At this time we do not know if such G_s is associated with intracellular membranes or soluble. Previous studies in which one of us (RI) had participated indicated the presence of soluble G_s activity in liver (Bhat *et al*, 1981). More recent data have indicated the presence of intracellular membrane pools of α_i-subunits in neutrophils measured by immunoblotting (Retrosen *et al*, 1988), and of unidentified G proteins in pancreas by pertussis toxin labeling (Audigier *et al*, 1988).

Translocation of G_s to and from the cell membrane does provide a simple explanation for the observation of decreased G_s activity in heterologously desensitized cell membranes. The mechanism whereby hormones stimulate such G_s translocation is unknown. It remains to be seen whether this is a more generally occurring phenomenon.

Lowered Responsiveness of the Effector

A surprising aspect of our studies on the desensitization processes in the chick hepatocyte system was the observation that addition of purified G_s to desensitized membranes resulted in partial but never full restoration of hormone- or G_s-stimulated adenylyl cyclase activities to levels observed with control cell membranes. Over a wide range of added G_s concentrations, a 10-15% difference remained between control and desensitized cell membranes. Further, we had found that 8-Br-cAMP treatment of chick hepatocytes resulted in a small (15-20%) loss of NaF-stimulated activity, and that this decrease also could not be restored by the addition of purified G_s. These observations led us to consider the possibility that the effector component of the system was altered such that its responsiveness to G_s is diminished. Since this was a cAMP-dependent phenomenon, protein kinase A phosphorylation of the adenylyl cyclase would appear to be a probable mechanism.

Preliminary experiments in our laboratory indicate that addition of the purified catalytic subunit of protein kinase A and ATP to control hepatocyte membranes, but not heterologously desensitized membranes, results in decreased forskolin or NaF-stimulated adenylyl cyclase activity. This effect appears to be more pronounced in membranes made from *kin*$^-$ S49 cells, which lack an active protein kinase A and therefore presumably all protein kinase A-mediated phosphorylation. Numerous experiments in our laboratory have indicated that the α-

and $\beta\gamma$-subunits of G proteins are not phosphorylated by protein kinase A. Since NaF stimulation of adenylyl cyclase involves only G_s and adenylyl cyclase, the results of these cell-free protein kinase A treatment experiments support our conclusion from the G_s addback experiments in hepatocyte membranes that adenylyl cyclase is a functional substrate for protein kinase A. Further, these results suggest that protein kinase A phosphorylation of adenylyl cyclase results in lowered adenylyl cyclase activity in the presence of NaF. Such a decrease could arise from two distinct types of alterations, as depicted in Figure 3.

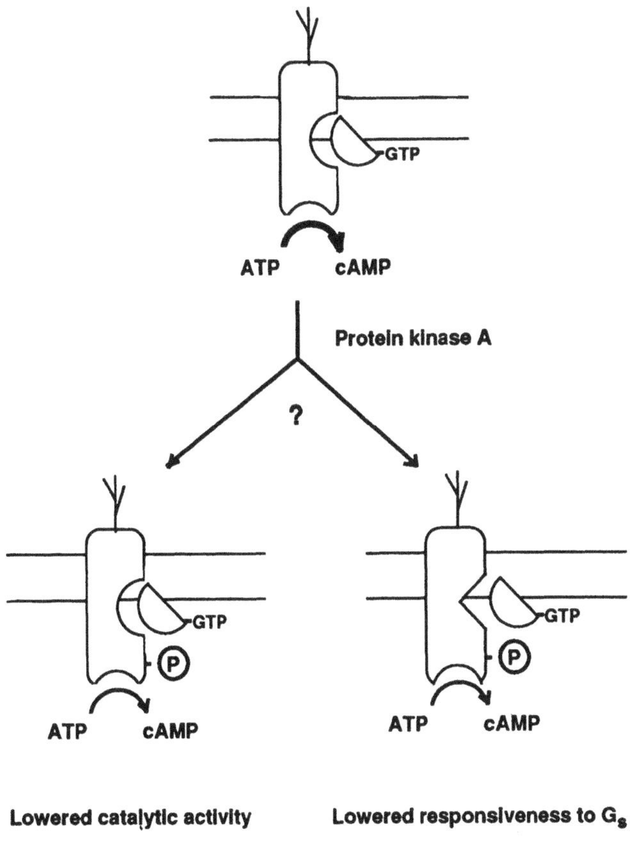

Figure 3. Reduced activity of adenylyl cyclase following protein kinase A phosphorylation may result from lowered catalytic activity or from a reduced ability of adenylyl cyclase to respond to stimulation by G_s.

The first alternative mechanism involves a reduction in the intrinsic catalytic rate of adenylyl cyclase by phosphorylation. In this model, there is no change in the affinity of α_s for the enzyme. In the second alternative, phosphorylation of adenylyl cyclase results in a lowered affinity of the enzyme for α_s, and hence a lowered responsiveness to stimulation. Just such a decreased affinity of α_s and adenylyl cyclase has been recently reported in heterologous desensitization in the rat reticulocyte system by Yamashita *et al* (1989). These two alternatives remain to be experimentally tested in our system. A combination of these two alternatives is also possible, and cannot be ruled out at this stage. Irrespective of the details of the mechanism by which phosphorylation lowers cAMP production by the catalyst, it seems quite certain that changes at the level of the effector will an important role in the development of heterologous desensitization.

Conclusions and Future Directions

Our studies on the chick hepatocyte system indicate that at least four of the five desensitization processes outlined in Figure 1 occur in the hepatocyte. The one process for which we have not found evidence is the multihormonal aspect of heterologous desensitization. In G_s addback experiments to glucagon-desensitized hepatocyte membranes, in addition to the partial restoration of NaF-stimulated adenylyl cyclase activity previously described, we also observed parallel changes in activity stimulated by a heterologous hormone, vasoactive intestinal peptide. VIP stimulation could be partially restored by addition of exogenous purified G_s, but the extent of desensitization remaining was identical to that seen for NaF-stimulated activity. If there were some heterologous alteration at the level of the receptors, we would expect that there would be some component of the total observed heterologous desensitization that could not be accounted for by changes at the level of G_s or of adenylyl cyclase. Such changes do not occur for VIP stimulation, but we cannot rule out the possibility that other hormone receptors may be regulated in this fashion in the hepatocyte.

Our observations in the hepatocyte is in contrast to our observations in the MDCK cell system, where it was found that addition of purified G_s to desensitized cell membranes resulted in full restoration of NaF-stimulated adenylyl cyclase activity, but not hormone-stimulated activity. In the MDCK system, it appears that while there is cAMP-dependent multihormonal desensitization, the responsiveness of the catalyst is unaltered. Since both the hepatocytes and the MDCK cells contain protein kinase A, it is not readily apparent as to why each process occurs in one cell type and not the other.

An extension of this line of reasoning leads to a general question as to why some processes of desensitization as listed in Figure 1 occur in some cell types but not others. The simplistic answer would be that some cells lack the machinery required to achieve certain forms of desensitization. This is not a credible answer to this problem, however, as demonstrated above for protein kinase A. Hence, there probably exist as yet unidentified conditions which influence the cell in choosing certain desensitization pathways. Whether such selection is made and altered under appropriate physiological pressure would be an appropriate issue to be addressed in future research. Answers to such questions will undoubtedly be useful in developing a molecular understanding of how cells alter their capacity to respond to external signals.

Acknowledgments

This work was supported by National Institutes of Health grants CA 44948 and DK 38761. RTP is the recipient of an National Science Foundation Predoctoral Fellowship.

References

Audigier, Y, Nigam, SK and Blobel, G (1988) Identification of a G protein in rough endoplasmic reticulum of canine pancreas. J. Biol. Chem. 263: 16352-16357.

Bhat, MK, Iyengar, R, Abramowitz, J, Bordelon-Riser, ME and Birnbaumer, L (1981) Naturally soluble component(s) that confer(s) guanine nucleotide and fluoride sensitivity to adenylate cyclase. Proc. Natl. Acad. Sci. USA 77: 3836-3840.

Clark, RB, Kunkel, MW, Friedman, J, Goka, T and Johnson, JA (1988) Activation of cAMP-dependent protein kinase is required for heterologous desensitization of adenylyl cyclase in S49 wild-type lymphoma cells. Proc. Natl. Acad. Sci. USA 85: 1442-1446.

Green, DA and Clark, RB (1981) Adenylate cyclase coupling proteins are not essential for agonist-specific desensitization of lymphoma cells. J. Biol. Chem. 256: 2105-2108.

Lefkowitz, RJ and Caron, MG (1988) Adrenergic receptors: models for the study of receptors coupled to guanine nucleotide regulatory proteins. J. Biol. Chem. 263: 4993-4996.

Premont, RT and Iyengar, R (1988) Glucagon-induced desensitization of adenylyl cyclase in primary cultures of chick hepatocytes: evidence for multiple pathways. J. Biol. Chem. 263: 16087-16095.

Premont, RT and Iyengar, R (1989) Heterologous desensitization of the liver adenylyl cyclase: analysis of the role of G proteins. Endocrinol 125: in press.

Premont, RT and Iyengar, R (1989) Adenylyl cyclase and its regulation by G_s. In: Iyengar R and Birnbaumer L (eds) G Proteins. Academic Press, Orlando, Fl. (in press).

Retrosen, D, Gallin, JI, Spiegel, AM and Malech, HL (1988) Subcellular localization of $G_i\alpha$ in human neutrophils. J. Biol. Chem. 263: 10958-10964.

Rich, KA, Codina, J, Floyd, G, Sekura, R, Hildebrandt, JD and Iyengar, R (1984) Glucagon-induced heterologous desensitization of the MDCK adenylyl cyclase: increases in the apparent levels of the inhibitory regulator (N_i). J. Biol. Chem. 259: 7893-7901.

Rich, KA and Iyengar, R (1989) Desensitization of the β-adrenergic receptor of the S49 murine lymphoma cell: mechanism of receptor loss and recovery. J. Mol. Endocrinol. in press.

Shear, M, Insel, PA, Melmon, KL and Coffino, P (1976) Agonist-specific refractoriness induced by isoproterenol: studies with mutant cells. J. Biol. Chem. 251: 7572-7576.

Su, Y-F, Harden, TK, and Perkins, JP (1980) Catecholamine-specific desensitization of adenylate cyclase: evidence for a multistep process. J. Biol. Chem. 255: 7410-7419.

Yamashita, A., Kurokawa, T, Une, Y and Ishibashi, S (1989) Reduction in the stability of the G_s-catalytic unit complex of adenylate cyclase in isoproterenol-induced heterologous desensitization. Eur. J. Pharmacol. 159: 247-256.

DISCUSSION

Downes: In the situation where you are using expression of the bovine brain mRNA and you have responses with bombesin and CCK did you look at the paradigm of augmentation injected Go in those situations ?

Iyengar: We do not want to do these experiments with the crude RNA, because, there are at least three papers now in the literature that say that if you inject total RNA you get both G-proteins as well as receptors, and so the idea has been that in collaboration with the receptor cloners to get some cloned messages and start looking at that. We have not done that and we are going to do it only for the cloned messengers, because those are going to be easy to interpret.

Downes: One additional piece of information: The human erythrocytes which contain so many different G-proteins, I can tell you they also contain a phospholipase C, which is remarkable in its insensitivity to guanine nucleotides.

Iyengar: All I can tell you is that none of the human erythrocyte proteins have worked in this assay in our hands, and I do not know why; we have α_1-adrenergic expression from cloned messages, which is always truly pertussis insensitive and it might be the one place where we can test, but I am acquainted with the ideas that there will be multiple setting in the insensitive pathways, so I would rather not adress the issue but if I have to I will.

Simon: Most of the receptor cloners have found that if they put in some receptor they can easily use the endogenous Xenopus G-proteins, it just surprised me that the endogenous G-proteins have not interfered with your assays at all, or at least in the augmentation assay, I wondered what other kind of experiments one could do to make sure that what you are seeing is really totally independent of the endogenous systems.

Iyengar: The augmentation assay is not independent at all of the endogenous system; in fact if you take the total response, 75 % of the contribution, on average, comes from the endogenous stimulation. What you can do is to add on to it, that is what one can do. Now in the case of the α-subunit because we inject α's that are stochiometrically liganded with GTPτS we do not have to worry about the endogenous G-proteins because there is no agonist or free GTPτS to activate.

Dumont: When you make these injections in the oocyte how do you exclude direct effects on the chloride channels;of course there is EGTA, but lets say Ca is necessary and there is a direct effect, how then would you exclude that?

Iyengar: The only way I can really exclude a direct effect, is going to be doing the experiment I want to have it done before I sent this out for publication, that is that we desensitize this pathway and show that at that stage when we have desensitization at levels subsequent to IP3 production that the αo does not work.

Dumont: I mean, you never checked intracellular calcium or IP3 generation or things like that, to show you that the whole chain is there.

Iyengar: The calcium is all intracellular, these are EGTA injections into the cell.

Dumont: I agree, but you never check, you put in the αo and then you measure the chloride and then you assume that you have activation of the PIP2 cascade and increase intracellular calcium and that activates the channel. What I am asking you is do you prove the other steps, I mean do you prove you have IP3, do you prove you increase intracellular calcium or not ?

Iyengar: We can prove that we are increasing an IP3 sensitive calcium pool. We will probably do the experiments doing direct measurements of IP3, but I would like to avoid that because everybody who does phospholipase C measurements tells me that there are several phospholipase C's and they all use PIP2. So we are going to look for selectivity for the calcium release that works the best. I think that at some stage we will actually measure IP3 but sofar we have not; now coming back to whether these are direct effects, we are going back to the same logic of doing sequential injections and I am willing to accept that Go regulates chloride channels; it does not bother me but the most reasonable explanation of it right now seems to be that it is an IP3 mediated effect. When we saw the βτ effects, first we had considered quite extensively the possibility that things subsequent to IP3 production would be modulated by βτ and that would have been just as exciting for me because it was a new system and we were mapping out the pathways; it is just that there has been no evidence to say that.

Boyer: I want to ask you do you have a control to rule out the possibility that GTPτS is coming out of Go, because we have done some experiments before on reconstitution with G-proteins in turkey erythrocyte membranes, and sometimes you get increases in activity that you could account for with GTPτS leaking of our protein. Even after passing the protein through a G25 column.

Iyengar: G25 is clearly inadequate, you need to be able to actually bind the protein and wash off your free GTPτS quantitatively. G25 is a poor way of separating these things. The way we do it is: The α-subunit is attached to the hydroxylapitite column, and you wash it until you can measure by counting no more GTPτS. But that is not the issue, the nice thing about Xenopus oocytes is that because the calcium release seems to involve some kind of GTP dependent step that is blocked both by GDP and GTPτS, just injections of GTPτS do not always mimic this response. We do not appear to carry over GTP S, and that is not a particular worry for me.

Boyer: Maybe you could inject GDPβS.

Iyengar: You can not do that because you are not going to get your calcium release.

Clark: Liver and HL60 cells both have Gi2 and Gi3 as you said the phase of response is pertussis toxin insensitive, so really the bottom line is do HL60 cells have Go.

Iyengar: You know looking for G-proteins by antibodies only goes so far, if you look at membranes of many cells you will find very little Gi1 in them , in most of the peripheral tissues, but that does not mean that it does not exist, you need to be able to purify these G-proteins; the kind of blot that I showed you is with a highly purified preparation and even there we are pushing the limits of sensitivity, and to put this in perspective, there is about a fifth Gi1 as there is Gs and it never bothered me or anybody else if there was just 15 ng Gs per μg of G-protein that we could activate; we diluted a thousand fold into our assays generally to do the cyc-reconstitution, so I think that any comparison between the amount of protein and the activity certainly is not warranted from our experience with Gs.

Clark: Going to your desensitization, I was not sure about your first and second slide, your second slide showed that there was a transient nature of the NaF desensitization, which peaked, or bottomed, by 5 min and was back again by ten, but your first slide appeared to be showing that you got desensitization at 30 minutes.

Iyengar: The first slide was a fixed study where you expose to hormone for 30 minutes, the second slide you just had a brief exposure to hormone for two minutes, and it was washed off and transferred to medium that contained no hormone, but that does not mean that there is no glucagon bound to the cell surface; there is, we know that. And so when the ambient concentration is approaching zero, what you can do is, you still continue your homologous desensitization proces, but your heterologous is a blip, it just comes on and goes of, it just illustrates the rapidity at which it occurs.

THE MECHANISMS OF LUTEINIZING HORMONE-INDUCED ACTIVATION AND DESENSITIZATION OF ADENYLATE CYCLASE

B.A. Cooke, D.R.E. Abayasekera & M.P.Rose

Department of Biochemistry, Royal Free Hospital School of Medicine, Rowland Hill Street, London NW3 2PF, UK

1. INTRODUCTION

It is well established that LH interacts with its plasma membrane receptor to activate the adenylate cyclase system to form cyclic AMP followed by a subsequent increase in steroidogenesis. Repeated administration of the hormone leads to a desensitization of that same hormonal response and eventually to loss of receptors (down regulation). Recent evidence suggests that LH, may also directly activate other transducing mechanisms e.g. leading to activation of phospholipase A_2 and formation of arachidonic metabolites. Increases in intracellular calcium and possibly activation of protein kinase C also occur. Work in our laboratory is focused on the possible roles of these transduction mechanisms in the activation and desensitization of adenylate cyclase and the subsequent formation of the male androgen, testosterone.

2. DESENSITIZATION OF ADENYLATE CYCLASE

In addition to the stimulatory actions of LH and hCG on steroidogenesis in Leydig and ovarian cells, these hormones also cause a refractoriness or desensitization of

NATO ASI Series, Vol. H 44
Activation and Desensitization of Transducing Pathways
Edited by T. M. Konijn, M. D. Houslay, P. J. M. Van Haastert

that same steroidogenic response. This may involve a loss of LH receptors (down regulation), an uncoupling of the LH receptor from the adenylate cyclase, an increase in the metabolism of cyclic AMP due to an increased phosphodiesterase activity and a decrease in the activities in some of the enzymes in the pathways of steroidogenesis (see Rommerts & Cooke (1988) for references). Our work has concentrated mainly on the desensitization of LH-stimulated cyclic AMP production. Evidence from in vivo experiments after administration of hCG suggested that a lesion develops between the LH receptor and the adenylate cyclase in rat testis cells and ovarian cells (see Rommerts & Cooke (1988) for references). Further work carried out in vitro with tumour Leydig cells demonstrated that LH produced a decrease in LH stimulation of adenylate cyclase in intact cells. This loss of response persisted in plasma membranes prepared from the desensitized cells (Dix et al., 1982). The response of the intact cells to cholera toxin and of the plasma membranes to fluoride and p(NH)ppG were not decreased. These studies suggested that the lesion occurred between the LH receptor and the stimulatory G protein (G_s) whereas the coupling between G_s and the adenylate cyclase catalytic unit was intact. Such a conclusion is in agreement with that of other workers investigating the catecholamine-induced desensitization of S49 lymphoma cells (Iyengar et al., 1981; Green & Clark 1981). This lesion may be necessary before internalization of the receptor can occur.

Interactions between different transducing systems i.e. adenylate cyclase and protein kinase C, have been reported in both mouse and rat Leydig cells (Themmen et al., 1986; Mukhopadhyay & Schumacher, 1985; Dix et al., 1987) and in many other cell types (Nishizuka, 1988). Treatment with the phorbol ester 12-0-

tetradecanolyphorbol-13-acetate (TPA) can mimic hormone induced desensitisation and has been demonstrated both in tumour Leydig cells (Dix et al., 1987; Rebois & Patel, 1985) and in other cell types, e.g. avian erythrocytes (Sibley et al., 1984). However, differences between hormone- and TPA- induced desensitisation have also been noted (Dix et al., 1987; see also Dix et al for refs.).

In further studies, using testis Leydig cells, we have now found that although both LH and TPA induced desensitisation of adenylate cyclase, there were marked differences between the effects of these compounds upon adenylate cyclase activity subsequently stimulated by cholera toxin. These effects may have been due to either a modification of G_s such that activity is increased or to an inhibitory effect upon the G_i protein. These differential effects of LH and TPA were therefore investigated further using pertussis toxin (which ADP-ribosylates and hence inactivates G_i) to examine whether this protein is inhibited by TPA (Platts et al., 1988). Cyclic AMP production from cultured rat testis Leydig cells was stimulated by both LH and cholera toxin. Pretreatment with LH for 1 hour inhibited, i.e., desensitised the ability of the cells to respond subsequently to further LH. In agreement with previous results (Habberfield et al., 1987), it was found that in LH treated cells the 'basal' production of cyclic AMP was increased and the addition of further LH had little effect. When basal levels were subtracted, it was found that the response to cholera toxin by LH-desensitised cells was not significantly different from that found in the control i.e., LH did not desensitise the response to cholera toxin.

Pretreatment of the cells with TPA also caused an inhibition of subsequent LH stimulated cyclic AMP production. However, TPA desensitised cells were found to have an enhanced response to cholera toxin compared to control cells. These effects were not observed in cells treated with phorbol-12,13-didecanoate, a phorbol ester which does not activate protein kinase C.

To examine whether this effect was due to an interaction of TPA with the G_i protein, cells were incubated in the presence of pertussis toxin before being desensitised. Pertussis toxin potentiated cyclic AMP production in both basal and stimulated cells. Pertussis toxin also potentiated LH and cholera toxin stimulated cyclic AMP production in LH desensitised cells, although the effect of LH upon pertussis toxin treated LH desensitised cells was still found to be less than the effect of LH upon pertussis treated non-desensitised cells. Thus inactivation of G_i did not completely reverse LH induced desensitisation.

The effects of TPA were not modified by culture with pertussis toxin. The fact that pertussis toxin was not able to potentiate further the enhanced response to cholera toxin implied that TPA and pertussis toxin have a common site of action. It is concluded that the testis Leydig cell differs from the tumour Leydig cell (Dix et al., 1987) in that TPA enhances the cells response to cholera toxin. That this effect is not potentiated further by pertussis toxin demonstrates that the difference between LH and TPA induced desensitisation is probably due to a modification of G_i by TPA but not LH. Phosphorylation of G_i and attenuation of G_i activity in platelets and S49 lymphoma cells by TPA has been demonstrated (Katada et al.,1985; Jakobs et al., 1985; Bell & Brunton, 1986). The difference between LH and TPA induced desensitisation may then be that LH brings about homologous desensitisation, whereas TPA desensitises in a heterologous manner i.e. it inhibits G protein function generally (Sibley & Lefkowitz, 1985).

3. EFFECTS OF PROTEIN KINASE C INHIBITORS

The effects of PMA on Leydig cell adenylate cyclase activity suggested that the desensitization involves protein kinase C. Further studies have therefore been carried out (Rose & Band, 1988) on the effects of inhibitors of protein kinase C. Two compounds which are involved in the synthesis of sphingolipids- sphingosine and psychosine, have been found to inhibit protein kinase C (Hannun et al., 1986). Their effects on cyclic AMP formation in Leydig cells have been investigated.

It was found that preincubation of Leydig cells with either inhibitor resulted in **inhibition** of LH-stimulated cyclic AMP production. Maximum inhibition occurred with 1uM suggesting that Leydig cells are very sensitive to these protein kinase C inhibitors. Recently it has also been shown that sphingosine and psychosine inhibit luteal cell function (Sender Baum & Ahren, 1988);

The present results indicate that both stimulation (with PMA) and inhibition (with sphingosine and psychosine) of protein kinase C lead to inhibition of adenylate cyclase activity. This apparent paradox may be explained by the fact that the effect of the inhibitors was found to dependent upon the time of addition with respect to LH (Rose & Band 1988) i.e. they only inhibited during the first 15 min of stimulation with LH. This implies that the initial activation of the adenylate cyclase may be dependent upon the activation of protein kinase C. Continued stimulation under conditions which lead to desensitization may cause additional protein kinase C mediated phosphorylation leading to uncoupling of the adenylate cyclase. It is also possible that sphingosine and psychosine have other effects in addition to inhibiting protein kinase C. Winicov & Gershengorn (1988) have reported that sphingosine inhibits thyrotropin releasing hormone binding to pituitary cells by a mechanism not involving protein kinase C. However, high concentrations were required (100uM) and psychosine had no effect. We have now investigated the effect of sphingosine on the binding of hCG to Leydig cells. We were unable to detect any effect of sphingosine on the numbers of hCG binding sites or on the affinity constant. It has also been reported that sphingosine inhibited insulin- and phorbol ester-stimulated uptake of 2-deoxyglucose in 3T3 cells (Nelson & Murray, 1986). We have previously shown that the glucose transporter protein is present in Leydig cells and is regulated by LH (Amrolia et al., 1986). We have carried out similar studies in the presence of sphingosine and found that 2-deoxyglucose uptake was unaffected.

4. RELEASE AND METABOLISM OF ARACHIDONIC ACID

It is well established that in many cell types protein kinase C is activated by diacylglycerol which is released together with inositol 1,4,5, trisphosphate by the action of phospholipase C on polyphosphoinositides. This pathway has not yet been demonstrated in Leydig cells. Nor has the direct action of LH on protein kinase C although this enzyme is present and LH releasing hormone causes a translocation from the cytosol to membrane fractions in Leydig cells (Nikula et al., 1988). Arachidonic acid is released by the action of phospholipase A_2 on phospholipids and/or by hydolysis of diacylglycerol. In pituitary cells it has been demonstrated that arachidonic acid can activate protein kinase C (see Nishizuka, 1988). It was therefore of interest to investigate the possible release and metabolism of arachidonic acid in the presence of LH in Leydig cells.

Our previous studies have shown that inhibitors of the lipoxygenase (but not the cyclooxygenase) pathway of arachidonic acid metabolism inhibited LH, LH releasing hormone (LHRH) and dibutyryl cyclic AMP-stimulated steroidogenesis in purified testis and tumour Leydig cells (Dix et al., 1984; Sullivan & Cooke, 1985). In our current studies we have shown in Leydig cells prelabelled with [^{14}-C] arachidonic acid, LH caused a rapid (within 30 sec) release of arachidonic acid from phospholipids (Chaudry et al., 1988).

In order to clarifiy the role of phospholipase A_2 in the action of LH the effects of exogenous phospholipase A_2 and the effects of inhibitors of this enzyme (dexamethasone and quinacrine) have now been investigated (Abayasekara et al., 1989). It was found that phospholipase A_2 stimulated both basal and submaximal LH-induced testosterone production without affecting cyclic AMP levels. No effect on maximum stimulated testosterone was found. Both dexamethasone and quinacrine, two structurally unrelated phospholipase A_2 inhibitors, caused a dose dependent inhibition of LH-induced testosterone output but had no effect on cyclic AMP production. No direct inhibitory effects of dexamethasone and quinacrine were found on the steroidogenic enzymes involved in cholesterol side chain cleavage of on cell viability. These studies provide further evidence for an additional transducing system for LH involving activation of phospholipase A_2 which is independent of cyclic AMP.

These results indicate the arachidonic acid released from phospholipids by phospholipase A_2 is not involved in the control of cyclic AMP formation. However recent preliminary results from our laboratory indicate that arachidonic acid is a potent inhibitor of cyclic AMP production in Leydig cells. Didolkar & Sundaram

(1987) have reported that arachidonic acid has a dose dependent biphasic effect on cyclic AMP and testosterone production in rat Leydig cells. These authors also showed that phospholipase C addition had similar biphasic effects. It is possible that there is a time- and dose- related release of arachidonic acid via phospholipase A_2 and C pathways with consequent stimulatory and inhibitory effects of arachidonic acid and/or metabolites. These effects may involve protein kinase C and other pathways both on cyclic AMP and cyclic AMP independent pathways. It is also relevant that LHRH has been shown to release arachidonic acid from pituitary cells and that this effect can be mimiced by phorbol esters, again implicating protein kinase C in the release of arachidonic acid (Naor & Catt, 1981).

5. THE EFFECT OF DEGLYCOSYLATED LH ON CYCLIC AMP AND TESTOSTERONE PRODUCTION

Previously it had been shown the deglycosylated LH (dGLH) and hCG (dGhCG) had little of no effect on cyclic AMP and testosterone production in testis interstitial cells, but rather functioned as competitive antagonists, inhibiting the effects of LH and hCG (Berman et al., 1985). These deglycosylated hormones are being currently investigated in our laboratory to investigate their effects on LH transducing and desensitization mechanisms.

It was found that although dGLH did inhibit LH-stimulated cyclic AMP production, dGLH itself produced a small but significant and dose-dependent increase in cyclic AMP levels. DGLH was found to be 50-100% as potent as LH in the stimulation of testosterone production. However when dGLH was added in the presence of LH, testosterone was inhibited.

Pretreatment of the cells with both dGLH and LH inhibited cyclic AMP production in response to a further dose of LH. It was thought that this effect of dGLH might be due to residual hormone bound to the receptor, thus blocking the effect of LH. However acid washing of the cells to remove receptor bound hormone did not alter the ability of either LH or dGLH to desensitize cyclic AMP production.

It is concluded that dGLH is a weak agonist for LH in terms of its ability to stimulate cyclic AMP production but is nearly equipotent as LH in stimulating testosterone production. They were also mutually antagonistic. DGLH is similar in its action on desensitization to LH. One explanation for these results is that dGLH is more effective in stimulating the cyclic AMP independent pathway(s) of LH than the intact hormone. If these involved protein kinase C activation and subsequent phosphorylation of the LH receptor, then dGLH would more effectively uncouple the LH receptor from G_s and cyclic AMP formation would be lower than with LH.

6. CONCLUSIONS

The results of these and other studies demonstrate that although cyclic AMP is a second messenger for LH, other transducing systems are involved in the action of this trophic hormone. In both the LH-induced activation and desensitization of adenylate cyclase, protein kinase C has been implicated. This is based on the effects of inhibitors (spingosine and psychosine) and stimulators (phorbol esters) of this enzyme. Additional effects of phorbol esters, not detected with LH, were found including the inactivation of G_i.

Inhibition of PLA_2 was found to inhibit LH-induced steroidogenesis but not cyclic AMP production. LH was also found to cause a transient increase in the release of arachidonic acid. These results are consistent with our previous work showing that arachidonic acid metabolites formed via the lipoxygenase pathway are involved in controlling steroidogenesis.

Although the PLA_2 inhibition did not inhibit LH-induced cyclic AMP production, the addition of arachidonic acid to the Leydig cells markedly inhibited the formation of this nucleotide. It has been demonstrated in pituitary cells that arachidonic acid is an activator of protein kinase C. It is possible therefore that similar effects are operating in the Leydig cells and that the proposed modulation of adenylate cyclase is dependent on the dose and time of exposure to LH and subsequent release of arachidonic acid.

Finally it was shown that, in agreement with previous work that deglycosylation of LH markedly decreases its ability to stimulate adenylate cyclase. However dGLH was still effective in stimulating steroidogenesis and causing desensitization of LH stimulated adenylate cyclase. These results again implicate non-cyclic AMP dependent pathways in the action of LH.

Acknowledgments

We are grateful to the MRC and the AFRC for financial support.

REFERENCES

Abayasekara, D.R.E., Band A. & Cooke, B.A. (1989) J. Endocrinol. Suppl. 121, abs. 189

Amrolia, P., Sullivan, M.H.F., Garside, D., Baldwin, S.A. & Cooke, B.A. (1988) Biochem. J. 249, 925-928

Bell, J.D. & Brunton, L.L. (1986) J. Biol. Chem. 261, 12036-12041

Berman, M.I., Anand-Srvastava, M.R. (1985) Mol. Cell. Endocrinol 42, 49-57

Chaudry, L., Schulster, D. & Cooke, B.A. (1989) Biochem. Soc. Trans. in press abs. 211 (629th Meeting of the Biochemical Society)

Didolkar, A.K. & Sundaram, K. (1987) Life Sciences 41, 471-477.

Dix, C.J., Schumacher, M., and Cooke, B.A. (1982) Biochem. J. 202, 739-745

Dix, C.J., Habberfield, A.D., Sullivan, M.H.F., and Cooke, B.A. (1984) Biochem. J. 219, 529-537.

Dix, C.J., Habberfield, A.D. & Cooke, B.A. (1987) Biochem. J. 243, 373-377

Green, D.A. & Clark, R.B. (1981) J. Biol. Chem. 256, 2105-2108

Habberfield, A.D., Dix, C.J. & Cooke, B.A. (1987) Molec. Cell. Endocrinol. 51, 153-161

Hannun, Y.A., Loomis, C.R., Merril, Jr., A.H. & Bell, R.M. (1986) J. Biol. Chem. 261, 12604-09

Iyengar, R., Vhat, M.K., Riser, M.E., and Birnbaumer, L. (1981) J. Biol. Chem. 256, 4810-4815

Jakobs, K.H., Bauer, S. & Watanabe, Y. (1985) Eur. J. Biochem. 151, 431-437

Katada, T., Gilman, A.G., Watanabe, Y., Bauer, S. & Jacobs, K.H. (1985) Eur. J. Biochem. 151, 431-438

Mukhopadhyay, A.K., & Schumacher, M. (1985) FEBS Lett. 187, 56-60

Naer, Z & Catt, K.J. (1981) J. Biol. Chem. 256, 2226-2229

Nelson, D.H. & Murray, D.K. (1986) Biochem. Biophys. Res. Comm. 138, 463-467

Nikula, H., Naor, Z., Parvinen, M. & Huhtaniemi, I. (1987) Molec.

Cell. Endocrinol. 49, 39-50

Nishizuka, Y. (1988) Nature 334, 661-665

Platts, E.A., Schulster, D. & Cooke, B.A. (1988) Biochem J. 253, 895-899

Rebois, R.V. & Patel, J. (1985) J. Biol. Chem. 260, 8026-8031

Rommerts, F.F.G. & Cooke, B.A. (1988) New Comprehensive Biochemistry: Hormones and their actions, Vol II (pp 163-180) Eds B.A.Cooke, R.J. King H.J. van der Molen. Elsevier.

Rose, M.P. & Band, A. (1989) Biochem. Soc. Trans. in press. Abs 105 (629th Meeting of the Biochemical Society)

Sender Baum, M.G. & Ahren, K.E.B. (1988) Molec. Cell. Endocrinol 60, 127-135

Sibley, D.R., Nambi, P., Peters, J.R. & Lefkowitz, R.J. (1984) Biochem. Biophys. Res. Comm. 121, 973-979

Sibley, D.R. & Lefkowitz, R.J. (1985) Nature 317, 124-129

Sullivan, M.H.F. & Cooke, B.A. 1985) Biochem.J. 232, 55-59

Themmen, A.P.N., Hoogerbrugger, J.W., Rommerts, F.F.G. & Van der Molen, H.J. (1986) J. Endocrinol. 108, 431-440

Winicov, I. & Gershengorn, M.C. (1988) J. Biol. Chem 263, 12179-12182

Winicov, I. & Gershengorn, M.C. (1988) J. Biol. Chem. 263, 12179-12182

DISCUSSION

Clark: First of all, we are talking the other night about sphingosine and what it might do other than inhibiting protein kinase C which is what Bell has proposed and demonstrated in an in vitro model system, using protein kinase C incorporated into liposomes. Have you looked at whether sphingosine inhibits adenylate cyclase itself in a cell-free preparation?

Cooke: No, we haven't done those studies yet. We have however carried out other control experiments. We looked at the effect of sphingosine on the binding of LH to its receptor, because it has been shown that TRH binding to its receptor in the pituitary is inhibited by sphingosine at high concentrations (100μM) and this is not through the protein kinase C pathway. However we can find no inhibition of binding of LH to Leydig cell LH receptors. Also there is a glucose transport protein system in Leydig cells which can be stimulated by LH and it has also been shown that sphingosine will inhibit glucose transport in adipocytes. However we can find no effect of sphingosine on glucose transport in Leydig cells. Those are the two controls we have carried out so far. We have shown that protein kinase

C can be inhibited in these cells by sphingosine, which is what you would expect. I agree it is the non-specific effect that one is to be concerned about it.

Clark: With sphingosine we found there was a progressive inhibition of the cyclase in a cell-free system. It does not just block hormone stimulation, it is more complicated in that it blocks forskolin stimulation as well. A lot of the effects seems to be due to the compound acting as a detergent as opposed to the specific inhibitory effect on protein kinase C.

Cooke: We have checked cell viability, using very sensitive tests and we do not find any decrease in viability within the range we are working in (1-50μM).

Clark: When you start getting up to 100μM you start seeing things. A lot of the studies have been done with sphingosines on second messenger systems with concentrations higher than 50μM. At those concentrations it makes holes in the cell and acts like a detergent. What I was wondering as well is that you seemed to have ruled out cAMP. It can be involved in induction since dibutyryl increases testosterone production. Do you have any clue, do you think it might be the arachidonic acids cascade phospholipase C activation. Is that what you thought to account for the difference in the dose response curves. Secondarily have you measured cAMP kinase activation ratios with the low concentrations of hormones to see if the kinase is activated even when you cannot measure cAMP levels.

Cooke: Yes, we have measured protein kinase A and it is activated with concentrations of LH when we cannot detect changes in cyclic AMP. With respect to the phospholipase A2 all our indications are that this pathway is involved in steroidogenesis through the leukotrienes. The difficulty has been to get very specific inhibitors of lipoxygenases and we now have these and we are now carrying out these studies. All the studies we have done so far indicate that the lipoxygenase pathway is involved. Also if you inhibit phospholipase A2 with dexamethasone then you inhibit steroidogenesis but not cAMP production which indicates that this pathway is independent of cAMP.

Clark: Can I just take on that last point. Have you added leukotrienes to intact cells and looked for steroidogenesis or do you invisage some kind of intracellular mechanism?

Cooke: We added 5HPETE and that does shift the dose response curve to LH to the left. With all the compounds we have tried we have never had an enhancement of LH maximum stimulated effects. It has always been a shift at submaximal levels.

Simon: What is the time scale in terms of when you first begin to see an increase in steroidogenesis after the application of LH?

Cooke: The shortest time scale is about 3 or 4 minutes. In the adrenal cell it is much faster, you can see an increase in steroidogenesis within seconds.

Simon: Would the arachidonic acid pathway be fast enough?

Cooke: That is very fast. The release of arachidonic acid is maximal within 30 seconds.

Chabre: What is the time scale of the desensitization? You said that you normally obtain desensitization within 30 minutes. Have you looked at shorter times?

Cooke: Yes, in fact desensitization is difficult to separate from stimulation. My feeling is that both mechanisms are operating almost stimultaneously.

Chabre: If you preincubate for one or two minutes with the hormone and then wash it out, would you see desensitization?

Cooke: Yes.

Snaar-Jagalska: Can you stimulate or inhibit adenylate cyclase in vitro?

Cooke: You don't get LH-induced desensitization as far as I know with plasma membranes, but if you preincubate intact cells with LH or phorbol esters and then prepare plasma membranes, the plasma membranes remain desensitized.

Snaar-Jagalska: What are the effects of pertussis toxin on this regulation of adenylate cyclase in vitro?

Cooke: As I explained pertussis toxin enhances the effect of LH, but doesn't completely prevent desensitization.

Dumont: Is it true that there is no proof that LH in rat Leydig cells stimulates the IP3 cascade?

Cooke: I think it is difficult to conclude convincingly that there is no effect. There has been a long series of papers from Farees and Davies and their colleagues, mainly on rat luteal cells, showing that you do get PI turnover after stimulation with LH. But what I am saying is that the actual data for IP3 formation is not very convincing and other workers do not find any. It is only in Farees's laboratory that they claim to see an increase in IP3.

Dumont: I would like to follow on that point and state that in the dog thyroid we have never been able to find an IP3 response to TSH. Other people have reported a small increase but that was with a thousand times physiological levels of TSH. On the other hand, in the human thyroid there are the two pathways. In the dog TSH stimulates cAMP. There is also a stimulation of PI turnover, but it is not accompanied by release of IP3 or activation of IP2 cascade.

Iyengar: Your TPA effects may not be via Gi. Caron and Lefkowitz have shown that cyclase is phosphorylated by C kinase. You may have increased sensitivity to Gs; would that explain your paradox? Instead of inactivating Gi?

Cooke: All our studies show that we get a homologous desensitization and not heterologous desensitization. In other words if we use other ligands such as cholera toxin or forskolin the cells do not desensitize at least over short periods. So it is not the Gs coupling to the adenylate cyclase which is defective but rather the receptor coupling to the Gs.

Iyengar: Do you get your enhanced sensitivity at early times?

Cooke: To cholera toxin, not to LH. The disturbing part is that phorbol esters are used so extensively and many workers have suggested that they mimic desensitization and are comparable with the natural hormone, but here is a case where TPA is clearly doing something else. Also we have shown that if we take Leydig tumor cells and desensitize with phorbol esters, and prepare the plasma membranes, they are no longer desensitized, whereas with LH desensitized cells the membranes remain desensitized. So I think phorbol esters are doing other things.

Clementi: Do you have desensitization before internalization of

the receptor? Is phosphorylation relevant for internalization and desensitization of your system? How many cytoplasmic receptors do you have normally in the cell?
What is the percentage of internalized receptors?

Cooke: These are all studies I would like to do but I do not think I can give you any definite answers. All I can say is that the internalization is very rapid. If we inhibit recycling with monesin, within 10 minutes all the surface receptors have disappeared. So internalisation and recycling is a very rapid process in which I would like to think that desensitization (uncoupling) comes first and then you get internalization. But we really do not have proof that this happens. We do have some evidence that you can get internalization of the receptor in the absence of LH. It may be that internalization goes on slowly in the absence of the hormone and that it speeds up in the presence of the hormone.

Clementi: When you use monesin you probably also stop the release of receptors from the cytoplasmic membranes. Also this is a rapid process.

Cooke: Yes, I take your point. We do not think that there is new synthesis. But that doesn't exclude stores of the receptors that may be mobilized and inserted into the plasma membrane.

Clark: I want to reemphasize the point you made about phorbol esters being tricky and certainly superficially this does not reflect desensitization as is caused by the hormone. In S49 cells diacylglycerol mimics the effect of TPA on Gi whatever that effect may be directly or indirectly. The EC50 for this effect with TPA is about 2 - 3 nM. The effects on the receptor show an EC50 of about 50-100nM, this is almost 20-50 fold higher. So, either there are different protein kinase C responding in very different ways or something else is happening. When you blast the protein kinase C to the extent it starts phosphorylate anything and can show phosphorylation of the receptors. But again it is a 50 fold higher concentration. The immediate rapid effect is on Gi, which I think will interest us in the future studies more than anything because it indicates there is a regulation of cyclase at the tranducer level not at the receptor level which may be more physiological. So it is important to use diacylglycerol or something that will mimick the phorbol esters.

Iyengar: You should really say coupling rather than Gi, because you don't really measure Gi activity, you measure increase of cAMP levels or cyclase.

Clark: Well, we measure either somatostatin inhibition or GPNHP inhibition of forskolin stimulation in cell-free assays.

ROLE OF PROTEIN KINASES IN THE DESENSITIZATION-SENSITIZATION OF β-ADRENERGIC RECEPTOR AND PROSTAGLANDIN RECEPTOR STIMULATION OF ADENYLYL CYCLASE

Richard B. Clark, Mark W. Kunkel, John A. Johnson, Richard Goldstein, and Jacqueline Friedman
Graduate School of Biomedical Sciences
University of Texas Health Science Center at Houston
P. O. Box 20334, Astrodome Station
Houston, Texas

Introduction

Desensitization of hormonal stimulation of adenylyl cyclase in the S49 wild type (WT) lymphoma cells is complex. At least five types of desensitization of the β-adrenergic receptor response have been identified (fig 1) at the receptor level. These include: (1) cAMP-dependent protein kinase (cA•PK)-mediated heterologous desensitization of hormonal stimulation which occurs rapidly in response to low concentrations (EC_{50} of about 10 nM) of either epinephrine or PGE_1, or any drug which increases cAMP (Clark *et al.* 1988); (2) homologous desensitization associated with the sequestration and internalization of the β-adrenergic receptor which occurs rapidly in response to high concentrations of epinephrine (EC_{50} about 200 nM), is dependent on substantial agonist occupation of the receptor, and is independent of G_s, cAMP or cA•PK (Green *et al.* 1981, Clark *et al.* 1985) and has been suggested to involve multiple phosphorylations by β-adrenergic receptor kinase (Benovic *et al.* 1987); (3) a protein kinase C (PKC)-mediated heterologous desensitization which occurs in response to relatively high concentrations of the phorbol ester PMA with an EC_{50} of about 100 nM (Johnson *et al.* 1986); (4) a slow homologous down regulation of the β-adrenergic receptor, characterized by the loss of antagonist binding, which proceeds with a half-life of about 2 hr in the WT and 6 hr in the cyc^- mutant (Clark *et al.* 1985); and (5) a slow homologous desensitization of the β-adrenergic receptor which occurs in response to treatment of cells with low concentrations of epinephrine (3-5 nM) for prolonged periods of time (half-life about 2-4 hr) and which may be, in part, mediated by cA•PK (Butcher *et al.* 1987).

In addition to these types of desensitization a very rapid PKC-mediated heterologous sensitization of hormonal stimulation has been characterized (number 6 in figure 1) which

NATO ASI Series, Vol. H 44
Activation and Desensitization of Transducing Pathways
Edited by T. M. Konijn, M. D. Houslay, P. J. M. Van Haastert

occurs in S49 lymphoma cells in response to low concentrations of PMA (EC_{50} about 2-5 nM). Functional assays of adenylyl cyclase indicate that G_i function was impaired (Johnson *et al.* 1986) and evidence from other systems seemed to indicate that G_i was phosphorylated. Recently David Manning (personal communication) has discovered that in human platelets it is G_z, not G_i, which is phosphorylated. In this model we suggest that the phosphorylated form acts as an inhibitor of G_i.

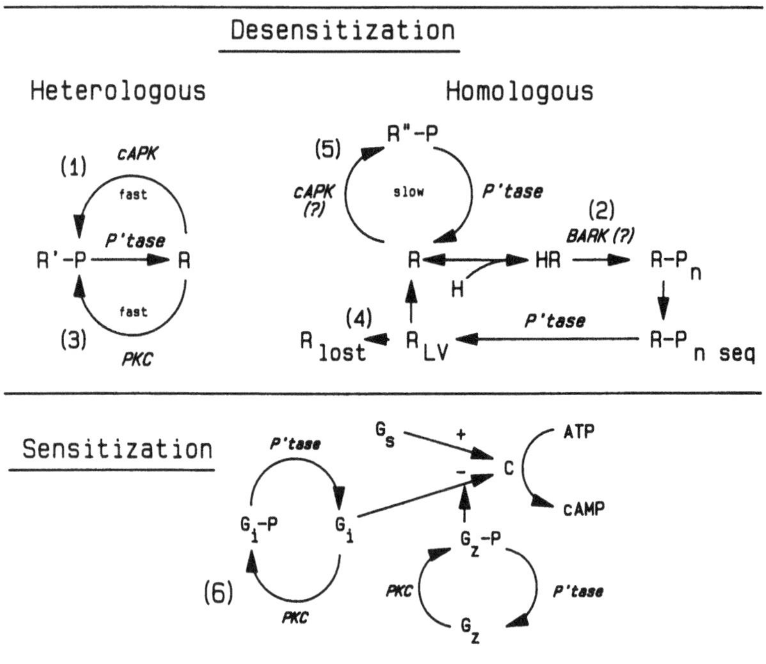

Figure 1. Mechanisms of desensitization and sensitization of the β-adrenergic receptor in S49 lymphoma cells Abbreviations are: H, stimulatory hormone; R, native β-adrenergic receptor; R'-P, R"-P, $R\text{-}P_n$, different phosphorylated states of the βAR; R_{seq}, sequestered βAR; R_{LV}, internalized βAR in light vesicles; cAPK, cAMP-dependent protein kinase; PKC, Ca^{2+}/phospholipid-dependent protein kinase; BARK, βAR kinase; p-tase, phosphatase; G_s, stimulatory GTP binding protein; G_i, inhibitory G protein; G_z, GTP binding protein-function unknown; C, adenylyl cyclase catalytic subunit. Numbers in parentheses refer to the different types of desensitization and are defined in the text.

The molecular mechanisms of these modes of desensitization are either not known or are poorly understood. In this paper we will focus on the evidence we and others have obtained which strongly indicates that cA•PK mediates the physiologically relevant rapid heterologous desensitization of hormonal stimulation of adenylyl cyclase in the S49 WT lymphoma cell and a homologous desensitization of the β-adrenergic receptor response in L cells. In addition we will review the evidence that PKC mediates a heterologous desensitization of hormonal stimulation and that BARK (Beta Adrenergic Receptor Kinase) mediates a homologous desensitization of the β-adrenergic receptor in the S49 WT lymphoma cells.

Krebs and Beavo (1979) outlined the evidence needed to establish the role of phosphorylation-dephosphorylation in the physiological regulation of the function of an enzyme. The essence of these criteria was that changes in function of the enzyme under investigation must be correlated with the phosphorylation-dephosphorylation of the enzyme both *in vitro* (cell-free) and *in vivo* (the intact cell). The activity of the kinases and phosphatases and concentrations of modulators of the enzyme in the cell must be consistent with the functional effects on the enzymes, and these concentrations of the kinases/phosphatases/modulators must be effective in the cell-free systems as well.

For the past several years our laboratory has studied cA•PK-mediated desensitization using the Krebs-Beavo criteria. We discovered the role of cA•PK in the heterologous desensitization of hormonal stimulation of adenylyl cyclase in these cells using predominantly the genetic-biochemical approach made possible through the use of somatic cell mutants of the S49 wild type lymphoma cells. The path which led us to an understanding of its role in desensitization was indirect. In fact, our early studies of desensitization in the S49 lymphoma cells appeared to conclusively eliminate a role for cAMP and cA•PK based on the demonstration that treatment of intact cells with high non-physiological concentrations of hormones caused the homologous desensitization of adenylyl cyclase in membranes of WT S49 lymphoma cells and the cyc^- and kin^- mutants. Since the cyc^- lack the α subunit of G_s, and the kin^- lack cA•PK, we concluded that the homologous desensitization was independent of G_s, cAMP, and cA•PK. We found that the EC_{50} for this epinephrine-induced homologous desensitization was about 200 nM in both the WT and the cyc^-, and that it was correlated with the sequestration-internalization of the β-adrenergic receptor (Clark *et al.* 1985).

It was at this time that my colleagues Drs. Roger Barber, Tom Goka, and R. W. Butcher in Houston found that epinephrine-induced desensitization of the β-adrenergic receptor stimulation of cAMP accumulation in the intact S49 WT cell occurred at concentrations of 10-20 nM epinephrine, well below the levels required to cause the epinephrine-induced homologous desensitization of adenylyl cyclase and the sequestration/internalization of the β-adrenergic receptor (Clark *et al.* 1985, Barber *et al.* 1984, Barber 1986). In conflict with these results we could not observe desensitization of the β-adrenergic receptor stimulation of adenylyl cyclase in membrane preparations following treatment of the WT cells with 20 nM epinephrine.

Mg^{2+} at high levels obscures the measurement of desensitization in cell-free assays of adenylyl cyclase

The explanation of our inability to measure desensitization following treatment with the more physiological concentrations of epinephrine came from the realization that the concentrations of Mg^{2+} we used in our standard assays of adenylyl cyclase, 5-10 mM, obscured the desensitization which was caused by these low concentrations of epinephrine. We first discovered the Mg^{2+} effect in studies of phorbol ester-induced desensitization and sensitization of adenylyl cyclase. Treatment of cells with 0.5 μM PMA for 5 min causes a 50-70% desensitization of hormonal stimulation of adenylyl cyclase which is progressively obscured as the free concentration of Mg^{2+} is increased above 1.0 mM. (Clark *et al.* 1987). This observation led us back to the problem of our inability to measure any desensitization of adenylyl cyclase following the pretreatments with the physiological concentrations of

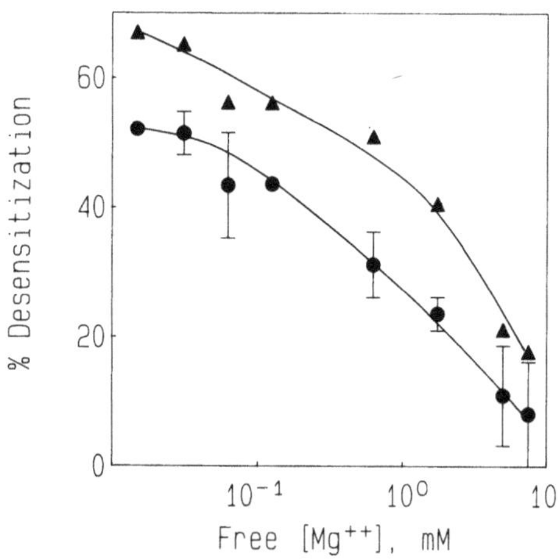

Figure 2. Effect of Mg^{2+} concentration on the measurement of epinephrine-induced desensitization of Wild Type S49 Lymphoma cells and DDT_1MF-2 Smooth Muscle Cells S49 WT (●) or DDT_1MF-2 (▲) cells were treated for 15 min at 37° C with either 50 nM epinephrine in 0.1 mM ascorbate/1 mM thiourea or ascorbate/thiourea alone. Membranes were prepared as previously described (Clark *et al.* 1987). The membranes were assayed for 50 *nM* epinephrine-stimulated adenylyl cyclase activity at various concentrations of free Mg^{2+}. The percent desensitization was calculated as (1 - [activity of treated/activity of control]) x 100.

epinephrine. Not surprisingly we found that Mg^{2+} also largely obscured the desensitization caused by these low concentrations of epinephrine. An experiment demonstrating the effect of Mg^{2+} on our ability to measure the epinephrine-induced desensitization of the WT lymphoma cells and the DDT_1 MF-2 smooth muscle cell line is shown in fig 2.

In summary, our results demonstrate that the characteristics of adenylyl cyclase activities assayed at low free Mg^{2+} levels better reflect the activities of adenylyl cyclase in the intact cell; namely, i) low basal activities; ii) greater fold-stimulation and inhibition of adenylyl cyclase by hormones; and iii) greater desensitization. Grubbs *et al.* have estimated that the free Mg^{2+} in S49 cells is about 0.2-0.4 mM which is in good agreement with the levels which we have found to be optimum.

Evidence that cA•PK mediates the heterologous desensitization of adenylyl cyclase in cells

After discovering the importance of using physiological levels of free Mg^{2+} (0.2-0.4 mM, see ref 9 and 10) in the cell-free assay of the desensitization of adenylyl cyclase, we re-investigated the characteristics of epinephrine-induced desensitization in the S49 cyc^- and the kin^- mutants. We found that these mutant lines, unlike the WT, did not undergo significant desensitization in response to treatment with 20 nM epinephrine (Clark *et al.* 1988) These observations suggested for the first time that cAMP activation of cA•PK was involved in desensitization of the β-adrenergic receptor in the S49 WT lymphoma cells.

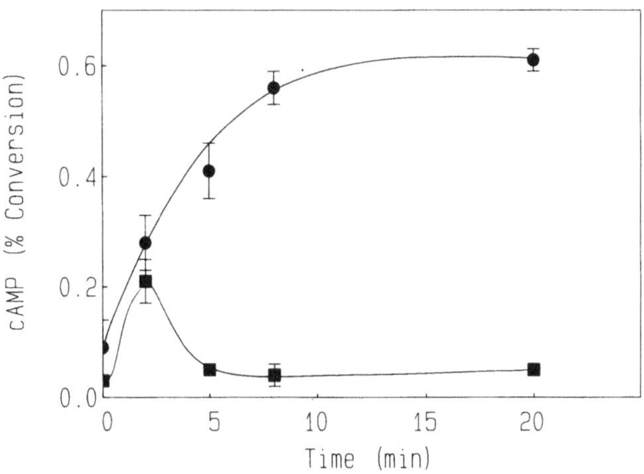

Figure 3. Time course of 10 nM epinephrine-induced cAMP accumulation in S49 WT and kin^- cells Intact WT (■) or kin^- (●) were treated with 10 nM epinephrine for various times and the cAMP accumulation was measured as previously described (Clark *et al.* 1988).

The following observations are consistent with the conclusion that cA•PK mediated the desensitization. First, we have shown that treatment of the WT cells with either dibutyryl cAMP or forskolin quickly desensitized both epinephrine and PGE_1 stimulations of adenylyl cyclase without affecting forskolin or NaF activation. Second, dibutyryl cAMP evoked a similar heterologous desensitization of hormonal stimulation of adenylyl cyclase in the cyc^- (measured after reconstitution of cyc^- membranes with cholate extracts of G_s), but not in the kin^-. Third, comparisons of the time course of cAMP accumulation in the WT versus the kin^- (fig 3) in response to treatment with 10 nM epinephrine demonstrated that the WT rapidly desensitized, but the kin^- did not. Fourth, epinephrine caused the homologous desensitization of the kin^- and the cyc^- cell lines with an EC_{50} of about 50 nM and 200 nM respectively, but there was no heterologous desensitization. Finally, the characteristics of the cA•PK-mediated desensitization, namely about a 2-3 fold decrease in the K_{act} for epinephrine stimulation of adenylyl cyclase, were similar whether the desensitization was caused by incubation of cells with forskolin, epinephrine, or PGE_1 (fig 4).

These observations constitute our evidence that cA•PK mediates the heterologous desensitization of hormonal stimulation in the intact S49 lymphoma cell. What remains is to correlate the functional changes in the epinephrine and PGE_1 stimulations of adenylyl cyclase with phosphorylation-dephosphorylation of the WT receptors after [^{32}P]-phosphate labeling of the cells.

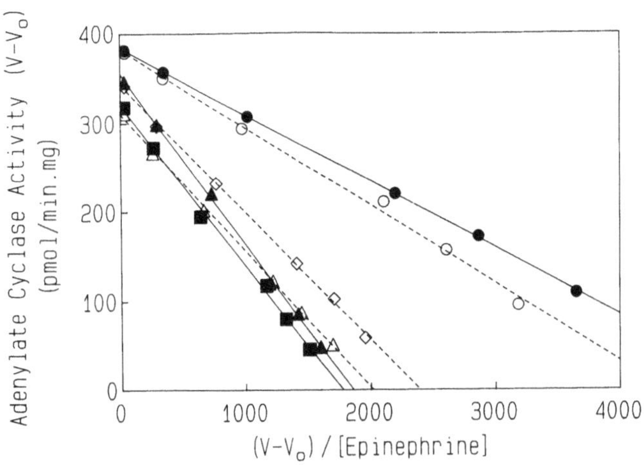

Figure 4. Desensitization in membranes from S49 WT cells pretreated with epinephrine, PGE_1, or Forskolin S49 WT cells were treated for 10 min at 37° C with ascorbate-thiourea (● ○), 20 nM epinephrine (△), 50 nM epinephrine (▲), 10 nM PGE_1 (◇), or 10 μM forskolin (■). Membranes were prepared and assayed for adenylyl cyclase activity at various concentrations of epinephrine. The Eadie-Hofstee transform of the data is shown.

Cell-free cA•PK-mediated desensitization of adenylyl cyclase in S49 WT and kin$^-$ lymphoma cell membrane preparations

Our previous inability to achieve cell-free desensitization of the β-adrenergic receptor response in plasma membrane fractions of the S49 lymphoma cells hindered progress in elucidating the mechanism of desensitization. Recently we successfully developed a cell-free system for the rapid heterologous desensitization of adenylyl cyclase in plasma membranes which mimics the intact cell process (Kunkel *et al.* 1989). The system is simple. Fig 5 shows the result of a typical experiment in which S49 WT membranes were incubated for 5 min with various concentrations of highly purified catalytic subunit of cA•PK (cA•PK_c), washed and then hormonal stimulation of adenylyl cyclase was measured under conditions of low free Mg^{2+}. Both epinephrine and PGE_1 stimulations were reduced 50-60% by the treatment while forskolin stimulation was unaffected, similar to the results following intact cell desensitization. Near maximal reduction of hormonal stimulation was achieved with a cA•PK_c concentration of 10 μg/ml which closely approximates intracellular levels of the kinase.

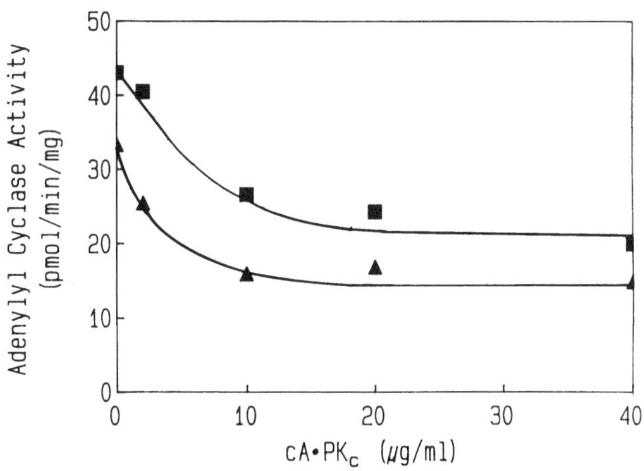

Figure 5. Concentration effects of cA•PK_c treatment of S49 WT membranes S49 WT membranes were treated at 30° C for 5 min with various concentrations of cA•PK_c as previously described. The washed membranes were assayed for adenylyl cyclase activity stimulated by 50 nM epinephrine (▲) or 2 μM PGE_1 (■).

In retrospect it seems that previous attempts to achieve cell-free desensitization did not succeed for a variety of reasons. First, early attempts were geared towards achieving the homologous desensitization in cell-free preparations, which, assuming it involves a sequestrations and internalization, via clathrin coated pits, of the β-adrenergic receptor, is unlikely to occur in small membrane vesicles. Second, all of the early studies were performed under high Mg^{2+} adenylyl cyclase assay conditions which greatly diminished our ability to measure heterologous desensitization. Third, most assays of desensitization involved use of a single high concentration of epinephrine. Since the characteristic of the S49 heterologous desensitization of the β-adrenergic receptor is a right-shift in the response to epinephrine (an increase in the K_{act} for epinephrine with no change in V_{max}), assay at a single high hormone concentration would not detect any change in β-adrenergic receptor function.

Perhaps the most exciting result achieved with the cell-free system was the heterologous desensitization of the kin^- adenylyl cyclase (Kunkel *et al.* 1989). The intact kin^- cells are incapable of undergoing the heterologous cA•PK-mediated desensitization since they lack any measurable cA•PK activity. While this is the only defect known in these cells, conclusions based on the intact cell result had to be tempered by the fact that other processes important to desensitization might have been abnormal in the mutant as an indirect consequence of the lack of cA•PK. Since the cell-free reconstitution of cA•PK with the kin^- membranes causes a desensitization of hormonal stimulation similar to the WT, it seems clear that the lack of heterologous desensitization in the intact kin^- cells is directly attributable to the absence of the cA•PK.

The cell-free heterologous desensitization of the kin^- membranes with $cA•PK_c$, however, is not identical to the WT. Kinase treatment of the WT membranes does not alter forskolin or NaF stimulations, whereas it consistently decreases forskolin stimulation of the kin^- by 20 to 30%. This decrease does not appear to be caused by the alteration of G_s, since cholate extracts of G_s from kin^- membranes incubated with the $cA•PK_c$ reconstitute the adenylyl cyclase activities of cyc^- membranes as well as extracts from the untreated controls. The decrease in forskolin stimulation could not be attributed to a modification of G_i since G_i-mediated inhibition is unaffected. We are left with the possibility that in the kin^- the catalyst is phosphorylated and its activity reduced by the $cA•PK_c$ treatment.

Localization of the domain of the β-adrenergic receptor altered by the $cA•PK_c$-mediated desensitization

Only by quite heroic efforts will the domain of the β-adrenergic receptor modified by cA•PK in intact cells be demonstrated by biochemical means. While this must ultimately be accomplished, the use of β-adrenergic receptor mutants in the study of desensitization, an approach pioneered by Dixon, Sigal and Strader at Merck Sharp and Dohme, offers an alternative (Strader *et al.* 1989). In collaboration with this group we have begun an exploration of the domains of the β-adrenergic receptor likely to be involved in the cA•PK$_c$-mediated desensitization (manuscript submitted). Cloning of the hamster β-adrenergic receptor by Dixon *et al.* (1986) revealed that there are two consensus sites for cA•PK in the primary sequence, one in the third intracellular loop of which ser 262 would be the predicted phosphorylation site, and another in the C-terminal domain of which ser 346 or ser 347 would be predicted phosphorylation sites.

We have examined the cA•PK-mediated desensitization of the native hamster β-adrenergic receptor and several mutant constructs expressed in L cells. These cells are ideal since the parent cell line does not have any detectable levels of the β-adrenergic receptor, but does have a good response to PGE_1 stimulation. Also, the expression of the plasmids containing the wild type and mutant receptors is stable, allowing long term studies. Three mutants have been studied in detail, one with the third loop consensus site deleted (residues 259-262), another with the C-terminal consensus site deleted (residues 343-348), and a truncation mutant (T(354)βAR) with residues from 354 to the COOH terminus deleted.

	Intact Cell				Cell-free	
	Epi treatment		PGE_1 treatment		$cAPK_c$ Treatment	
Mutant	Epi	PGE_1	Epi	PGE_1	Epi	PGE_1
βAR	+++	0	+++	++	+++	0
D(343-348)βAR	+++	0	++	++	+++	0
T(354)βAR	++	0	++	++	+++	0
D(259-262βAR	0	0	0	++	0	0

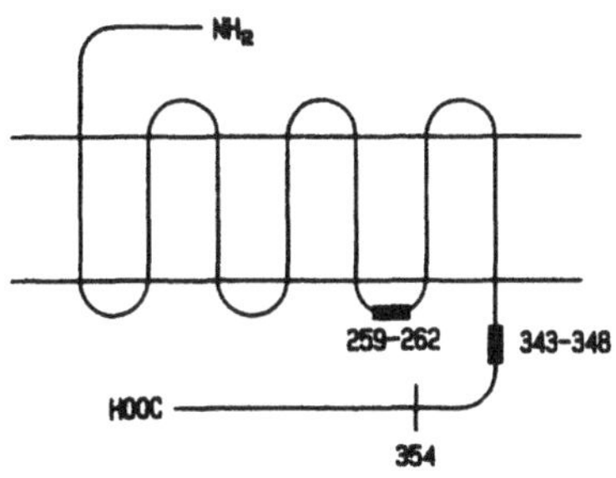

Figure 6. Summary of cA•PK-induced desensitization in β-adrenergic receptor mutants of the mouse L cell Intact L$_c$ cells expressing the wild type βAR or various mutants were treated for 7 minutes at 37^0 C with 50 nM epinephrine (Epi treatment) or 2 μM PGE_1 (PGE_1 treatment). Membranes were isolated on sucrose density gradients and adenylyl cyclase responses to concentrations near the EC_{50} were assayed. (Epi = 50 nM epinephrine, PGE_1 = 2 μM PGE_1). Cell-free cA•PK$_c$ treatments (cA•PK$_c$ at 25 μg/ml) were performed on sucrose density gradient purified membranes from βAR or mutant L cells. After 5 min at 30^0 C adenylyl cyclase responses to epinephrine and PGE_1 were measured. +++ indicates 45-55% desensitization, ++ indicates 25-35% desensitization, and 0 indicates no significant desensitization.

Preliminary results with these mutants indicate that the crucial domain for the cA•PK-mediated desensitization of the hamster β-adrenergic receptor is that containing ser 262 on the third intracellular loop. Deletion of the locus containing this serine residue prevents the cA•PK-mediated desensitization both in intact L cells, and in the cell-free protocol using L-cell membranes and purified cA•PK_c as discussed above. In contrast, deletion of the cA•PK consensus site on the C-terminal chain of the β-adrenergic receptor has no effect on the cA•PK-mediated desensitization of epinephrine stimulation, nor does a truncation of the C-terminal domain from residue 354 to residue 418. These results are summarized in fig 6.

Of further interest, the cA•PK-mediated desensitization affects only the β-adrenergic receptor response in the L cells; the PGE_1 stimulation is not altered. This illustrates an important point. Our results with the cA•PK-mediated desensitization of the S49 lymphoma cell and the DDT_1 MF-2 smooth muscle cell line led us to believe that the action of cA•PK on hormonal stimulation would be heterologous, affecting all stimulatory responses in mammalian cell lines. This is obviously not the case since in L cells the desensitization response can be, depending on the stimulatory hormone used to induce the desensitization, either heterologous or homologous. These descriptive terms, originally used by John Perkins' group to distinguish between desensitizations caused by two stimulatory receptors which affect either only their own stimulation of adenylyl cyclase (homologous) or that of both hormones (heterologous), have been of great value. New terms, such as hemi-heterologous suggest themselves for cases such as the L cell, but such an approach to nomenclature might well become burdensome as the mechanisms of desensitization are elucidated and become better understood on a molecular basis. Perhaps cA•PK-mediated desensitization should be called precisely that and additional terms used to further define the cA•PK-mediated effect. The rapid desensitization of the S49 WT adenylyl cyclase is clearly a heterologous receptor level cA•PK-mediated desensitization, in the L-cell it is, in part, a different process, while in the turkey erythrocyte or in the kin^-, where the cA•PK appears to alter both NaF- and β-adrenergic receptor-stimulated activities, there is yet another level of effects to be considered and described.

Reversal of cA•PK-mediated desensitization by cA•PK in the presence of high concentrations of ADP

After several unsuccessful attempts at reversing the cA•PK-mediated desensitization with nonspecific phosphatases such as alkaline phosphatase, we decided that it would be useful to attempt reversal of the cA•PK-mediated effect by incubation of membranes with ADP and cA•PK_c. This technique has been shown to be successful in studies of soluble enzymes (Flockhart 1983) and has the benefit that it retains the site-specificity of the kinase.

Preliminary studies with this technique indicate that the cA•PK-induced desensitization of the WT adenylyl cyclase can be reversed, whereas the homologous desensitization of the kin$^-$ cannot. We would like to suggest that this technique might be of general use to gain insight into the nature of the kinase which causes the intracellular phosphorylation of receptors.

Summary and Discussion

Considerable progress has been made in defining the physiological role of protein kinases in desensitization using the combination of genetics and biochemistry. The power of the genetic approach is especially useful in the intact cell studies allowing direct answers to otherwise seemingly intractable problems. If a cell lacks a kinase which has been proposed to mediate a type of desensitization, or if its activity can be selectively manipulated by genetic techniques, then its role can be more easily characterized. Similarly the construction of mutant receptors and chimeras and their expression in cells allows a parallel approach to the same questions addressed by the somatic cell mutants. These approaches have allowed us to determine the role of cA•PK in the S49 cells as well as the domain of the β-adrenergic receptor which is likely phosphorylated.

It is important at this point in our studies on the role of kinases in desensitization of the mammalian β-adrenergic receptor to step back and objectively review the extent to which the criteria outlined by Krebs and Beavo have been fulfilled. It is obvious that the cell-free and intact cell correlations between functional modifications and phosphorylation have not yet been accomplished for the cA•PK-mediated desensitization. Thus while we have gathered considerable evidence from the genetic approaches that cA•PK likely mediates the heterologous and homologous desensitization in the S49 lymphoma cells and the L cells respectively, there is more to be done to fulfill the criteria for a physiologically meaningful phosphorylation-dephosphorylation.

Since the receptor may be phosphorylated by multiple kinases it is important that we not just measure phosphorylation of the β-adrenergic receptor but rather define exactly which amino acid is phosphorylated. For example, Strasser *et al.* (Strasser *et al.* 1986) have nicely demonstrated that the β-adrenergic receptor in the S49 WT, kin$^-$ and cyc$^-$ is phosphorylated. However the phosphorylation was measured after treatment of the intact cells with high concentrations of isoproterenol. Since these concentrations cause heterologous and homologous desensitization of β-adrenergic receptor stimulation of adenylyl cyclase, sequestration and internalization, and down regulation, it is not clear which of these processes are correlated with the phosphorylation. Obviously the phosphorylation of the β-adrenergic receptor observed by this group in the kin$^-$ and the cyc$^-$ cannot be attributed to heterologous desensitization. We believe that the complexity of the desensitization demands that in the future the specific domains and even the precise amino acid(s) modified be defined. To do otherwise could result in a meaningless correlation. A real stumbling block in establishing the Krebs-Beavo criteria has been the difficulty of purifying femtomolar amounts of the bar from small numbers of cells and then doing the necessary biochemistry to establish the specific amino acid residues phosphorylated. Perhaps this problem will be solved as the techniques of immunology are applied to the isolation of the β-adrenergic receptor and used in conjunction with standard chromatographic techniques.

Two accomplishments which we report here should enable us to achieve the correlation of phosphorylation/dephosphorylation with function. First, we have a system for cell-free cA•PK-mediated desensitization which mimics the intact cell desensitization. Second, we have established conditions in the intact S49 cell and the L cell where cA•PK-mediated desensitization of the β-adrenergic receptor can be studied in the absence of homologous desensitization; namely, with forskolin or PGE_1 treatment. We also have conditions for producing homologous desensitization in the absence of the cA•PK-mediated event by treating the cyc^- or kin^- S49 cells or the D(259-262)βAR L cell with high concentrations of epinephrine.

We have not determined which phosphatase catalyzes the reversal of the cA•PK-mediated desensitization in the S49 cells. While use of the $cA{\cdot}PK_c$/ADP reversal technique demonstrated that the functional effects can be reversed, the restoration of function must ultimately be correlated with the loss of a specific phosphorylated domain of the β-adrenergic receptor with the appropriate phosphatase.

There are similar deficiencies in our understanding of the role of protein kinase C-mediated desensitization. Since the phorbol esters seem to specifically activate PKC we can be reasonably assured that PKC is involved in this desensitization of the β-adrenergic receptor in the intact cell. However, there is no evidence concerning the specific domains of the β-adrenergic receptor modified, nor do we yet know whether PKC treatment of membranes will mimic the intact cell effects of the phorbol esters on the β-adrenergic receptor. The PKC story is further complicated by the fact that the active phorbol esters also cause the sensitization of hormonal stimulation and the loss of G_i-mediated inhibition of adenylyl cyclase.

In an effort to gain more insight into the mechanisms of PKC-mediated sensitization and desensitization we have evaluated several drugs which have been reported to be specific inhibitors of PKC. We have found them either to be ineffective or to have multiple, non-specific effects when used on intact S49 cells (John Johnson, unpublished results). For example, while we have found that the drug sphingosine appeared to block epinephrine-induced desensitization and PMA-induced sensitizations of cAMP accumulations, in depth characterization of sphingosine's action revealed multiple effects which could explain the observations but which do not involve the inhibition of PKC. These effects included a sphingosine-induced detergent-like permeabilization of the plasma membranes, inhibition of hormone-stimulated adenylyl cyclase activity, and an inhibition of the cellular mechanisms for the elimination of cAMP, presumably phosphodiesterases. These nonspecific effects preclude its use as a specific inhibitor of PKC in the S49 cell lines.

Lefkowitz's group has suggested that BARK, a kinase which recognizes only the agonist-occupied form of the β-adrenergic receptor, phosphorylates the β-adrenergic receptor on the C-terminal chain causing the homologous desensitization and the internalization of the β-adrenergic receptor (Benovic *et al.* 1986). More recently they have elaborated on their model and have suggested that an arrestin-like protein is necessary to cause the uncoupling of adenylyl cyclase after phosphorylation by BARK since pure BARK has no significant effects on β-adrenergic receptor function even though it phosphorylates multiple residues of the receptor (Benovic *et al.* 1987). The evidence that BARK is involved in a physiologically meaningful desensitization process is almost totally dependent upon the *in vitro* data which demonstrates that BARK phosphorylation of pure lung β-adrenergic receptors in reconstituted liposomes is dependent on a high level of β-adrenergic receptor occupancy. It has not yet been proven whether treatment of membranes with BARK and/or arrestin causes any functional effects on adenylyl cyclase activities in the manner used in the studies with cA•PK . To summarize, there is no conclusive evidence which supports a role for BARK in the intact cells. What is needed is either i) a specific permeable activator; ii)an inhibitor of BARK; or iii) a BARK mutant analogous to the S49 kin^- mutant. It has been suggested that heparin is specific for BARK but unfortunately heparin is not permeable (Lohse *et al.* 1989) and its specificity remains to be proven.

References

Barber R (1986) Discrimination between intact cell desensitization and agonist affinity changes. Mol. and Cell. Endo. 46:263-270

Barber R, Goka TJ, Butcher WR (1984) Effects of desensitization on the responses of WI-38 and S49 cells to hormones. Mol. Cell Endocrinol. 36:29-35

Benovic JL, Kuhn H, Weyand I, Codina J, Caron MG, and Lefkowitz RJ (1987) Functional desensitization of the isolated β-adrenergic receptor by the β-adrenergic receptor kinase: Potential role of an analog of the retinal protein arrestin (48-kDa protein). Proc. Natl. Acad. Sci. USA 84:8879-8882

Benovic JL, Strasser RH, Caron MG, and Lefkowitz RJ (1986) β-adrenergic receptor kinase: Identification of a novel protein kinase that phosphorylates the agonist-occupied form of the receptor. Proc. Natl. Acad. Sci. USA 83:2797-2801

Butcher RW, Goka TJ, and Barber R (1987) Desensitization and the control of cellular cAMP levels. In: Cabot, MC and McKeehan W (eds) Mechanisms of Signal Transduction by Hormones and Growth Factors, Proceedings of the Second International Symposium on Cellular Endocrinology, Lake Placid, NY. Liss, New York, p 21-30

Clark RB, Friedman J, Johnson JA, and Kunkel MW (1987) β-adrenergic receptor desensitization of wild-type but not cyc^- lymphoma cells unmasked by submillimolar Mg^{2+}. FASEB J 1:289-297

Clark RB, Friedman J, Prashad N, and Ruoho AE (1985) Epinephrine-induced sequestration of the β-adrenergic receptor in cultured S49 WT and cyc^- lymphoma cells. J Cyclic Nucleotide and Protein Phosphorylation Res. 10:97-119

Clark RB, Kunkel MW, Friedman J, Goka TJ, and Johnson JA (1988) Activation of cAMP-dependent protein kinase is required for heterologous desensitization of adenylyl cyclase in S49 wild-type lymphoma cells. Proc. Natl. Acad. Sci. USA 85:1442-1446

Dixon RAF, Kobilka BK, Strader DJ, Benovic JL, Dohlman HG, Frielle T, Bolanowski MA, Bennett CD, Rands E, Diehl RE, Mumford RA, Slater EE, Sigal IS, Caron MG, Lefkowitz RJ, and Strader CD (1986) Cloning of the gene and cDNA for mammalian β-adrenergic receptor and homology with rhodopsin. Nature 321:75-79

Flockhart DA (1983) Removal of phosphate form protein by the reverse reaction. Meth. Enz. 99:14-20

Green DA, Friedman J, and Clark RB (1981) Epinephrine desensitization of adenylate cyclase from cyc^- and S49 cultured lymphoma cells. J Cyclic Nucleotide Res. 7:161-172

Grubbs RD, Collins SD, and Maguire ME (1984) Differential compartmentation of magnesium and calcium in murine S49 lymphoma cells J. Biol. Chem. 259:12184-12192

Johnson JA, Goka TJ, and Clark RB (1986) Phorbol ester-induced augmentation and inhibition of epinephrine-stimulated adenylate cyclase in S49 lymphoma cells. J. Cyclic Nucleotide Res. 11:199-215

Krebs EG and Beavo JA (1979) Phosphorylation-dephosphorylation of enzymes. Ann. Rev. Bioc. 48:923-959

Kunkel MW, Friedman J, Shenolikar S, and Clark RB (1989) Cell-free heterologous desensitization of adenylyl cyclase in S49 lymphoma cell membranes mediated by cAMP-dependent protein kinase. FASEB J in press

Lohse MJ, Lefkowitz RJ, Caron MG, and Benovic JL (1989) Inhibition of β-adrenergic receptor kinase prevents rapid homologous desensitization of β_2-adrenergic receptors. Proc. Natl. Acad. Sci. USA 86:3011-3015

Strader CD, Sigal IS, and Dixon RAF (1989) Structural basis of β-adrenergic receptor function. FASEB J 3:1825-1832

Strasser RH, Sibley DR, and Lefkowitz RJ (1986) A novel catecholamine activated adenosine cyclic 3',5'-phosphate independent pathway for β-adrenergic receptor phosphorylation in wild-type and mutant S49 lymphoma cells: Mechanism of homologous desensitization of adenylate cyclase. Biochemistry 25:1371-1377

DISCUSSION

Simon: You seem to be able to account so nicely for the desensitization with the cAMP effect, and to some extent the phosphorylation. Why do you have to invoke phosphorylation of the G-protein as a mechanism of desensi- tization?

Clark: Phosphorylation of the G-protein?

Simon: Yes, I thought that in one of your earlier slides, you were worried about G_i being phosphorylated.

Clark: That is with regard to the phorbol ester-PKC induced augmentation (sensitization) of hormonal stimulation which is accompanied by the loss of G_i-mediated inhibition of AC. There is good evidence that PMA treatment impairs G_i function. Whether the sensitization caused by PMA is caused by direct phosphorylation of G_i is not known. Phosphorylation of G_i could result in the loss of its inhibitory effect on adenylyl cyclase and the augmentation of hormonal stimulation. Resolution of whether G_i phospho- rylation by PKC plays a role in PMA-induced sensitization should be resolved in the next couple of years. That is the only example I know of where modification of a G-protein in S49 cells is implicated in sensiti-zation or desensitization. The effect of the A-kinase is almost certainly at the receptor level, since the heterologous desensitization occurs in the cyc^- with addition of dibutyryl cAMP. Reconstitution studies indicate that there is no modification of Gsα following activation of either PKC or cAPK. Further, there is no obvious cAPK concensus site in the primary structure of the G-protein α-subunits. Modification of the βτ subunits remains a possibility.
We must keep in mind that the PKC inhibition of G_i function could be indirect; e.g., by stimulation of the ADP-ribosylation of G_i. ADP-ribosylation of G-proteins has been indirectly linked to what we might think of as adaptation and desensitization. Its involvement in the PMA-induced sensitization has not been eliminated; in fact, we have shown that the augmentation of hormonal stimulation by PMA is very similar (and not additive) to that caused by pertussis toxin.

Houslay: Is C-kinase induced desensitization as extensive as A-kinase desensitization?

Clark: Yes, in fact we usually observe greater desensitization of epi stimulation following PMA activation of PKC. Within 5 to 10 minutes, PMA causes about a 3-4 fold increase in the K_{act} for either epi or PGE_1 activation of adenylyl cyclase. The cAPK-mediated desensitization causes about a 2.0 fold increase in the K_{act} for hormonal stimulation.

Houslay: So if you go to maximal desensitization with A-kinase and then treat cells with TPA, do you observe additional desensitization?

Clark: We have not done those additivity experiments. We want to do these experiments both with intact cells and with cell-free preparations. We have been trying to observe the C-kinase effects in a cell-free system, but it is turning out to be a lot more intractable problem than the cAPK-mediated desensitization. The consensus sites for the C-kinase on the hamster β-adrenergic receptor are identical to those for cAPK namely serines 261,262 on the third cytoplasmic loop and serines 345 and 346 on the C-terminal tail. It is interesting that the functional effects of PKC and cAPK activation on the βR activation are very similar, if we factor out the PKC effect on G_i.

Houslay: So you have not done the deletion studies then, with phorbol ester?

Clark: Those experiments are being done right now (see comments below).

Devreotes: You said that under conditions where you get a nice desensi- tization, a heterologous cAMP kinase mediated desensitization, you cannot detect receptor phosphorylation, in vitro or in vivo?

Clark: We have not yet been able to observe phosphorylation of the βR in intact cells with concentrations of epinephrine below 50 nM or with the cell-free desensitization protocol, although I think the phosphorylation is occurring. We simply are not good enough yet at resolving the βR from the background labeling. We are keeping an open mind, since it is conceivable that something other than the βR is phosphorylated.

Devreotes: The deletion mutant would suggest that it is the receptor. When Caron and Lefkowitz get phosphorylation of the βR in mammalian cells in response to high concentrations of catecholamines, do they measure approximately 3 moles PO_4 per mole of βR or something like that?

Clark: They have done a lot of different phosphorylations, do you mean in intact cells or cell-free?

Devreotes: In intact cells.

Clark: Strasser et al. Biochem. 25:1371 (1986) have shown that the β-receptor is phosphorylated in the S49 wild type, cyc^- and the kin^- lymphoma cells to an approximate stoichiometry of 1.0 mol PO_4/mol of βR following treatment with 10 μM isoproterenol.

Devreotes: Right, so my question is that if you can see 1.0 mole PO4 per mole βR, what is the amount that you cannot see? I mean, is it much less than a mole per mole that you are getting.

Clark: Again, the problem may simply be that we are not obtaining sufficient purification of the βR on the alprenolol affinity column since we would predict a stoichiometry of 1 mol PO4/mol βR. Another problem is the estimation-calculation of stoichiometry. I think anyone in this room would agree that the measurement of the stoichiometry of labeling of the βR is at best an approximation at this point. There are a lot of calculations, assumptions, and measurements involved in estimating stoichiometry (i.e. levels of the βR, assumption that βR population is homogeneous, spec. act. of the ATP pool, measuring low levels of counts in the βR).

Mulle: Is there any possibility of cross desensitization between hormones like glucagon and epinephrine in your system?

Clark: There are no glucagon receptors in the S49 lymphoma cells. If there were, and if they increased adenylyl cyclase, there would likely be heterologous desensitization mediated by cAPK. If you mean cross desensitization from, for example, activation of phospholipase C; we do not have direct evidence for that in our S49 cells. It is all indirect, that is through activation of protein kinase C with phorbol esters. I wish we knew of a hormone that activates phospholipase C in the S49 lymphoma cell or L-cell. We could then determine whether a physiological activation of PCK regulates adenylyl cyclase. I know of only one study (Abou-Samra et al. (1987) JBC 262: 1129-1136) and that shows a hormonal (vasopressin) augmentation (sensitization) of cAMP which is nearly identical to the augmentation caused by phorbol esters. I think that this result is real important and needs to be duplicated in other model systems.

Corbin: When you delete the cyclic AMP kinase concensus sites in L-cells, do you still have protein kinase C effects?

Clark: We do not yet have a definitive answer. The prediction is that we should not see the PMA-induced desensitization of the β-receptor, since the consensus sites for PKC on the third loop of the βR, ser 261, is just one residue removed from the cAPK consensus site (ser 262). The concensus site for cAPK and PKC on the C-terminal tail is the same (ser 346). Of course we should still observe the PMA-induced augmentation or sensitization of hormonal stimulation in all of the mutants since it likely involves G_i. Preliminary results do seem to indicate that the third loop concensus site is the critical one for PKC-mediated desensitization and that the sensitization is unaffected in the various mutants.

DIRECT INVOLVEMENT OF THE CYCLIC NUCLEOTIDE BINDING SITES IN THE CYCLIC-NUCLEOTIDE-INDUCED CHARGE SHIFT OF PROTEIN KINASES

Lynn Wolfe[*], Terry A. Woodford[$], Sharron H. Francis[*$], and Jackie D. Corbin[*$]

[*]Department of Molecular Physiology and Biophysics and [$]the Howard Hughes Medical Institute, Vanderbilt University, Nashville, TN 37232-0295

SUMMARY

The holoenzymes of cAMP- and cGMP-dependent protein kinases (cAK and cGK) exist as cyclic nucleotide-free and -bound forms (C.E. Cobb, A.H. Beth, and J.D. Corbin (1987) J. Biol. Chem. 262, 16566-74; L. Wolfe, S.H. Francis, L.R. Landiss, and J.D. Corbin (1987) J. Biol. Chem. 262, 16906-13), which can be resolved by DEAE chromatography due to a cyclic nucleotide-induced increase in surface electronegativity. In the present study, the isolated type I and type II regulatory subunits (R_I and R_{II}) of cAK also exhibited an electronegative charge-shift in the presence of cAMP, implying that the catalytic subunit was not involved in the apparent change in conformation resulting in the charge-shift of the holoenzyme. A mutant bovine R_I that contained the cAMP binding sites, but was missing regions involved in dimerization and interaction with the catalytic subunit, was expressed in E. coli and purified. This truncated R_I also demonstrated an electronegative charge-shift following cAMP binding, indicating that the binding sites themselves were directly involved in the shift. The increase in surface electronegativity of the cyclic nucleotide binding sites may either be a reflection of an overall conformational change due to cyclic nucleotide binding or it could be directly involved in regulating function.

[1]The abbreviations used are: cAK, cAMP-dependent protein kinase; cGK, cGMP-dependent protein kinase; HPLC, high performance liquid chromatography; R_I and R_{II}, types I and II regulatory subunits of cAMP-dependent protein kinase; C, catalytic subunit of cAMP-dependent protein kinase.

NATO ASI Series, Vol. H 44
Activation and Desensitization of Transducing Pathways
Edited by T. M. Konijn, M. D. Houslay, P. J. M. Van Haastert

INTRODUCTION

The cAMP-dependent protein kinase (cAK)[1] and cGMP-dependent protein kinase (cGK) are thought to be homologous proteins, with overall similarities in molecular structure and mechanism of activation (Lincoln and Corbin, 1977; Gill, 1977; Takio et al., 1984). The cAK is a tetramer composed of two catalytic (C) and two regulatory (R) subunits. The two main isozymic forms of cAK, types I and II, have the same C subunits, but different R subunits (R_I and R_{II})(Corbin, Keely, Park, 1975; Nimmo, Cohen, 1977; Carlson, Bechtel, Graves, 1979; Doskeland, Ogreid, 1981). The cGK also has two R and two C domains, but since R and C are joined by peptide bond, this enzyme is a dimer (Gill et al., 1976; Lincoln et al., 1976; Lincoln, Dills, Corbin, 1977). There are also at least two isozymic forms of this enzyme (Wolfe, Corbin, Francis, 1989). The cAK and cGK dimerize in the amino-terminal region of the R components. The cAK and cGK R components each have two cyclic nucleotide binding sites which differ in cyclic nucleotide affinity and analog selectivity (Rannels, Corbin 1980; Rannels, Corbin, 1981; Corbin et al., 1982; Mackenzie, 1982; Corbin et al., 1986; Doskeland et al., 1987). The slowly-exchanging binding site is designated site 1 for each of these enzymes, and the rapidly-exchanging site is designated site 2[2]. The cAK and cGK each has an inhibitory domain that is believed to lie between the dimerization domain and the two cyclic nucleotide binding domains in the primary sequence (Takio et al., 1984; Corbin et al., 1978; Flockhart, Corbin, 1982; Weldon, Taylor, 1985; Hofmann, Gensheimer, Gobel, 1985). This region interacts with the respective C component to suppress C activity. However, the inhibitory function is blocked when the cyclic nucleotide binding domains are occupied.

The two intrasubunit cyclic nucleotide binding sites of the regulatory components of cAK and cGK are relatively specific for the respective cyclic nucleotide (Corbin et al., 1986). The binding sites have apparently

[2]The two intrasubunit binding sites of cyclic nucleotide-dependent protein kinase are denoted as the slowly exchanging site and the rapidly exchanging site or as site 1 and site 2, respectively. These sites have also been referred to as sites I and II (Mackenzie, 1982) and sites B and A (Doskeland et al., 1987), respectively.

evolved by gene duplication, and are homologous with the bacterial cAMP-binding protein termed CAP (Weber et al., 1987). Based on the crystallographic structure of CAP, studies of cyclic nucleotide analogs, molecular modeling, sequence homologies and mutagenesis of cyclic nucleotide binding sites of the protein kinases, certain predictions and conclusions can be drawn about the eucaryotic cyclic nucleotide binding sites (Weber et al., 1987; Weber et al., 1989). The consensus binding site is comprised of a sequence of ~120 amino acids, which contains three α-helices and eight β-strands that form a β-barrel structure. Substitution of glycine in a critical turn between two β-strands in cAMP-binding sites causes large decreases in binding affinity, presumably by disrupting the β-barrel (Woodford et al., 1989). Cyclic nucleotides bind in the syn conformation, with the equatorial exocyclic phosphate oxygen forming an ionic interaction with an arginine residue and the 2'OH forming a hydrogen bond with a glutamic acid residue. A threonine residue that is present only in cGMP-binding sites is hydrogen-bonded to the 2-NH_2 group of cGMP (Weber et al., 1989). This position apparently allows for discrimination between cAMP and cGMP binding. Based on results of cylic nucleotide analog studies, different isozymes of cAK and cGK differ in the structures of the binding sites surrounding the purine moiety.

Both cAK and cGK can exist as ternary complexes containing regulatory components, catalytic components and cyclic nucleotides. The type II isozyme of cAK (Cobb, Beth, Corbin, 1987) and type Iα cGK (Wolfe et al., 1987) have been shown to bind more tightly to DEAE resins when 50% of their cyclic nucleotide binding sites are filled. This permits the separation of cyclic nucleotide-free and cyclic nucleotide-bound enzyme forms. The cAMP-bound form of cAK has cAMP equally distributed between sites 1 and 2, is inactive, and has a higher Hill coefficient than does the cAMP-free enzyme. The K_a for cAMP is similar for the cAMP-free and -bound forms. The cGMP-bound form of cGK shows cGMP saturation of only sites 1, is 50% activated, and has a lower K_a for cGMP than does the cGMP-free enzyme. The Hill coefficients are similar for the cGMP-free and -bound enzyme forms. The higher Hill coefficient for the cAMP-bound cAK and the lower K_a for cGMP for the cGMP-bound cGK indicate that both of these enzymes are "primed" for activation by small increases in cyclic nucleotide concentrations, albeit through different mechanisms.

The cyclic nucleotide-induced charge shifts of cAK and cGK suggest that these enzymes may undergo similar cyclic nucleotide-induced conformational changes. In addition to suggesting similar domain substructure and primary sequence homology, this provides further evidence for a close evolutionary link among the cyclic nucleotide-dependent protein kinases. The cyclic nucleotide-induced conformational change, which increases surface electronegativity, may be a key event in the overall activation process for the protein kinases.

The present studies are designed to identify the domain(s) of the holoenzyme that are responsible for the charge-shift. The cAK is used for the studies because, unlike the cGK, the cAK has separate R and C subunits, an advantage that allows for isolation and independent study of R and C. Since a conformational change resulting from the binding of cyclic nucleotides could occur in the R subunit, which contains the cAMP binding sites, isolated bovine cAK R_I and R_{II} and a recombinant truncated mutant form of R_I lacking the dimerization and inhibitory domains are used.

EXPERIMENTAL PROCEDURES

Purification of R subunits

R_I was purified from bovine lung and R_{II} was purified from bovine heart as described (Rannels, Corbin, 1983). Tissues were obtained fresh at a local slaughterhouse and transported to the laboratory on ice, where all of the manipulations were done at 4^oC. Ten kg of tissue (heart or lung) was ground and homogenized in 4 volumes of 10 mM KH_2PO_4/ 1 mM EDTA, pH 6.8 (buffer A) in a Waring Blender (three 30-second bursts at high speed). The homogenate was centrifuged at 12,000 x g for 30 minutes at 4^oC, and the supernatant was filtered through several layers of glass wool and the pellets were discarded. The supernatant was then mixed batchwise with 10:1 (volume:volume) DE-52 cellulose (approximately 4 liters of DE-52) and stirred occasionally for 2 hours. After decanting the supernatant, the DE-52 was washed on a Buchner funnel under vacuum with buffer A until the eluate was colorless (~ 40 liters). The resin was then packed into a glass column (75 x 7.5 cm). The enzymes were eluted by washing the column with 6 liters of 300 mM NaCl in buffer A. Fractions (500 ml) were collected and assayed for [^{3}H]cAMP binding activity (see below). Active fractions were precipitated with 400 g/l ammonium sulfate and then centrifuged at 12,000 x g for 20 minutes. The pellets were resuspended in 100 ml of buffer A. The pooled sample was dialyzed overnight against several changes of buffer A and centrifuged at 12,000 x g for 20 minutes to remove any remaining insoluble material. The sample was applied to a 1-ml affinity column, which was 8-(aminohexyl)-amino-cAMP agarose for R_{II} and 6-(aminoethyl)-amino-cAMP agarose for R_I. The respective columns were washed sequentially with 10 ml each of buffer A, buffer A containing 2 M NaCl, buffer A, buffer A containing 10 mM 5'-AMP, and buffer A. Each column was then eluted with four 2-ml overnight elutions of 10 mM cGMP. The pooled elutions were loaded onto a DEAE-Sephacel column (0.9 x 5 cm) equilibrated in buffer A, and the column was washed extensively with buffer A for R_I or buffer A containing 50 mM NaCl for R_{II} (2 liters) to remove the cGMP from the enzyme. Different 150-ml linear NaCl gradients were used to elute R_I (0 to 100 mM) and R_{II} (50 to 300 mM) from the DEAE-Sephacel column. Fractions (2-ml) were collected and assayed for [^{3}H]cAMP binding activity. Both R_I and R_{II} preparations were judged to be homogeneous based on analysis by SDS PAGE.

Preparation and Purification of Recombinant Truncated R_I

A pBR322 plasmid containing the entire coding region of bovine testis R_I plus flanking sequence was kindly provided by Dr. G.S. McKnight (Department of Pharmacology, University of Washington). The R_I cDNA was subcloned into pUC13 using a modification of the procedure described by Saraswat *et al.* (1986). Following partial digestion of the plasmid with NcoI and NarI, the generated fragments were "blunt-ended" with DNA polymerase I (Klenow fragment). A 1155 base pair R_I cDNA fragment was gel-purified, then ligated to Hinc II-linearized phosphatase-treated pUC13.

Expression of the R_I cDNA was done in *E.coli* strain 222, a double mutant which lacks adenylate cyclase and has a defective CAP protein that binds DNA in the absence of cAMP (Puskas *et al.*, 1983). Following transformation of *E.coli* strain 222 with the recombinant pUC13 plasmid, bacterial colonies were detected by ampicillin resistance, insertional inactivation of the lac Z gene and photolabeling of expressed R_I protein with [^{32}P]8-N_3-cAMP following the transfer of bacterial colonies to nylon membranes.

Full-length R_I was expressed by the majority of bacterial clones selected. However, one clone expressed a truncated form of R_I from which the first 109 amino acids, as determined by DNA sequence analysis, were deleted. As with the full-length R_I, the truncated R_I, designated R_{tr}, was a fusion protein and containing an additional 11 amino acids at the amino terminus (Thr-Met-Ile-Thr-Pro-Ser-Leu-Gly-Cys-Arg-Ser).

For isolation of R_{tr}, Luria broth (2 liters) containing 25 mg/ml ampicillin was inoculated with 100 ml exponentially-growing bacteria (A_{600}= 0.6) and grown overnight at 37°C. The cells were then pelleted by centrifugation at 12,000 x g for 20 minutes at 4°C, and resuspended in 40 ml buffer A. After freezing at -70°C and thawing, the cells were lysed by a single passage through a French pressure cell at 15,000 psi. The lysate was centrifuged at 12,000 x g for 20 minutes at 4°C and the R_{tr} was then purified from the supernatant using the above procedure for R_I beginning with the affinity column step.

The cAMP binding sites of R_{tr} were functionally identical to those of the full-length R_I. R_{tr} was found to have a binding stoichiometry of 2 mol [^{3}H]cAMP/ mol subunit, and dissociation of [^{3}H]cAMP gave a characteristic biphasic curve. R_{tr} would not, however, recombine with the catalytic subunit at any concentration.

High Performance Liquid Chromatography (HPLC)

All HPLC was done using a BioRad Biosil TSK-DEAE-5-PW column (75 x 7.5 mm) connected to a Beckman 344 HPLC system with a 421 controller, a 165 variable wavelength detector, and two 112 analytical solvent delivery modules.

[^{3}H]cAMP Binding Assay

[^{3}H]cAMP binding was measured as described by Sugden and Corbin (1976). The [^{3}H]cAMP binding mix contained 50 mM KH_2PO_4 at pH 6.8, 1 mM EDTA, 2 M NaCl, 0.5 mg/ml histone IIA, 1 uM cAMP, and 0.3 uM [^{3}H]cAMP (15

Ci/mmol). Fifty ul of [^{3}H]cAMP binding mix was added to 50 ul of each HPLC-DEAE fraction and the samples were incubated for 30 minutes at 30°C to allow the binding to reach equilibrium. The reaction was stopped by adding 1 ml of ice-cold buffer A (see above), and bound [^{3}H]cAMP was separated from the free form by filtration through a 0.45 um nitrocellolose filter. The filter was rinsed 4 times with 1 ml ice cold buffer A, dried, and radioactivity was determined by scintillation counting The activity was expressed as picomoles of [^{3}H]cAMP bound per ml of enzyme sample.

Determination of [^{3}H]cAMP Retained During Chromatography

An aliquot of each HPLC-DEAE fraction was added to 4 ml aqueous scintillation fluid and counted. The specific activity of the stock [^{3}H]cAMP was used to calculate the amount of [^{3}H]cAMP retained per ml. Free [^{3}H]cAMP eluted very early in the gradient and was well-separated from the enzymes and protein-bound [^{3}H]cAMP.

[^{3}H]cAMP Dissociation Assay

NaCl and histone IIa were added to 75 mM and 0.75 mg/ml final concentrations, respectively, to forty ul of HPLC-DEAE fractions containing protein-bound [^{3}H]cAMP and precipitated with 2 ml of saturated aqueous ammonium sulfate for determination of total [^{3}H]cAMP bound (B_0). Five ul of 10 mM cAMP in 3 M NaCl and 30 mg/ml histone IIA was then added to 200 ul of each fraction and the samples were placed at 30°C. Forty ul aliquots were taken at various times and added to 2 ml saturated ammonium sulfate. The samples were then filtered through 0.45 um nitrocellulose filters prewetted with 1 ml deionized water. The filters were rinsed twice with 2 ml saturated ammonium sulfate, placed into vials containing 1 ml 2% SDS, capped, and shaken. Four ml of aqueous scintillation fluid was then added and the vials were capped, shaken, and counted. The data were plotted as the natural logarithm of the ratio of bound [^{3}H]cAMP at a given time to B_0 versus time.

Protein Determination

Protein determination was by the method of Bradford (1976) using bovine serum albumin fraction V as standard.

MATERIALS

[^{3}H]cAMP was obtained from Amersham. 8-(aminohexyl)-amino-cAMP agarose, 6-(aminoethyl)-amino-cAMP agarose, and DEAE-Sephacel were from Pharmacia. DE-52 was obtained from Whatman, and restriction enzymes were purchased from New England Biolabs and Boehringer Mannheim.

RESULTS

When purified cyclic nucleotide-free R_I was chromatographed on HPLC-DEAE, a single peak of [^{3}H]cAMP binding activity was observed (Fig. 1, panel A). However, when 250 nM [^{3}H]cAMP was added to the R_I immediately prior to chromatography, a second peak of binding activity eluting at higher ionic strength was observed, and most of the [^{3}H]cAMP retained during chromatography eluted with this second peak (Fig. 1, panel B). Thus, just as for the cAK holoenzyme (Cobb, Beth, Corbin, 1987), the second peak of R_I appeared to be a cAMP-bound form and the first peak was

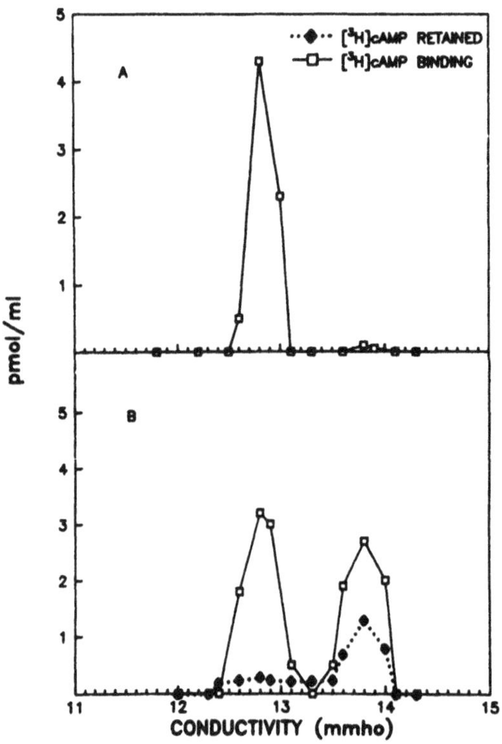

Fig. 1. Effect of [^{3}H]cAMP preincubation on HPLC-DEAE profile of bovine lung R_I. Purified R_I alone (panel A) or mixed with 250 nM (panel B) [^{3}H]cAMP in a final volume of 250 ul were immediately injected onto a BioRad TSK-DEAE-5-PW HPLC column equilibrated in 10 mM KH_2PO_4 (pH 6.8), 1 mM EDTA, and maintained at 2.4°C. The column was eluted with a linear 60-ml gradient from 0 to 100 mM NaCl, and 1-ml fractions were collected. Purification of R_I, assay of [^{3}H]cAMP binding activity, and determination of [^{3}H]cAMP retained during chromatography were done as described in "Experimental Procedures".

relatively cAMP-free. It was calculated that the [^{3}H]cAMP coeluting with the second peak was sufficient to saturate 53% of the cAMP-binding sites. Some of the retained [^{3}H]cAMP was associated with the first peak of binding activity, but it was not contained within the distinct peak associated with the binding activity. The level of retained [^{3}H]cAMP in the first peak was too low to accurately determine whether it was free or protein-bound. It was possible that this represented [^{3}H]cAMP bound to one site of the R dimer, analogous to that which was found for the first peak of cGK holoenzyme (Wolfe _et al._, 1987). A second possibility was that this [^{3}H]cAMP was bound to two sites of the R dimer, but that the conformational change which increased surface electronegativity did not occur. A third possibility was that a portion of this [^{3}H]cAMP was the result of dissociation of the [^{3}H]cAMP bound to the second peak of R_I during chromatography. R_{II} gave results similar to those of R_I (not shown), with the first peak being 25% saturated and the second peak being 57% saturated with [^{3}H]cAMP. These results indicated that the R subunits alone could undergo the cyclic nucleotide-induced charge-shift in the absence of the C subunit.

Since the R subunits contain several functional domains (dimerization, C subunit interaction, and cAMP binding), it was necessary to determine which domains were involved in the charge-shift. A recombinant truncated mutant of R_I (R_{tr}) was used to study the involvement of the different domains of R in the charge-shift. Since R_{tr} was missing the first amino-terminal 109 amino acids, which includes the dimerization domain and the majority of the C subunit-interaction (inhibitory) domain, it contained essentially just the two cAMP binding domains (Fig. 2).

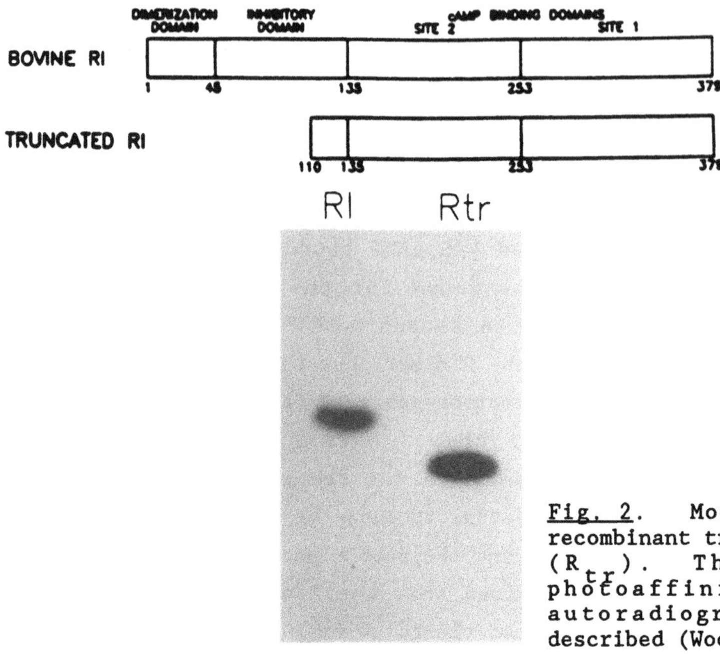

Fig. 2. Model and SDS PAGE of recombinant truncated mutant of R_I (R_{tr}). The 8-N_3[^{32}P]cAMP photoaffinity labeling and autoradiography were done as described (Woodford et al., 1989).

R_{tr} was also found to undergo a shift in elution from HPLC-DEAE when cAMP was bound to it. Fig. 3 shows HPLC-DEAE profiles of R_{tr} in the

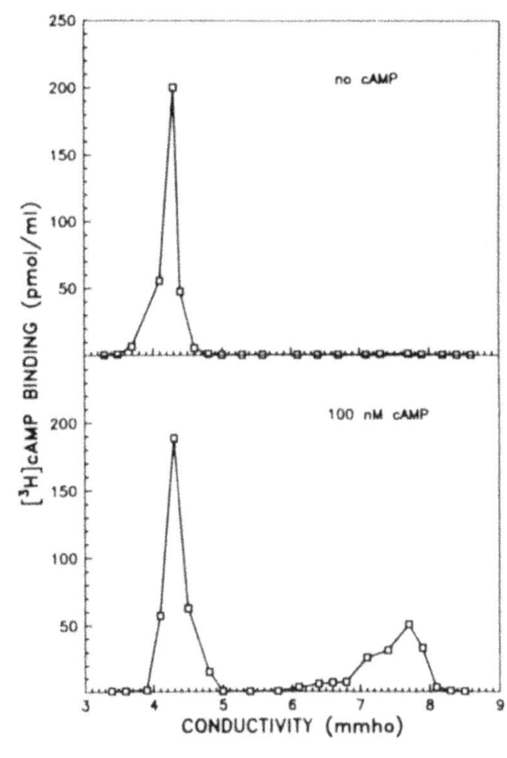

Fig. 3. Effect of [^{3}H]cAMP preincubation on HPLC-DEAE profiles of R_{tr}. Samples of purified R_{tr} with (bottom panel) and without (top panel) 100 nM [^{3}H]cAMP in a final volume of 250 ul were injected onto a BioRad TSK-DEAE-5-PW HPLC column (75 x 7.5 mm) equilibrated in 10 mM KH_2PO_4 (pH 6.8), 1 mM EDTA, and maintained at 2.4^{o}C. A linear 45-ml gradient from 0 to 75 mM NaCl was applied to the column and 1 ml fractions were collected. Preparation of R_{tr} and assay of [^{3}H]cAMP binding were done as described under "Experimental Procedures."

absence (top panel) and presence (bottom panel) of 100 nM [^{3}H]cAMP. Just as for R_I, R_{tr} eluted in one peak when cAMP was absent and in two peaks when cAMP was present. The majority of [^{3}H]cAMP again eluted with the second peak (not shown), indicating that the second peak was the cAMP-bound form of R_{tr}. 25% of the total binding sites in the first peak of R_{tr} contained [^{3}H]cAMP, while 51% of the binding sites in the second peak were saturated with [^{3}H]cAMP. These results indicated that the dimerization and inhibitory domains of the R were not necessary for the charge-shift to occur, and that the cAMP binding sites alone were sufficient.

Dissociation curves of the [^{3}H]cAMP bound to the second peak of R_{tr} (Fig. 4), R_I, and R_{II} (not shown) were biphasic, indicating that both sites

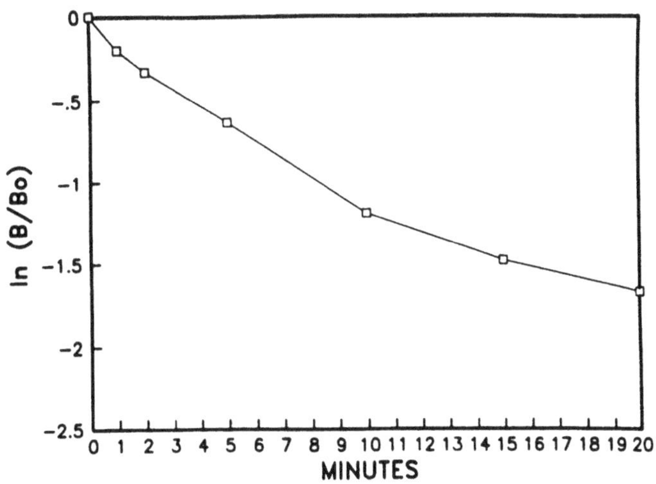

Fig. 4. Dissociation behavior of [^{3}H]cAMP bound to cAMP-bound form of R_{tr}. NaCl and histone IIb were added to 40 ul of peak fractions from the HPLC-DEAE cAMP-bound form of R_{tr} (Figure 3, lower panel) at a final concentration of 75 mM and 0.75 mg/ml, respectively. Samples were then precipitated with 2 ml saturated ammonium sulfate for determination of [^{3}H]cAMP bound at T=0 (B_o). Ten ul of 10 mM cAMP in 3 M NaCl and 30 mg/ml histone IIb were then added to 240 ul of each fraction, and the samples were placed at 30°C. Forty ul aliquots were taken at various times and added to 2 ml saturated ammonium sulfate for determination of bound [^{3}H]cAMP as described under "Experimental Procedures".

1 and 2 contained bound [^{3}H]cAMP. In the case of R_{tr}, the antilogarithm of the y-intercept (-0.70) of the line formed by the dissociation from site 1 gave an estimated value of 0.50, indicating that approximately 50% of the

[^{3}H]cAMP was bound to site 1, and approximately 50% was bound to site 2. When a similar analysis was done for R_I, 58% of the [^{3}H]cAMP was bound to site 1; and for R_{II}, 54% was bound to site 1.

DISCUSSION

The increased affinity of the cyclic nucleotide-bound enzymes for positively charged DEAE resins suggests that the net surface negative charge of these enzymes is increased with the binding of cyclic nucleotide. This is likely achieved by a conformational change in the binding sites which either exposes regions of negative charge or masks regions of positive charge on the surface of the proteins. Results presented here suggest that the catalytic subunit has little, if any, role in the cAMP-induced charge-shift that occurs with the holoenzyme on ion exchange columns. Since the R subunit contains the cAMP binding sites, it was thought that this subunit alone could undergo a charge-shift resulting from a conformational change induced by the binding of cAMP. The observation that the truncated R subunit, R_{tr}, also binds more tightly to DEAE when cAMP is bound to it suggests that the cAMP binding sites themselves are indeed involved in the charge-shift. Because the cAK and cGK are structurally and functionally homologous, it is probable that the cGMP-induced charge-shift of the cGK occurs within the cGMP binding sites in a manner similar to that of the cAK.

It has been proposed that fewer than 10 charged residues on the surface of a protein determine its chromatographic behavior (Regnier, 1987). The change in net surface electronegativity could occur due to the bound negatively-charged cAMP itself, to a conformational change induced in the cAMP binding site(s) due to cAMP binding, or to translation of the cAMP binding effects through the protein structure following cAMP binding. Since the cyclic phosphate of cAMP carries a negative charge, it would appear that binding of two cAMP molecules to the R subunit would greatly increase the overall negative charge of the complex. However, three-dimensional computer models of the cAMP binding sites of the R subunits based on the crystal structure of the E. coli CAP protein (Weber et al., 1987) suggest that the cyclic phosphate would not be near the surface of the R subunit-cAMP complex. It seems unlikely, therefore, that the negative charge of the cAMP molecule is the sole factor involved in the charge-shift, and that a conformational change in the cAMP binding site(s) is a more likely prospect. Structural analyses suggested that the negative charge of the cyclic phosphate of cAMP or cGMP interacts with a highly-conserved arginine residue within the binding sites of either cAK or cGK (Weber et al., 1987). When cAMP or cGMP binds, the charge on this arginine presumably becomes neutralized by the negative charge on the cyclic phosphate. This mechanism could be involved in the charge-shift, although computer modeling does not support a surface location of this arginine residue (Weber et al., 1987).

Although either the negative charge of the cyclic phosphate itself, or the masking of the positive charge of the arginine with the negative charge of the cyclic phosphate may play a role in the charge-shift, these are probably not the main factors involved. Since the cAMP-bound R subunits are each about 50% saturated with cAMP, the contribution of the arginine residues in the binding sites and the negative charges of the cAMP molecules would theoretically be the same for all R isoforms. However, the degree of the charge-shift is different for each of the R subunits (1.0 mmho for R_I, 1.3 mmho for R_{II}, and 3.2 mmho for R_{tr}), so there are obviously other factors that contribute to the charge-shift. The amino termini of R_I and R_{II} show the most divergence in primary sequence. The R_{tr} is missing this region, which could account for these differences. Exposure of negative charge or masking of positive charge must therefore occur somewhere downstream from Ser 110 due to structural change(s) induced by cAMP binding to sites 1 and 2.

It is necessary to consider at least two factors other than electrostatic forces when analyzing chromatographic behavior. These are (i) change in degree of hydration which may occur with a change in conformation and the associated changes in surface polarity, and (ii) changes in hydrophobic interactions between the enzymes and the support matrix of the DEAE resin. However, the charge-shift occurs on DEAE linked to either cellulose, Sephacel, or HPLC resin, indicating that interaction with the resin matrices is probably not important in the charge-shift.

Although the mechanism is as yet unknown, it appears that the cyclic nucleotide-induced charge-shift reflects a cyclic nucleotide-induced structural change that has been well-conserved through evolution. It could therefore be important in the physiological function of these enzymes, and may be involved in translating cyclic nucleotide binding into enzyme activation.

ACKNOWLEDGEMENTS

The authors would like to thank Dr. G. Stanley McKnight for providing the cDNA for bovine testis R_I, Dr. Fred Regnier for helpful discussions, and Alfreda Beasly-Leach for excellent technical assistance.

REFERENCES

Bradford MM (1976) A rapid and sensitive method for the quantitation of microgram quantities of protein utilizing the principle of protein-dye binding. Anal Biochem 72:248-254

Carlson GM, Bechtel PJ, Graves DJ (1979) Chemical and regulatory properties of phosphorylase kinase and cyclic AMP-dependent protein kinase. Adv Enzymol 50:41-115

Cobb CE, Beth AH, Corbin JD (1987) Purification and characterization of an inactive form of cAMP-dependent protein kinase containing bound cAMP. J Biol Chem 262:16566-16574

Corbin JD, Doskeland SO (1983) Studies of two different intrachain cGMP-binding sites of cGMP-dependent protein kinase. J Biol Chem 258:11391-11397

Corbin JD, Keely SL, Park CR (1975) The distribution and dissociation of cyclic adenosine 3':5'-monophosphate -dependent protein kinases in adipose, cardiac, and other tissues. J Biol Chem 250:218-225

Corbin JD, Ogreid D, Miller JP, Suva RH, Jastorff B, Doskeland SO (1986) Studies of cGMP analog specificity and function of the two intrasubunit binding sites of cGMP-dependent protein kinases. J Biol Chem 261:12081-1214

Corbin JD, Rannels SR, Flockhart DA, Robinson-Steiner AM, Tigani MC, Doskeland SO, Suva RH, Miller JP (1982) Effect of cyclic nucleotide analogs on intrachain site 1 of protein kinase isozymes. Eur J Biochem 125:259-266

Corbin JD, Sugden PH, West L, Flockhart DA, Lincoln TM, McCarthy D (1978) Studies on the properties and mode of action of the purified regulatory subunit of bovine heart adenosine 3':5'-monophosphate-dependent protein kinase. J Biol Chem 253:3997-4003

Doskeland SO, Ogreid D (1981) Binding proteins for cyclic AMP in mammalian tissues. Int J Biochem 13:1-19

Doskeland SO, Vintermyr OK, Corbin JD, Ogreid D (1987) Studies on the interactions between the cyclic nucleotide-binding sites of cGMP-dependent protein kinase. J Biol Chem 262:3534-3540

Flockhart DA, Corbin JD (1982) Regulatory mechanisms in the control of protein kinases. CRC Crit Rev Biochem 12:133-186

Gill GN (1977) A hypothesis concerning the structure of cAMP- and cGMP

dependent protein kinase. J Cyc Nuc Res 3:153-162
Gill GN, Holdy KE, Walton GM, Kanstein CB (1976) Purification and characterization of 3':5'-cyclic GMP-dependent protein kinase. PNAS USA 73:3918-3922
Hofmann F, Gensheimer HP, Gobel C (1985) cGMP-dependent protein kinase. Autophosphorylation changes the characteristics of binding site 1. Eur J Biochem 147:361-365
Lincoln TM, Corbin JD (1977) Adenosine 3':5'-cyclic monophosphate dependent protein kinases: Possible homologous proteins. PNAS, USA 74:3239-3243
Lincoln TM, Hall CL, Park CR, Corbin JD (1976) Guanosine 3':5'-cyclic monophosphate binding proteins in rat tissues. PNAS USA 73:2559-2563
Lincoln TM, Dills WL, Corbin JD (1977) Purification and subunit composition of guanosine 3':5'-monophosphate-dependent protein kinase from bovine lung. J Biol Chem 252:4269-4275
Lincoln TM, Flockhart DA, Corbin JD (1978) Studies on the Structure and mechanism of activation of the guanosine 3':5'-monophosphate-dependent protein kinase. J Biol Chem 253:6002-6009
Mackenzie CW (1982) Bovine lung cyclic GMP-dependent protein kinase exhibits two types of specific cyclic GMP-binding sites. J Biol Chem 257:5589-5593
Monken CE, Gill GN (1980) Structural analysis of cGMP-dependent protein kinase using limited proteolysis. J Biol Chem 255:7067-7070
Nimmo HG, Cohen P (1977) Hormonal control of protein phosphorylation. Adv Cyc Nuc Res 8:145-266
Puskas R, Fredd N, Gazdar C, Peterkofsky A (1983) Methylglyoxal-mediated growth inhibition in an Escherichia coli cAMP receptor protein mutant. Arch Biochem Biophys 223:503-513
Rannels SR, Corbin JD (1980) Two different intrachain cAMP binding sites of cAMP-dependent protein kinases. J Biol Chem 255:7085-7088
Rannels SR, Corbin JD (1980) Studies of functional domains of the regulatory subunit from cAMP-dependent protein kinase isozyme I. J Cyc Nuc Res 6:203-215
Rannels SR, Corbin JD (1981) Studies on the function of the two intrachain cAMP binding sites of protein kinase. J Biol Chem 256:787-7876
Rannels SR, Beasley A, Corbin JD (1983) Regulatory subunits of bovine heart and rabbit skeletal muscle cAMP-dependent protein kinase isozymes. Meth Enzymol 99:51-60
Regnier FE (1987) The role of protein structure in chromatographic behavior. Science 238:319-323
Reimann EM (1986) Conversion of bovine cardiac adenosine cyclic 3',5'-phosphate dependent protein kinase to a heterodimer by removal of 45 residues at the N-terminus of the regulatory subunit. Biochemistry 25:119-125
Saraswat LD, Filutowicz M, Taylor SS (1986) Expression of the type I regulatory subunit of cAMP-dependent protein kinase in Escherichia coli. J Biol Chem 261:11091-11096
Sugden PH, Corbin JD (1976) Adenosine 3':5'-cyclic monophosphate-binding proteins in bovine and rat tissues. Biochem J 159:423-437
Takio K, Wade RD, Smith SB, Krebs EG, Walsh, KA, Titani K (1984) Guanosine cyclic 3':5'-phosphate dependent protein kinase, a chimeric protein homologous with two separate protein families. Biochemistry 23:4207-4218
Weber IT, Steitz TA, Bubis J, Taylor SS (1987) Predicted structures of cAMP binding domains of type I and II regulatory subunits of cAMP-dependent protein kinase. Biochemistry 26:343-351

Weber IT, Shabb JB, Corbin JD (to be published) Predicted structures of the cGMP-binding domains of the cGMP-dependent protein kinase: A key alanine/threonine difference in evolutionary divergence of cAMP and cGMP binding sites. Biochemistry

Weldon SL, Taylor SS (1985) Monoclonal antibodies as probes for functional domains in cAMP-dependent protein kinase II. J Biol Chem 260:4203-4209

Wolfe L, Corbin JD, Francis SH (to be published) Characterization of a novel isozyme of cGMP-dependent protein kinase from bovine aorta. J Biol Chem

Wolfe L, Francis SH, Landiss LR, Corbin JD (1987) Interconvertible cGMP-free and cGMP-bound forms of cGMP-dependent protein kinase in mammalian tissues. J Biol Chem 262:16906-16913

Woodford TA, Correll LA, McKnight GS, Corbin JD (to be published) Expression and characterization of mutant forms of the type I regulatory subunit of cAMP-dependent protein kinase: The effect of defective cAMP binding on holoenzyme activation. J Biol Chem

DISCUSSION

Houslay: Does the phosphodiesterase that you talked about also hydrolyze cAMP? cGMP actually augments cAMP-hydrolysis by certain phosphodiesterases.

Corbin: No, this enzyme is quite specific for cGMP, both at the catalytic site and at the binding site. Although it could be in the same family of enzymes, this phosphodiesterase is quite different from the cGMP-stimulated cAMP phosphodiesterase. What the physiological role for this enzyme is, I don't know. Our present hypothesis is that cGMP-binding does activate the enzyme. This would be in line with what Peter Van Haastert has seen for a *Dictyostelium* phosphodiesterase and also what Joe Beavo has seen for a liver cAMP specific phosphodiesterase. This is also consistent with some other observations that we have made. For example, we know that the catalytic site is in communication with the binding site since analogs such as methylxanthines, that are specific for the catalytic site, stimulate cGMP binding. We have not shown the reversal - that binding stimulates the catalytic site.

Clark: I wondered if you had a strategy for finding the cGMP-kinase substrates. Do you see amplification of the endogenous phosphorylated proteins or do you see some number of them phosphorylated in the hepatocyte, like Garrison's studies.

Corbin: My bias is, because of our experience with this kind of thing, that if we were going to do these experiments I would look for proteins with an eye on the physiology of the system. For example, if it is known that cAMP elevation in liver causes glycogen breakdown, one might test the known enzymes of glycogen degradation, such as phosphorylase and phosphorylase kinase, to see if the purified proteins are substrates for protein kinase, and if they are activated by phosphorylation. To look for ^{32}P-labeled proteins is a very difficult way to do it. The cGMP-dependent protein kinase has been implicated in changing the calcium levels of cells. Smooth muscle cells relax when cGMP-dependent protein kinase is activated. Smooth muscle relaxation causes blood vessels to dilate. There must be some protein involved in the lowering of calcium somehow. Whether there is an extrusion of calcium is not certain, but some protein is phosphorylated that causes a lowering of calcium in cells. Most substrates for protein kinases have been discovered, or arrived at, through work on the physiology of the system rather than by just looking for ^{32}P-labeled proteins.

Cooke: Is there some homology between the cAMP-binding protein and the CAP-protein? What happens to the CAP-protein when cAMP binds to it? Is there any homology with the binding of the cAMP-binding protein with the catalytic subunit and whatever binds with the CAP-protein?

Corbin: It is hard to answer that question, but the CAP-protein has only homology in the cAMP binding site. The DNA binding domain of CAP is very different from any part of the regulatory subunit or cGMP-dependent protein kinase. CAP binds to DNA only when cAMP is bound to it, and it turns on certain genes to influence sugar transport in bacteria. So there is no protein kinase involvement in the function of CAP, although the physiology is still a bit analogous. In mammals, when glucose is low glycogen break-down and gluconeogenesis are turned on by cAMP elevation. In bacteria the same sort of thing happens - when sugar is low, cAMP is elevated and sugar transport is increased. The bacterium goes about it by totally different mechanisms. Apparently there is no protein kinase involvement in the action of CAP.

Clementi: Is this cAMP binding domain somehow similar to the domain that could be present in the cAMP receptor of *Dictyostelium*?

Corbin: Peter Devreotes addressed that question before, and according to him there is no similarity in the amino acid sequence. The two proteins could be related by convergent evolution, where for example an arginine residue of either protein could bind to the negative charge on the phosphate, or the glutamic acid residue of either protein could be used for hydrogen bonding to the ribose OH. Since the amino acid sequence homology is not present, the two proteins are apparently not related by divergent evolution. You have to wonder, too, whether or not the cGMP-binding phosphodiesterase could be related to one or the other of these two proteins. The binding site seems very different, based on biochemical properties, from the binding sites of the protein kinases. So it does not seem that there is homology between this protein and the protein kinase. Whether or not the phosphodiesterase could be homologous to the *Dictyostelium* receptor remains to be seen. Peter may have already compared the *Dictyostelium* receptor sequence to some known sequences of phosphodiesterases. Whether or not the catalytic domain could be homologous with the receptor binding site should also be considered. There is some published sequence on the catalytic domains of at least some phosphodiesterases, and they appear to be all in the same family. There may be at least three families of cAMP- and cGMP receptors: the *Dictyostelium* receptor family, the protein kinase receptor family and the phosphodiesterase receptor family.

Inositol phospholipid-coupled systems

HETEROGENEITY AMONGST MYO-INOSITOL DERIVATIVES: METABOLISM OF SECOND MESSENGERS AND SYNTHESIS OF CO-SIGNALS

C. P. Downes, L. R. Stephens[1] and P. T. Hawkins[2]. Smith Kline and French Research Limited The Frythe WELWYN Herts AL6 9AR England

Introduction

Myo-inositol is one of nine possible isomers of hexahydroxy cyclohexane in which the 2-hydroxyl is axial to the plane of the 6 membered ring with the remaining hydroxyl groups all being equatorial (Cosgrove, 1980). This configuration ensures that simple derivatives of myo-inositol, such as inositol phosphates, can occur in a variety of isomeric forms e.g. there are 20 possible inositol trisphosphate isomers. Recent studies have uncovered a bewildering complexity of cellular inositol metabolism, much of which is concerned with metabolism of the Ca^{2+}-mobilising second messenger, inositol 1,4,5-trisphosphate ($Ins1,4,5P_3$ [3]). On the other hand, certain inositol

1 Present address: AFRC Institute of Animal Physiology, Babraham, Cambridge CB2 4AT

2 Present address: MRC Molecular Neurobiology Unit, MRC Centre, Hills Road, Cambridge CB2 2QH

3 Abbreviations: these are as recommended in Biochem. J. (1989) 258, 1-2; Ins, myo-inositol; InsP-$InsP_6$, inositol mono-hexakisphosphate, unspecified isomeric configuration. The positions of the phosphates on a given inositol phosphate are denoted by numbering from the position of the phosphate in D-Ins1P.

NATO ASI Series, Vol. H 44
Activation and Desensitization of Transducing Pathways
Edited by T. M. Konijn, M. D. Houslay, P. J. M. Van Haastert

phosphates isomers appear to be involved with the synthesis of inositol polyphosphates such as inositol pentakis- and hexakisphosphate ($InsP_5$ and $InsP_6$) whose function and metabolic relationships to the fast signalling events triggered by receptor stimulated cleavage of phosphatidylinositol 4,5-bisphosphate ($PtdIns4,5P_2$) are not yet understood (see Downes, 1988 and references therein). In addition a novel inositol phospholipid (phosphatidylinositol 3-phosphate, PtdIns3P) has recently been described (Whitman et al, 1988; Stephens et al 1989b) whose synthesis may be regulated by the association of a PtdIns 3-kinase with activated tyrosine kinases such as the receptor for platelet derived growth factor (PDGF; Whitman et al, 1987)

Contemporary studies of the formation and degradation of $PtdIns4,5P_2$-derived second messengers thus require sophisticated separation and identification techniques in order to verify the precise structures of the compounds detected and to understand the metabolic relationships amongst them. Moreover, workers in this field must also have in mind the possibility that additional signal molecules may be generated during the complex metabolism of inositides (see e.g., Morris et al, 1987). In this paper we briefly describe progress in the identification of a number of novel myo-inositol metabolites and discuss our current understanding of the metabolic pathways involved in their formation.

$Ins1,4,5P_3$ metabolism

Metabolism of this ubiquitous second messenger was initially thought to procede by sequential removal of first the 5-, then the 4- and finally the 1-phosphate. However, with the introduction of anion exchange hplc techniques for the separation of cellular inositol phosphates, this idea was soon rejected. It is now clear that $Ins1,4,5P_3$ is metabolised by two main routes each of which generates a distinct fingerprint of products (Downes et al, 1982; Irvine et al, 1986; Majerus et al, 1988). The 5-phosphomonoesterase pathway involves 3 enzymes and generates $Ins1,4P_2$ and Ins4P as intermediates. Dephosphorylation of the latter 2 compounds is potently inhibited by lithium. By contrast the 3-kinase pathway generates $Ins1,3,4,5P_4$, $Ins1,3,4P_3$, $Ins3,4P_2$, Ins3P and Ins1P as intermediates and is blocked by lithium at the level of dephosphorylation of $Ins1,3,4P_3$, Ins3P and Ins1P (Ackerman et al, 1987; Majerus et al, 1988).

The purpose of the above pathways is primarily the inactivation of Ins1,4,5P_3 as neither metabolite can mimic the former's Ca^{2+}-mobilising activity. Secondly these reactions are an important source of Ins in cells undergoing active stimulation of receptors coupled to inositol phospholipid hydrolysis. Finally, and more speculatively, they may provide cells with a co-signal because there is now some experimental support for the idea that Ins1,3,4,5P_4 may promote Ca^{2+} entry into cells either by opening a channel in the plasma membrane or by facilitating Ca^{2+} transport between Ins1,4,5P_3-sensitive and insensitive stores (Morris et al, 1987).

The existence of the above pathways has been convincingly demonstrated in tissue homogenates, but support for their occurrence in intact cells is largely limited to co-chromatography of cellular inositol phosphates with standards of defined structure. It will be important to establish the structure of inositol containing compounds present in cell extracts by more rigorous techniques particularly as additional pathways, concerned with the metabolism of inositol polyphosphates such as InsP_5 and InsP_6, appear to coexist with the pathways of Ins1,4,5P_3 metabolism in many cells.

Methods for structural elucidation of inositol phosphate isomers

InsP_6 is a major phosphoric ester in many plant seeds where it may function as a phosphorus reserve and InsP_5 occurs at high concentration in avian erythrocytes, apparently bound tightly to haemoglobin (Cosgrove, 1980), but it is now clear that they are probably also ubiquitous components of mammalian cells (Szwergold et al, 1987). In order to determine the metabolic relationship of these compounds to the products of Ins1,4,5P_3 metabolism it was essential to rigorously identify inositol phosphate species present in a variety of cells. We have used chick erythrocytes as a cell type which is specialised for the synthesis of InsP_5 and compared its inositol phosphate content with that found in mammalian cells such as murine macrophages and 1321N1 human astrocytoma cells.

The analytical methods are essentially adaptations of those developed by Ballou and his colleagues to establish the head group structure of PtdIns4,5P_2 (Grado and Ballou, 1961). They comprise a variety of anion exchange hplc systems for the separation of InsP, $InsP_2$, $InsP_3$ and $InsP_4$ isomers; chemical or enzyme catalysed dephophorylation of unknown inositol phosphates; and periodate oxidation, reduction and dephosphorylation of unknown inositol phosphates to give linear polyols that can be identified by co-chromatography with known standards. Two novel techniques that greatly augment the existing methods for inositol phosphate analysis have also been introduced. The first is a simple hplc system for the separation of linear polyols that is capable of separating all such compounds that can result from periodate oxidation of $InsP_2$s, $InsP_3$s and $InsP_4$s. The second is the use of commercially available bacterial or yeast L-iditol dehydrogenase to define the enantiomeric configuration of trace quantities of any radiolabelled inositol phosphate that yields an iditol, altritol, glucitol or arabitol after periodate oxidation, reduction and dephosphorylation (Stephens et al, 1988b).

Cellular $InsP_4$ species and the synthesis of $InsP_5$

Use of the above battery of techniques has revealed at least three species of $InsP_4$ in both mammalian and avian cells; $Ins1,3,4,5P_4$, $Ins1,3,4,6P_4$ and $Ins3,4,5,6P_4$ (Stephens et al 1988a; 1988b). The only significant route accounting for the metabolism of $Ins1,3,4,5P_4$ in cellular homogenates was its dephosphorylation to $Ins1,3,4P_3$. $Ins3,4,5,6P_4$, however, is the substrate for a widely distributed 1-kinase (Stephens et al, 1988c) that converts it to $Ins1,3,4,5,6P_5$, the species of $InsP_5$ found in avian erythrocytes (there are 6 possible isomers of $InsP_5$), suggesting it may be an important precursor of this molecule. $Ins1,3,4,6P_4$ can also be converted to the same $InsP_5$ isomer, at least in rat brain homogenates, by a distinct 5-kinase (Stephens et al, 1988a).

The latter $InsP_4$ isomer was originally described as the product of an $Ins1,3,4P_3$ 6-kinase found in the cytosolic fractions of hepatocytes and adrenal glomerulosa cells (Shears et al, 1987; Balla et al, 1987). We have confirmed that a similar enzyme is present in rat brain homogenates which are thus capable of synthesising $InsP_5$ from $Ins1,4,5P_3$ with $Ins1,3,4,5P_4$, Ins $1,3,4P_3$ and $Ins1,3,4,6P_4$ as intermediates (Stephens et al, 1988a). This suggests there may be two distinct routes for the synthesis of $InsP_5$:

one would allow crosstalk between $InsP_5$ synthesis and the fast signalling events initiated by $PtdInsP_2$ hydrolysis; the other involves $Ins3,4,5,6P_4$ as its immediate precursor, but the pathway leading to the latter's formation is currently unknown. A summary of these findings is illustrated in figure 1.

The pathways depicted in figure 1 have been demonstrated in tissue homogenates, but it is important to know if they operate effectively in intact cells. This question has been approached by labelling chick erythrocytes with ^{3}H-inositol and then briefly with ^{32}P-orthophosphate. The individual, dual-labelled $InsP_4$ isomers and the $InsP_5$ fraction were purified by hplc

Figure 1 Complexities of inositol phosphate metabolism

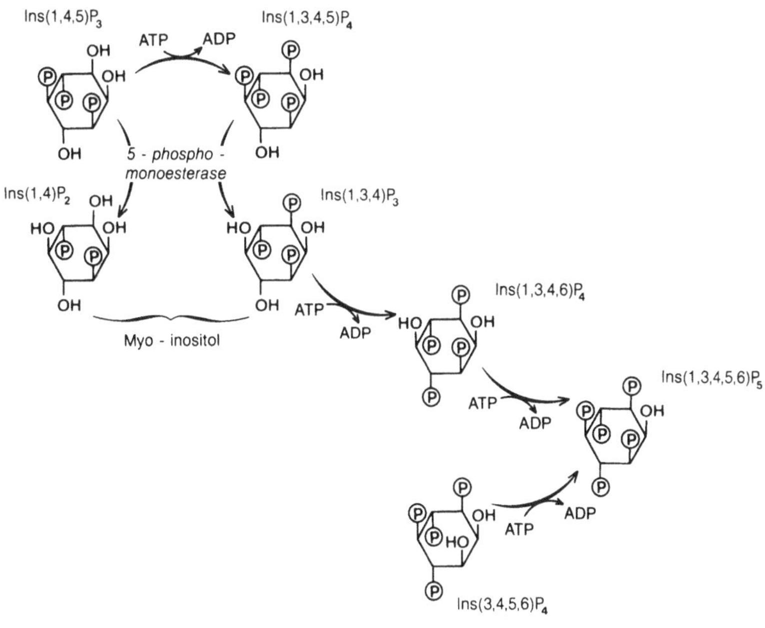

$Ins1,4,5P_3$ is metabolised by 5-phosphomonoesterase and 3-kinase reactions and the products are dephosphorylated to replenish the cellular pool of Ins as described in the text. $InsP_5$ can be synthesised by phosphorylation of $Ins3,4,5,6P_4$ whose metabolic origin is unknown, or by a series of phosphorylation and dephosphorylation reactions starting from $Ins1,4,5P_3$. The relative importance of each of these pathways to the synthesis of $InsP_5$ in intact cells is discussed in the text.

and the distribution of the ^{32}P amongst the individual phosphate groups of each compound was determined by carrying out a series of selective dephosphorylations and determining the ^{32}P content of identified dephosphorylation products (Stephens and Downes, 1989). This information, together with the absolute specific radioactivities of the individual phosphate groups in each compound reveals that Ins3,4,5,6P_4 is indeed the major percursor of $InsP_5$ in unstimulated chick erythrocytes, but surprisingly indicates that Ins1,3,4P_3 is unlikely to be a major precursor of Ins1,3,4,6P_4 in vivo. There is clearly still much to be learned concerning the pathways of inositol polyphosphate metabolism and their regulation in intact cells.

How many $InsP_3$ isomers are present in cells?

Important questions that remain in studies of $InsP_5$ metabolism are whether ultimately its synthesis depends upon interconversions of inositol phosphates derived from an inositol phopholipid or directly from Ins and what are the identities of the inositol phosphates that act as intermediates in its synthesis and/or degradation? Accommodation of these pathways in cells which are acutely sensitive to the intracellular concentration of Ins1,4,5P_3 suggests that other $InsP_3$ isomers might be characteristic metabolites of $InsP_5$. We, therefore, carried out a detailed, quantitative survey of the inositol trisphosphates that could be identified in ^{3}H-inositol labelled chick erythrocytes (Stephens et al, 1989a). The results of these experiments are summarised in Table 1.

Chick erythrocytes contain 6 distinct $InsP_3$ species, two of which have been unambiguously established to be Ins1,4,5P_3 and Ins1,3,4P_3. The remaining isomers, present in unstimulated cells in the proportions indicated in table 1, may be metabolites or precursors of $InsP_5$ as they are major dephosphorylation products when Ins3,4,5,6P_4 or Ins1,3,4,6P_4 are incubated with soluble enzymes present in avian erythrocyte lysates and/or rat brain homogenates. Two of these $InsP_3$s co-migrated with Ins1,4,5P_3 on a high resolution anion exchange hplc column suggesting that, if present in other cells, they may contribute to background radioactivity in the fractions assumed to contain only Ins1,4,5P_3. However, since most mammalian cells contain considerably lower levels of $InsP_5$ than chick erythrocytes, it is conceivable that the $InsP_3$ isomers identified in these studies will be correspondingly less prominent "contaminants" of the Ins1,4,5P_3 fraction in mammalian cells.

Table 1. Inositol trisphosphates in chick erythrocytes: see Stephens et al (1989b) and the text for details.

$InsP_3$ isomer	Peak designation	Probable origin	% of total $InsP_3$
1,3,4	"Ins1,3,4,P_3	Ins1,3,4,5P_4	34
1,3,6		Ins1,3,4,6P_4	3
1,4,5		Ptd Ins4,5P_2	36
3,4,6	"Ins1,4,5P_3"	Ins1,3,4,6P_4	10
1,4,6		Ins1,3,4,6P_4	4
3,4,5	Peak III	Ins3,4,5,6P_4	13

PtdIns 3-kinase and mitogenic signalling

The most recently discovered and, perhaps, the most surprising branch-point in inositide metabolism is the occurence of two distinct PtdInsP isomers, PtdIns4P and PtdIns3P (Whitman et al, 1988; Stephens et al, 1989b). The discovery arose from independent studies: in this laboratory on the possibility that a novel inositol phospholipid might be a precursor for one of the unexpected inositol phosphates described above; and in L.Cantley's laboratory on the molecular, kinetic and regulatory distinctions between PtdIns kinase species present in fibroblasts (Whitman et al, 1987).

Cantley and his colleagues defined two fibroblast PtdIns kinases and we have determined the head group structures of their products. Type II PtdIns kinase is strongly inhibited by adenosine, activated by non-ionic detergents and makes PtdIns4P (PtdIns 4-kinase). Type I PtdIns kinase is relatively much less abundant than PtdIns 4-kinase, is insensitive to adenosine, inactivated by detergents and makes PdIns3P (PtdIns 3-kinase). Perhaps the most important characteristic of PtdIns 3-kinase is its tendency to associate specifically with activated tyrosine kinases. Included in the latter are the middle T antigen/pp60^{c-src} complex in polyoma virus transformed fibroblasts and pp60^{v-src} in rous sarcoma virus transformed cells. A compelling piece of evidence in addition to those cited above, that PtdIns3-kinase is important in mitogenesis, is its association with the platelet derived growth factor (PDGF) receptor of NIH 3T3 cells within seconds of exposing the cells to PDGF. The specificity of these interactions is indicated by the fact that, although PtdIns 4-kinase represents some 95% of detectable PtdIns kinase activity, immunoprecipitates of the cited tyrosine kinases contain, almost exlusively, PtdIns 3-kinase. Moreover, by utilising Chinese hamster ovary cell lines expressing a variety of mutated forms of the PDGF receptor, Coughlin et al (1989) have demonstrated that an active tyrosine kinase and kinase insert domain are required for the interaction with PtdIns 3-kinase and that the kinase insert domain is necessary, but not sufficient for mitogenic activity of the receptor. Whether PtdIns3P or a metabolite can act as a mitogenic signal remains to be determined.

Conclusions

In this paper we have illustrated the point that myo-inositol is a molecular building block of extraordinary versatility. Several pathways of inositol phosphate and inositol phospholipid metabolism coexist within cells, apparently segregated by the positional specificity of the phosphotransferases that catalyse all of the reactions that we have discussed. Most of the participants in these pathways probably act as stepping stones towards functionally significant end-points. Whatever is the function of this burgeoning complexity, elucidation of all of the pathways of myo-inositol metabolism and their functions is an important challenge in biochemistry.

References

Ackerman K E, Gish B G, Honchar M P and Sherman W R (1987). Evidence that inositol 1-phosphate in brain of lithium treated rats results mainly from phosphatidylinositol metabolism. Biochem J 242: 517-524.

Balla T, Guillemette G, Baukal A J and Catt K J (1987). Production of inositol 1,3,4,6-tetrakisphosphate during angiotensin II action in bovine adrenal glomerulosa cells. Biochem Biophys Res Commun 148: 199-205.

Cosgrove D J (1980) Inositol phosphates: their chemistry, biochemistry and physiology. Elsevier, Amsterdam.

Coughlin S R, Escobedo J A and Williams L T (1989). Role of phosphatidylinositol kinase in PDGF receptor signal transduction. Science 243: 1191-1194.

Downes C P (1988) Inositol phosphates: a family of signal molecules? Trends Neurosci 11:336-338

Downes C P, Mussat M C and Michell R H (1982). The inositol trisphosphate phosphomonoesterase of the human erythrocyte membrance. Biochem J 203: 169-177.

Grado C and Ballou C E (1961). Myo-inositol phosphates formed by alkaline hydrolysis of beef brain phosphoinositide. J Biol Chem 235: 54-60.

Irvine R F, Letcher A J, Heslop J P and Berridge M J (1986). The inositol tris/tetrakisphosphate pathway - demonstration of Ins(1,4,5)P_3 3-kinase activity in animal tissue. Nature 320: 631-634.

Majerus P W, Connolly T M, Bansal V S, Inhorn R C, Ross T S and Lips D L (1988). Inositol phosphates: synthesis and degradation. J Biol Chem 263: 3051-3054.

Morris A P, Gallacher D P, Irvine R F and Peterson O H (1987). Synergism of inositol trisphosphate and tetrakisphosphate in activating Ca^{2+}-dependent K^+ channels. Nature 330: 653-655.

Shears S B, Parry J B, Tang E K Y, irvine R F, Michell R H and Kirk C J (1987). Metabolism of D-myo-inositol 1,3,4,5- tetrakisphosphate by rat liver, including the synthesis of a novel isomer of inositol tetrakisphosphate. Biochem J 246: 139-147.

Stephens L R and Downes C P (1989). Product, precursor relationships amongst inositol polyphosphates. Biochem J submitted.

Stephens L R, Hawkins P T, Barker C and Downes C P (1988a) synthesis of myo-inositol 1,3,4,5,6-pentakisphosphate from inositol phosphates generated by receptor activation. Biochem J 253: 721-733.

Stephens L R, Hawkins P T, Carter N, Chahwala S B, Morris A J, Whetton A D and Downes C P (1988a) L-myo-inositol (1,4,5,6)-tetrakisphosphate is present in both mammalian and avian cells. Biochem J 249: 271-282.

Stephens L R, Hawkins P T and Downes C P (1989a). An analysis of [^{3}H]-myo-inositol trisphosphates found in [^{3}H]-myo-inositol prelabelled avian erythrocytes. Biochem J submitted.

Stephens L, Hawkins P T and Downes C P (1989b). Metabolic and structural evidence for the occurrence of a third species of polyphosphoinositide in cells: D-phosphatidyl-myo-inositol 3-phosphate. Biochem J 259: 267-276.

Stephens L R, Hawkins P T, Morris A J and Downes C P (1988c) L-myo-inositol (1,4,5,6)-tetrakisphosphate (3-hydroxyl) kinase. Biochem J 249: 283-292.

Szwergold B S, Graham R A and Brown T R (1987). Observation of inositol pentakis- and hexakis-phosphates in mammalian tissues by ^{31}P NMR. Biochem Biophys Res Commun 149: 874-881.

Whitman M, Downes C P, Keeler M, Keller T and Cantley L (1988). Phosphatidylinositol kinase makes a novel phospholipid, Type I phosphatidylinositol 3-phosphate. Nature 332: 644-646.

Whitman M, Kaplan D R, Roberts T M and Cantley L (1987). Biochem J 247: 165-174.

DISCUSSION

Houslay: Did the kinases that were involved in making these higher inositol phosphates, have specific subcellular localizations?

Downes: Sofar as we can tell they are soluble proteins; in any kind of simple homogenization or in the case of avian erythrocytes just simple hypotonic lysis ,which is a fairly mild means of obtaining a membrane- and cytosol fraction . So I suspect they are true soluble proteins.

Houslay: Could I then follow up to something you eluded to which really was. If the cell is expending a lot of energy to make these compounds, I presume that this is not just to keep you off the street. Are there binding proteins on membranes for these or some of these.

Downes: In the first point, you suggested that the cell is expending a lot of energy on these pathways. It is an absolutely trivial amount of energy that is being expended. I am confident, for example, that the amount of energy expended generating IP5, as it turns over so slowly, is a tiny proportion of what is needed for synthesis and degradation of IP3.

Houslay: Yes, but remember it's expending energy making the enzymes that are doing this, I mean it is a lot of effort, instead of dephosphorylating the 1,4,5 to 1,4 and myo-inositol, why does it go the other way?

Downes: If we are talking about the distinction between the kinase that makes 1,3,4,5 and the phosphatase that degrades IP3 to Ins 1,4 P2 my speculation is very simple. You use a lot of energy to generate control of a key substance like IP3. I cannot imagine the amount of energy being used here is even comparable with the energy which is used to maintain membrane potential, calcium gradients and things of this kind. The control comes in because you have two enzyme activities with very different kinetic properties. One is a high Km high capacity phosphatase the other is a low Km ,probably relatively low capacity, but highly regulated, calcium sensitive kinase. If you think about that and translate it into your studies on phosphodiesterases, where you have one route to remove cAMP, what you do is incorporate in that single route a variety of different control steps by having different isozymes, which are regulated in different ways. You have two routes for metabolism of IP3 and generate control through that mechanism.

Houslay: Can I come back, has anyone looked for binding? For whether there is high affinity binding for any of the higher inositol phosphates on membranes?

Downes: With the exception of studies on labelled Ins 1,3,4,5 P4, nobody has looked systematically. We have one binding assay that we tried out, but it is not easy to make high specific activity IP5. We have got a strategy now so we think we can do that and generate the compounds at sufficient specific activity to do some basic hypothesis testing. If you make a membrane preparation from say brain, you can find, apparently specific Ip5 binding. That is all I can say, I have no idea whether it is in any way structurally specific. And what you need to be able to do in all those experiments, and that includes experiments for trying to elucidate the function of 1,3,4,5 tetrakisphosphate, is to use controls that basically use distinct isomers of these IP5's or IP4's, which, therefore, determine whether we are dealing with a specific binding site or not before we start claiming anything which binds is important. IP5 is known for example to bind with high affinity to hemoglobin in avian systems, that might be it's function. It clearly has the same effect as the binding of 2,3 DPG to the human hemoglobin i.e. changes it's oxygen binding affinity.

Iyengar: Can I follow up Miles' question; do you think all these metabolites other than Ins 1,4,5 P3 are totally unregulated by hormones and signal transduction and they work in intermediate metabolism in the cell, is that the line of thinking.

Downes: There is no evidence to argue anything else. Nobody has demonstrated a specific function for any of these compounds , but if you just think about it, I don't know how I want to control the level of I-1,4-bisphosphate when it's a product of something that is a second messenger. It is a bit like saying that AMP is a signal. I would find it impossible to imagine that they could be, it is because of the way they are made and where they fit into the metabolic pathway.

Iyengar/Houslay: But adenosine is an extracellular signal.

Downes: I would not exclude that possibility and obviously this has been a suggestion that has been made for IP5 and IP6.

Iyengar: Prostaglandins, coming from arachidonic, all those things produce signals.

Downes: Right, but most of the inositol phosphates are not themselves secreted from cells, they would need a very specific mechanism to get them out of cells.

Iyengar: Find a carrier.

Downes: That is fine, but I am not looking.

Dumont: Your second PIP, is it a substrate of the hormone regulated PLC, does it release Ins 1,3 P2 as a signal?

Downes: We have looked at the hormone sensitivity of the PI-3P, versus PI-4-P, and PI-4,5-bisphosphate pools in astrocytoma cells in response to normal PLC stimulus such as catachol. All I can say is, when PI-4,5-bisphosphate levels fall in response to catachol so does PI-4-P and so does PI-3-P. Obviously, there is a route to generate substantial amounts of I-1,3-bisphosphate via the complex metabolism of 1,4,5-IP3, one of the typical products via the kinase mechanism. So, it has been suggested by Williams who has been using fibroblasts transfected with mutated PDGF-receptors, and also by Cantley, that the function is to generate a set of novel lipids that are then substrates of PLC and generate different patterns of inositol phosphates. I again cannot imagine, that that is the function, simply because these things are present at levels that are 1 to 2 % the level of the conventional lipids. Unless the turnover is, let's say, 50 times the rate of the conventional lipids, they cannot possibly generate those products at a rate that can influence their levels compared to the rate at which they are made by direct metabolism of inositol phosphates. So my group is interested in the lipids themselves, as potential regulators, possibly in the lipid membrane domain.

Cooke: Is there a possibility that you have got PLC on the outer membrane; can you in fact release some of these inositol phosphates to the outside of the cell.

Downes: It is not beyond the realms of possibility, it is beyond the realms of any experiments I have carried out to figure out what the answer is. However, Michell's group found that hormone-stimulated accumulation of inositol phosphates in hepatocytes is essentially an intracellular response.

Cooke: There is no suggestion that you get inositol phosphates being generated outside the cell?

Downes: Some people have looked to that, certainly, if you stimulate hepatocytes by conventional mechanisms then to all intents and purposes the inositol phosphates that are generated are all intracellular. But it does not have to be that way in all situations, and certainly on a quantitative basis, if a portion of the PI-3-P broke down, but it all went outside, you really would not pick it up because it would only be a tiny percentage. You live with the idea that 99 % was intracellular, so I don't really think we have information that will tell us, whether that is happening or not.

Corbin: By analogy with the cyclase-phosphodiesterase system for the cAMP cascade; is it possible that PI-kinase stimulation would regulate IP3 levels?

Downes: We have considered that for a long time and it certainly would explain why there are active kinases al well as the phospho-monoesterases that break down PIP and PIP2; it seems they are pretty active in cells. The mono-ester phosphate groups of the conventional lipids turn over very rapidly; basically you cannot work out the turnover times, because it is too close to that of the τ-phosphate of ATP. One possibility is you could control the rate of PIP2 synthesis. Having said that, it is clear that a major arm of the control is at the level of the receptor, the G-protein and the phospholipase, just as is the cyclase, and that I do not think is very different. I think Boyer will show data on the turkey erythrocyte system, which suggests that the level of PIP2 in the membrane is not limiting over a period of a few minutes of stimulation. Perhaps the kinases exert longer term control, I mean there are many receptors, for example, that do not obviously desensitize over a period of maybe one to two hours, when looking at this kind of signalling system. And it could be that in those cases substrate supply becomes a major factor. One thing that Mike Berridge and I have suggested is that that might be a way in which lithium functions, by limiting the supply of inositol it actually prevents PI resynthesis. And if that were to translate into a reduced supply of PIP2 you may have a sort of stimulus dependent reduction in the supply of substrate for PLC. This would not happen unless you stimulated the cells, so it would only tend to inhibit cells that were already switched on.

Van Haastert: You have a whole set of inositol phosphates. Do you have any indication that the inositol phosphates are in different compartments of the cell?

Downes: There are very few compartments in avian erythrocytes. Since the same compounds are present in avian erythrocytes as in mammalian cells, just the proportions differ, and since the enzymes I have discussed are mainly in the soluble fraction, I think it is unlikely that that is going to be important. What is important is a metabolic segregation if you like. So I do not think they are kept apart obviously in the cell, though in a real cell the thing might be more complex.

ROLE OF GLYCOSYL-PHOSPHATIDYLINOSITOLS IN INSULIN SIGNALLING

Isabel Varela, Jose F. Alvarez, Jose Puerta, Rosa Clemente, Ana Guadaño, Matias Avila, Francisco Estevez, Susana Alemany and Jose M. Mato
Fundacion Jimenez Diaz and Instituto de Investigaciones Biomedicas del C.S.I.C., Reyes Catolicos 2, Madrid 28040, Spain

A. Introduction

Insulin is one of the best studied hormones. The main actions of insulin are to stimulate the synthesis of glycogen, lipids and proteins, through the modulation of the metabolic pathways implicated in these processes, and the stimulation of cell growth. The physiological effects of insulin include the stimulation of the uptake of glucose, amino acids and ions; the regulation of the state of serine-and threonine-phosphorylation of a variety of proteins and rate-limiting enzymes like glycogen phosphorylase, hormone-sensitive lipase and ATP citrate lyase; and the regulation of the expression of the genes for several regulatory enzymes (review by Denton, 1986). All these effects, whose chronology vary from seconds to hours, are initiated by the interaction of the hormone with the insulin receptor, an integral membrane glycoprotein composed of two α (Mr about 130 kDa) and two β (95 kDa) subunits. The α subunits bind insulin and are linked by disulphide bonds to each other and to the β-subunits. Following insulin binding, the β-subunits are rapidly autophosphorylated predominantly at tyrosine residues. Direct evidence, indicating that insulin action depends on the protein tyrosine kinase activity of the insulin receptor, has been obtained by site-directed mutagenesis of the insulin receptor cDNA as well as with monoclonal antibodies to the insulin receptor kinase domain (review by Rosen, 1987). How receptor autophosphorylation is coupled to the modulation of processes like the synthesis of glycogen, lipids and proteins is not well understood. One hypothesis suggests that the receptor kinase

NATO ASI Series, Vol. H 44
Activation and Desensitization of Transducing Pathways
Edited by T. M. Konijn, M. D. Houslay, P. J. M. Van Haastert

activity catalyses the phosphorylation of cellular protein substrates. A second hypothesis suggests that autophosphorylation of the receptor would lead to variations in its interactions with other membrane components. In both cases the interaction of insulin with its receptor would lead to the generation of biochemical signals involved in insulin action. Recently, evidence has been provided which suggests that inositol phospho-oligosaccharides might be one of the biochemical signals involved in insulin signalling (review by Mato, 1989).

B. Structure of the insulin-sensitive glycosyl-phosphatidylinositol

Insulin has been reported to promote the phosphodiesteratic hydrolysis of a novel glycosyl-phosphatidylinositol (glycosyl-PI) in a number of cells including intact murine myocytes (Saltiel et al. 1986), H35 hepatoma cells (Mato et al. 1987a), T lymphocytes (Gaulton et al. 1988) and CHO cells (Villalba et al. 1989). These insulin-sensitive glycosyl-PIs have features in common with the glycosyl-PI anchor of a number of membrane proteins (Cross, 1987). These similarities include (i) the finding that the insulin-sensitive glycosyl-PI is hydrolysed by the phosphatidylinositol specific phospholipase C (PI-PlC) from *Staphylococcus aureus*, *Bacillus cereus* or *Bacillus thuringiensis* (Saltiel et al. 1986; Mato et al. 1987a; Gaulton et al. 1988; Villalba et al. 1989); (ii) the observation that purified glycosyl-PI can be cleaved by nitrous acid deamination with generation of phosphatidylinositol, indicating the presence of inositol monophosphate linked to non-N-acetylated glucosamine (Mato et al. 1987a,b); and (iii) the finding that the diacylglycerol moiety of the phospholipid contains saturated fatty acids (palmitate and myristate)(Saltiel et al. 1986; Mato et al. 1987a). In addition, the polar head group of the insulin-sensitive glycosyl-PI isolated from rat liver membranes or H35 hepatoma cells has been shown to contain galactose (about four residues per residue of inositol) and up to three additional phosphates (Mato et al. 1987b; Merida et al. 1988), probably linked to a galactose residue (Merida et al. 1988). The purified glycosyl-PI is also sensitive to treatment with ß-galactosidase (Varela et al. 1989), indicating the presence of galactose residues with ß-glycosidic bonds. Another characteristic of this glycosyl-PI is the absence of amino acids and ethanolamine (Mato et al. 1987a,b). It should be noted however, that variations might exist between insulin sensitive glycosyl-PIs from different cell types. In fact the sensitivity to bacterial PI-PlC of insulin-sensitive glycosyl-PIs isolated from H35, T lymphocytes, and CHO is different (Mato et al. 1987a; Gaulton et al. 1989; Villalba et al. 1989) and in Rat-1 cells, the purified insulin sensitive glycosyl-PI is resistant to both treatment with PI-PlC and nitrous acid deamination (Alemany, Clemente & Mato, unpublished).

As with most glycolipids, the majority of the insulin-sensitive glycosyl-PI (which represents about 0.5% of the plasma membrane phospholipids) is present at the outer surface of rat hepatocytes (Alvarez et al. 1988) and adipocytes (Varela et al. 1989). This indicates, that in response to the addition of insulin the inositol phospho-oligosaccharide that forms the polar head group of glycosyl-PI is generated extracellularly. Similarly, Romero et al. (1988) have observed that the extracellular concentration of POS increased rapidly after the addition of insulin to BC3H1 myocytes. Furthermore, as the result of the phosphodiesteratic hydrolysis of glycosyl-PI diacylglycerol is also formed at the outer leaflet of the membrane bilayer. Since the rate of transbilayer movement of diacylglycerol is very slow, the physiological role, if any, of the diacylglycerol generated upon glycosyl-PI hydrolysis might be confined to the outer surface of the cell.

C. Insulin-dependent glycosyl-PI hydrolysis

As mentioned above the addition of insulin to a number of target cells stimulates the hydrolysis of glycosyl-PI. This process is dose-dependent and can be observed at physiological concentrations of the hormone (Mato et al. 1987a, Villalba et al. 1989). The time-course of glycosyl-PI hydrolysis is fast and is detected within 1 minute of the addition of insulin (Saltiel et al. 1986; Mato et al. 1987a; Gaulton et al. 1988; Villalba et al. 1989). The requirement of the insulin-receptor tyrosine kinase activity in insulin-dependent glycosyl-PI hydrolysis has been evaluated by analyzing glycosyl-PI hydrolysis in transfected CHO cell lines expressing either the wild type human insulin receptor or a mutant receptor that lacks kinase activity. Cells bearing normal human receptors hydrolysed up to 70% of the glycosyl-PI within 2 min of the addition of 0.1 nM insulin, whereas parental cells and cells expressing the mutant receptor hydrolysed only 20%-30% of the glycosyl-PI in response to 100 nm insulin (Villalba et al. 1989). These results strongly suggest that the receptor protein tyrosine kinase activity is necessary to transduce the effect of insulin on glycosyl-PI hydrolysis.

A number of experiments indicate that there is a correlation between the number of insulin receptors and the levels of glycosyl-PI. Thus, in parental CHO cells and in cells bearing the mutated human insulin receptor the amount of glycosyl-PI is about 3-fold lower than that found in cells bearing the normal insulin receptor (Villalba et al. 1989). A correlation of glycosyl-PI levels with the induction of insulin receptors has also been found in T lymphocytes (Gaulton et al. 1988), and there is a correlation of glycosyl-PI levels with the insulin-induced reduction of their own receptors (down regulation) in rat hepatocytes (Ruiz-Albusac, Mato & Blazquez, unpublished). The amount of glycosyl-PI in hepatocytes from rats treated with dexamethasone, which causes a reduction in the number of insulin receptors, is about 3-fold lower than in cells from control animals. Furthermore, the hydrolysis of glycosyl-PI and the synthesis of glycogen in response to insulin was also markedly reduced in hepatocytes isolated from rats treated with dexamethasone (Cabello, Mato & Feliu, unpublished).

The possibility that other receptors with tyrosine kinase activity might be coupled to glycosyl-PI hydrolysis has also been investigated by various laboratories. EGF (100 ng/ml) and IGF-I (50 ng/ml) have been reported to stimulate the hydrolysis of the insulin sensitive glycosyl-PI in BC3H1 myocytes (Farese et al. 1988)). In CHO cells, IGF-I (15 nM) has a small effect on glycosyl-PI hydrolysis (10%) which increases to 20% in cells transfected with the human IGF-I receptor and that exhibit about 4 times more receptors than the parental CHO cells (Villalba et al. 1989). Although suggestive, the small effect of IGF-I on glycosyl-PI hydrolysis in CHO raises the possibility that it may be mediated by the insulin receptor. In PC-12 cells, NGF (50 ng/ml) stimulates the hydrolysis of glycosyl-PI (Chan et al. 1989). Although tyrosine phosphorylation has been reported in PC-12 cells stimulated with NGF, there is no evidence for a tyrosine kinase domain in the NGF receptor. ACTH has also been reported to stimulate the turnover of a glycosyl-PI in calf adrenal glomerulosa cells (Cozza et al. 1988). Since there is no data about the hydrolysis by PI-PlC or sensitivity to nitrous acid deamination of this glycolipid in glomerulosa cells, the possibility exists that it might be a different type of lipid.

D. Insulin-like effects of inositol phospho-oligosaccharides

The results mentioned above indicate that insulin stimulates the generation at the outer cell surface of an inositol phospho-oligosaccharide (POS) which is the polar head group of a glycosyl-PI. In 1986 we reported that the addition of POS to isolated rat adipocytes had insulin-like effects on phospholipid methyltransferase (Kelly et al. 1986), the enzyme that converts phosphatidylethanolamine into phosphatidylcholine by three successive N-methylations being S-adenosyl-L-methionine the methyl donor. Since then, this or similar preparations of POS have been shown to have insulin-like effects on lipolysis (Kelly et al. 1987a), lipogenesis (Saltiel & Sorbara-Cazan, 1987) and pyruvate dehydrogenase (Gottschalk & Jarett, 1988) in adipocytes, and glycogen phosphorylase, pyruvate kinase, and cAMP levels (Alvarez et al. 1987) in hepatocytes. POS however has no effect on glucose transport in adipocytes (Kelly et al. 1987a; Saltiel & Sorbara-Cazan, 1987). It is interesting to note however, that all the biological effects of POS mentioned above were carried out with a molecule isolated from rat liver and, that in this tissue, insulin has no effect on glucose transport. Therefore, it would be important to know if POS isolated from adipocytes might facilitate glucose transport in this tissue. At this respect, it is interesting to note that glucose uptake has been stimulated by the addition to rat adipocytes of an inositol-phosphate oligosaccharide fraction isolated from Actovegin, a drug which is derived from calf blood (Obermaier-Kusser et al. 1989). Moreover, diacylglycerol has been shown to act, independently of protein kinase C, as an activator of glucose transport in adipocytes when applied outside the cell (Strälfors, 1988). This raises the interesting possibility that insulin generates two types of biochemical signals: POS, that mediates a number of biological effects of insulin and diacylglycerol that mediates the effects on glucose uptake. Finally, POS has also been found to inhibit glucose-induced insulin release in pancreatic islet cells without affecting glucose oxidation (Albor et al. 1989), its physiological significance however remains to be determined.

As mentioned above, the physiological effects of insulin include the regulation of the state of serine-and threonine-phosphorylation of a variety of proteins and rate-limiting enzymes like glycogen phosphorylase, hormone-sensitive lipase and ATP citrate lyase. If insulin action is dependent on the generation of POS, this molecule(s) might also mimic insulin effect on serine-and threonine-phosphorylation of key target enzymes and proteins. We have shown (Alemany et al. 1987; Kelly et al. 1987b) that in adipocytes POS faithfully copies the insulin-directed effects on phosphorylation/dephosphorylation of target proteins of the hormone. Thus POS has insulin-like effects on the phosphorylation of ATP citrate lyase, hormone sensitive lipase, glycogen phosphorylase, a 65 kD and a 50 kD phosphoprotein. In the case of ATP citrate lyase the site of phosphorylation stimulated by insulin and POS was also studied and found to be the same (Puerta, Alemany, Mato, unpublished). The conclusion of these results is that POS stimulates the same signalling pathways as does insulin. Since both cAMP-dependent and independent effects are mimicked by POS, this indicates that the hydrolytic production of POS is an early step in the insulin signalling mechanism.

POS has been found to stimulate the transport of amino acids in isolated rat hepatocytes through a mechanism that requires protein synthesis (Varela, Avila, Hue, Mato, unpublished), indicating that this molecule has also some of the long-term effects of insulin. At this respect, it is interesting to note that the incorporation of thymidine into DNA has been stimulated by the addition to cells of an oligosaccharide obtained from the conditioned media of reuber hepatoma cells (Witters & Watts, 1988). Whether this compound is related to POS however remains to be determined.

E. Mechanism of action of POS: intracellular vs extracellular action

Since POS has insulin-like effects when added to intact cells the question is whether POS acts intracellularly or through its interaction with a cell surface receptor. Evidence in favour of an intracellular action comes from the observation that POS affects the activity of several cellular enzymes such as cyclic nucleotide phosphodiesterase, pyruvate dehydrogenase, adenylate cyclase (Saltiel, 1987; Larner et al. 1988), cAMP-dependent protein kinase (Villalba et al. 1988) and casein kinase II (Alemany, Puerta, Mato, unpublished) in a similar manner to insulin when added to intact cells. The uptake of POS has been investigated by incubating isolated rat hepatocytes with [^{3}H-galactose]POS. Labelled POS was taken up by the cells at 37°C but not at 4°C which suggests the existence of an active uptake system. POS uptake was time-and dose-dependent, a maximal being observed within 1 and 5 minutes of the addition of the labelled molecule which agrees with the time-course of its biological actions. Moreover the uptake of POS was specific and could not be inhibited by the addition to the incubation media of a 20-fold excess of myo-inositol, galactose or inositol 1-phosphate. However, the presence of glucosamine in the assay media markedly inhibited the uptake of POS by rat hepatocytes (Alvarez, Estevez, Varela, Mato, unpublished). At this respect, it is interesting to note that the addition of glucosamine (but not of galactose or myo-inositol) has been found to inhibit the lipogenic activity (Machichao & Häring, personal communication) and the effect on protein phosphorylation (Alemany & Strälfors, personal communication) of both POS and insulin in rat adipocytes.

In conclusion, the present evidence supports a model where insulin promotes the hydrolysis of a glycosyl-PI at the outer surface of the cell with release of its polar head group and diacylglycerol. The inositol phospho-oligosaccharide that forms the polar head group enters the cell and modulates the state of phosphorylation of key target proteins.

ACKNOWLEDGEMENTS: Work at the authors' laboratory was supported by grants from Direccion General de Investigacion Cientifica y Tecnica (PM88-0011), Fondo de Investigaciones Sanitarias de la Seguridad Social (88/1489) and Europharma (01/88). IV was a post-doctoral fellow of the Ministerio de Educacion y Ciencia. JFA, JP and RC were fellows of the Fundacion Conchita Rabago de Jimenez Diaz. MA was a fellow of the Ministerio de Educacion y Ciencia. FE was a fellow of the Gobierno de Canarias.

REFERENCES

Albor A, Camara J, Valverde I, Mato JM, Malaisse WJ (1989) Inhibition of insulin release by a putative insulin-mediator in pancreatic islet cells. Med Sci Res 17:161-162

Alemany S, Mato JM, Strälfors P (1987) Phospho-dephospho-control by insulin is mimicked by a phospho-oligosaccharide in adipocytes. Nature 330:77-79

Alvarez JF, Cabello MA, Feliu JE, Mato JM (1987) A phospho-oligosaccharide mimics insulin action on glycogen phosphorylase and pyruvate kinase activities in isolated rat hepatocytes. Biochem Biophys Res Commun 147:765-771

Alvarez JF, Varela I, Ruiz-Albusac JM, Mato JM (1988) Localisation of the insulin-sensitive phosphatidylinositol glycan at the outer surface of the cell membrane. Biochem Biophys Res Commun 152:1455-1462

Chan BL, Chao MV, Saltiel AR (1989) Nerve growth factor stimulates the hydrolysis of glycosyl-phosphatidylinositol in PC-12 cells: A mechanism of protein kinase C regulation. Proc Natl Acad Sci USA 86:1756-1760

Cozza EN, Vila MC, Gomez-Sanchez CE, Farese RV (1988) ACTH stimulates turnover of the phosphatidylinositol-glycan. Biochem Biophys Res Commun 157:585-589

Cross GAM (1987) Eukaryotic protein modification and membrane attachment via phosphatidylinositol Cell 48:179-180

Denton RM (1986) Early events in insulin actions. "In:" Greengard P, Robison GA (eds) Advances in Cyclic Nucleotide Research. Raven Press, New York, vol 20, p 293

Farese RV, Nair GP, Stadaert ML, Cooper DR (1988) Epidermal growth factor and insulin-like growth factor-I stimulate the hydrolysis of the insulin-sensitive phosphatidylinositol-glycan in BC3H1 myocytes. Biochem Biophys Res Commun 156:1346-1352

Gaulton G, Kelly KL, Mato JM, Jarett L (1988) Regulation and function of an insulin-sensitive glycosyl-phosphatidylinositol during T lymphocyte activation. Cell 53:963-970

Gottschalk KW, Jarett L (1988) The insulinomimetic effects of the polar head group of an insulin-sensitive glycophospholipid on pyruvate dehydrogenase in both subcellular and whole cell assays. Arch Biochem Biophys 261:175-185

Kelly KL, Mato JM, Jarett L (1986) The polar head group of a novel insulin-sensitive glycophospholipid mimics insulin action on phospholipid methyltransferase. FEBS letters 209:238-242

Kelly KL, Mato JM, Merida I, Jarett L (1987a) Glucose transport and antilipolysis are differentially regulated by the polar head group of an insulin-sensitive glycophospholipid. Proc. Natl. Acad. Sci. USA 84:6404-6407

Kelly KL, Merida I, Wong EHA, DiCenzo D, Mato JM (1987b) A phospho-oligosaccharide mimics the effect of insulin to inhibit isoproterenol-dependent phosphorylation of phospholipid methyltransferase in isolated adipocytes. J Biol Chem 262:15282-15290

Larner J, Huang LC, Schwartz CFW, Oswald AS, Shen TY, Kinter M, Tang G, Zeller K (1988) Rat liver insulin mediator which stimulates pyruvate dehydrogenase phosphatase contains galactosamine and D-chiroinositol. Biochem Biophys Res Commun 151:1416-1426

Mato JM, Kelly KL, Abler A, Jarett L (1987a) Identification of a novel insulin-sensitive glycosphospholipid from H35 hepatoma cells. J Biol Chem 262:2131-2137

Mato JM, Kelly KL, Abler A, Jarett L, Corkey BE, Cashel JA, Zopf D (1987b) Partial structure of an insulin-sensitive glycophospholipid. Biochem Biophys Res Commun 152:1455-1462

Mato JM (1989) Insulin mediators revisited. Cellular Signalling 1:143-146

Merida I, Corrales FJ, Clemente R, Ruiz-Albusac JM, Villalba M, Mato JM (1988) Different phosphorylated forms of an insulin-sensitive glycosylphosphatidylinositol from rat hepatocytes. FEBS letters 236:251-255

Obermaier-Kusser B, Mühlbacher C, Mushack J, Seffer E, Ermel B, Machichao F, Schmidt F, Häring HU (1989) Further evidence for a two-step model of glucose-transport regulation. Inositol phosphate-oligosaccharides regulate glucose carrier activity. Biochem J In press

Romero G, Lutrell L, Rogol A, Zeller K, Hewlett E, Larner J (1988) Phosphatidylinositol-glycan anchors of membrane proteins: Potential precursors of insulin mediators. Science 240:509-511

Rosen OM (1987) After insulin binds. Science 237:1452-1458

Saltiel AR (1987) Insulin generates an enzyme modulator from hepatic plasma membranes: Regulation of adenosine 3',5'-monophosphate phosphodiesterase, pyruvate dehydrogenase, and adenylate cyclase. Endocrinology 120:967-972

Saltiel AR, Sorbara-Cazan LR (1987) Inositol glycans mimics the action of insulin on glucose utilization in rat adipocytes. Biochem Biophys Res Commun 149:1084-1092

Saltiel AR, Fox JA, Sherline P, Cuatrecasas P (1986) Insulin stimulates the hydrolysis of a novel membrane glycolipid causing the generation of cAMP phosphodiesterase modulators. Science 233:967-972

Strälfors P (1988) Insulin stimulation of glucose uptake can be mediated by diacylglycerol in adipocytes. Nature 335:554-556

Varela I, Alvarez JF, Ruiz-Albusac JM, Mato JM (1989) Asymmetric distribution of the phosphatidylinositol-linked phospho-oligosaccharide that mimics insulin action in the plasma membrane. Submitted

Villalba M, Kelly KL, Mato JM (1988) Inhibition of cyclic AMP-dependent protein kinase by the polar head group of an insulin-sensitive glycophospholipid. Biochim Biophys Acta 968:69-76

Villalba M, Alvarez JF, Russell D, Mato JM, Rosen O (1989) Hydrolysis of glycosyl-phosphatidylinositol in response to insulin is reduced in cells bearing kinase-deficient insulin receptors. Submitted

Witters LA, Watts TD (1988) An autocrine factor from reuber hepatoma cells that stimulates DNA synthesis and acetyl-CoA carboxylase. J Biol Chem 263:8027-8036

DISCUSSION

Downes: One possible advantage of having a mediator released from the outside of the cell is that it could interact with other cells; have you ever looked at POS on any noninsulin responsive cells.

Mato: That is an interesting possibility which we have not yet tested. We are planning to test the effect of POS on hepatocytes from insulin resistant animals.

Downes: All the studies I've seen including the one you have shown, shows characteristic rapid breakdown of the lipid and then a resynthesis, and if this is all happening at the outside of the cell how do the energy dependent steps get put in place so rapidly is there a precursor pool just waiting what exactly is going on there?

Mato: The time course of resynthesis varies with the cell. In hepatocytes there is a rapid breakdown but resynthesis is relatively slow and 10 min after the addition of insulin the levels of the glycolipid are still below the control value. In lymphocytes, however, resynthesis takes place within 1 min of the addition of the hormone. Lymphocytes are exceptionally rapid and in other systems, like in CHO cells, resynthesis is also relatively a slow process. The mechanism of resynthesis is not known.

Houslay: You showed that 80 % of these lipids was on the outside, can you exclude that the 20 % on the inside is not involved in the signalling.

Mato: It is difficult to exclude the possibility that the percent of glycolipid present on the inside of the cell is not involved in the signalling process. The fact that POS mimics insulin action when added to intact cells and the finding that cells release POS to the extracellular media upon insulin addition suggests that the pool of glycosyl-PI present on the cell surface plays an important role in insulin signalling.

Houslay: In your scheme there are three active sites that are important on the extracellular surface, one is the insulin receptor, the second presumably is the phospholipase and the third is the transport protein. Is it possible to show those sites on the outside. For example could you treat with a protease and block the transport.

Mato: There are two papers, one by Joe Larner and the other one by Alan Saltiel, showing that insulin activates the release of proteins which are attached to the membrane through a glycosyl-PI anchor. These papers indicate the presence of a phospholipase C whose active site is on the cell surface and that acts on glycosyl-PI. We have not done yet the experiments to show whether treatment with a protease could block transport.

Houslay: Is there energy dependent uptake of your POS?

Mato: Uptake of POS occurs at 37 °C but not at 4 °C. Moreover, metabolic poisoning of the cell with KCN also blocks POS uptake. These results suggest the existence of an energy dependent uptake of POS.

Corbin: Is the concentration that you add to your broken cell preparation or your intact cells consistent with the concentration you measure in a liver cell extract or an adipocyte extract?

Mato: We do not know what is the concentration of POS generated after insulin addition to adipocytes or hepatocytes. It is known, however, that the same concentration of POS is required to obtain effects with intact or broken cell preparations. We have measured uptake in hepatocytes at micromolar concentrations of POS in the incubation media.

Corbin: If you take fat cells and treat them with insulin and then make a fraction containing the POS and add that to a phosphodiesterase, would that work?

Mato: Experiments like that have been carried out by Alan Saltiel and Joe Larner with positive results.

Dumont: To follow up on that question I would like to know what are your structural requirements for POS action and so what are your negative controls?

Mato: Treatment of POS with nitrous acid, a procedure that breaks the glycosidic bond between the inositolphosphate and the molecule of glucosamine, results in inactivation of the molecule. Similarly, treatment with alkaline phosphatase abolishes the biological activity of POS. Moreover, simple sugars like myoinositol phosphate, glucosamine, etc. do not mimic insulin action under conditions where POS does.

Dumont: So in each experiment you use that as a negative control, for instance you say that if you remove the inositol that will not work, do you use that as a negative control?

Mato: You purify the glycolipid and at the same time you obtain a blank and treat both with phospholipase. Treatment of the sample containing the glycolipid yields an active molecule and the blank not. The biologically active molecule can be destroyed by treatment with nitrous acid or alkaline phosphatase. There is an correlation between biological activity and integrity of the phospho-oligosaccharide.

Clementi: Does diacylglycerol have some function that is on the outside; second is there any evidence that other hormones can activate this phospholipase too?

Mato: These are two interesting questions. Last year there was a paper by Peter Stralfors in Nature where he gave evidence that extracellularly added diacylglycerol stimulated glucose transport in adipocytes. In this paper there was not correlation between stimulation of glucose transport and activation of protein kinase C. As to your second question, Farese has shown that EGF and IGF-I stimulate the hydrolysis of glycosyl-PI in BC3H1 myocytes and we have detected a small hydrolysis of this lipid in CHO cells stimulated with IGF-I. In a recent publication by the group of Saltiel, NGF has also been shown to stimulate glycosyl-PI hydrolysis. I think that similarly to the situation with cAMP and the phosphoinositides, a variety of signals will induce glycosyl-PI hydrolysis.

Van Haastert: I have an other question if POS is liberated in the extracellular space then I would expect that POS is not only insulin dependent but also cell density dependent. And my question is whether insulin action is cell density dependent?

Mato: I do not have an answer to your question. You can argue that since POS is a small tail protruding from the cell surface compared to the proteins, when it is generated in response to insulin this molecule is not free to diffuse and might be transported very rapidly to the cell interior.

Devreotes: Just to follow that up, what if you apply insulin in the presence of alkaline phosphatase?

Mato: We have performed a similar experiment. Insulin was applied in the presence of β-galactosidase, to break the tail of the glycosyl-PI, and we observed that the hormone had no effect on pyruvate kinase activation by glucagon.

Iyengar: What about the tyrosine kinase activity in those experiments?

Mato: Insulin binding was not modified in those experiments.

Iyengar: When you treat your cells with β-galactosidase and then you get insulin binding do you then get tyrosine kinase activity?

Mato: That we have not measured.

Simon: I do not understand the DNA synthesis stimulation, do you grow the cells in minimal medium, so that there is serum and no insulin to start with?

Mato: The experiment I mentioned was carried out by the group of Lee Witters, so I do not remember the details. DNA synthesis in the control cells is measured in the absence of serum or in the presence of diluted media and is compared with the effect observed in cells treated with insulin or POS.

Simon: So presumably you could get the growth effects of insulin in minimal medium with POS?

Mato: As I remember the experiments, the authors did not measure actual growth. The only related experiment that we have carried out in my laboratory is the measurement of amino acid transport and the effect of insulin and POS. The stimulation by insulin of amino acid transport depends on protein synthesis and we observed the same dependency with POS.

AGONIST AND GUANINE NUCLEOTIDE REGULATION OF P_{2Y}-PURINERGIC RECEPTOR-LINKED PHOSPHOLIPASE C

J.L. Boyer, M.W. Martin, C.L. Cooper, G.L. Waldo, A.J. Morris, H.A. Brown, R.A. Jeffs, J.R. Hepler, C.P. Downes, and T.K. Harden
Department of Pharmacology
University of North Carolina School of Medicine
Chapel Hill, North Carolina 27599
USA

INTRODUCTION

Extracellular adenine nucleotides interact with cell surface receptors to produce a myriad of physiological effects in the central nervous system and peripheral tissues (Burnstock, 1978; Gordon, 1986; Burnstock and Kennedy, 1986; Fleetwood and Gordon, 1987). Burnstock proposed in 1978 that these responses are mediated by two major receptor types: those physiologically activated by adenosine, called P_1-purinergic receptors, and exhibiting a potency order of adenosine > AMP > ATP, and those activated by ATP or ADP, called P_2-purinergic receptors, and exhibiting the potency order of ATP > ADP > AMP > adenosine. Subsequent studies proved that at least two subtypes of P_1-purinergic receptors exist (A_1- and A_2-purinergic receptors) (Van Calker et al., 1979; Stiles, 1986; Williams, 1987). Subclassification of P_2-purinergic receptors proved more difficult, since in contrast to the receptors for adenosine, no good antagonists of P_2-purinergic receptors are available, and cell surface hydrolases readily metabolize ATP and ADP to adenosine, which can then produce physiological effects through P_1-purinergic receptors. However, the development of relatively non-hydrolyzable analogs of ATP helped resolve this issue, and the observation of differential effects of a large number of ATP and ADP analogs led Burnstock and Kennedy to propose in 1985 that subypes of P_2-purinergic receptors exist. P_{2X}-purinergic receptors exhibit the potency order of $Ap(CH_2)pp$ > $App(CH_2)p$ > ADP > 2-methylthio ATP

NATO ASI Series, Vol. H 44
Activation and Desensitization of Transducing Pathways
Edited by T. M. Konijn, M. D. Houslay, P. J. M. Van Haastert

(2MeSATP) and P_{2Y}-purinergic receptors exhibit the potency order of 2MeSATP > ATP > Ap(CH_2)pp = App(CH_2)p.

Biochemical mechanisms responsible for P_1-purinergic receptor action have been extensively scrutinized and principally consist of effects on cyclic AMP synthesis; A_1-adenosine receptors inhibit adenylate cyclase and A_2-adenosine receptors activate adenylate cyclase (Londos and Wolf, 1977; Stiles, 1986). Much less is known about the biochemical sequelae of P_2-purinergic receptor activation. However, the nascent understanding of the biochemical responses to P_{2Y}-purinergic receptor activation centers around the inositol phosphate/Ca^{++} signalling system. That is, it has been shown recently that a large class of neurotransmitter, hormone, and growth factor receptors produce their physiological effects through a phospholipase C-catalyzed breakdown of phosphatidylinositol 4,5-bisphosphate (PtdIns(4,5)P_2) to inositol 1,4,5-trisphosphate, which releases intracellular Ca^{++}, and to diacylglycerol, which activates protein kinase C (Berridge and Irvine, 1984; Kikkawa and Nishizuka, 1986; Berridge, 1987). In several target tissues, ADP and ATP analogs have been shown to increase inositol phosphate production or cytoplasmic Ca^{++} levels through what is apparently a P_{2Y}-purinergic receptor (Charest et al., 1985; Okajima et al., 1987; Pirotton et al., 1987; Boyer et al., 1989a).

Our interest in P_2-purinergic receptors evolved from a long-term interest in defining the mechanism(s) whereby hormone receptors activate phospholipase C. It has been known for fifteen years that GTP is necessary for hormonal regulation of adenylate cyclase, and during the last ten years it became clear that a guanine nucleotide regulatory protein (G-protein) coupled receptors to the adenylate cyclase catalyst. Similarly, direct evidence of a G-protein involvement in receptor regulation of phospholipase C was reported in 1985-1986 using cell-free preparations from a variety of target tissues (for review see Martin, 1989). Turkey erythrocyte membranes have proven particularly useful for study of the G-protein-regulated phospholipase C (Harden et al., 1987; Downes et al.,1988; Harden, 1989). In this review we summarize the salient features of this model system, and review in particular its regulation by P_{2Y}-purinergic receptors, and its properties during purinergic agonist-induced desensitization.

GUANINE NUCLEOTIDE-DEPENDENT REGULATION OF PHOSPHOLIPASE C BY P_{2Y}-PURINERGIC RECEPTORS

In contrast to mammalian erythrocytes, turkey erythrocytes possess a phosphatidylinositol synthase activity and therefore, [^{3}H]-labelled inositol can be used to label turkey erythrocyte phosphoinositides to very high specific activity. Membranes prepared from labelled erythrocytes exhibit little or no basal release of inositol phosphates, but respond to activators of G-proteins, e.g. GTPγS or AlF_4^-, with large increases in phospholipase C activity (Harden et al., 1987). In the absence of ATP the release of inositol phosphates is linear for only 5-7 minutes, whereas in the presence of ATP (> 300 μM) production is linear for up to 20 minutes (Harden et al., 1988). Analysis of phosphoinositide levels in GTPγS-stimulated membranes indicates that in the presence of ATP considerable interconversion of phosphatidylinositol to phosphatidylinositol 4-phosphate and interconversion of phosphatidylinositol 4-phosphate to PtdIns(4,5)P_2 occurs to maintain polyphosphoinositides at near initial levels. Several lines of evidence indicate that PtdIns(4,5)P_2 is the predominant substrate phosphoinositide for the activated phospholipase C (Harden et al., 1988). As is described in detail below, the effects of ATP on responsiveness involve more that support of interconversion of phosphoinositides.

The effects of ATP and analogs of ATP and ADP on phospholipase C activity can be most unambiguously defined under conditions of short-time assays, e.g. < 5 minutes, where substrate concentrations are not limiting and in the presence of a guanine nucleotide, e.g. GTP or low (< 1 μM) concentrations of GTPγS, that by itself causes little or no stimulation of the enzyme. In this situation, ATP and ADP analogs cause a marked concentration dependent stimulation of enzyme activity with an order of potency of 2MeSATP > ADPβS > ATPγS > ATP > App(NH)p = ADP > Ap(CH_2)pp > App(CH_2)p. Adenosine, AMP, phenylisopropyladenosine, and 5'-dideoxyadenosine have no effect on phospholipase C activity at very high concentrations (Boyer et al., 1989a). Taken together, these results are consistent with the receptor specificity described by Burnstock and Kennedy (1985) for a P_{2Y}-purinergic receptor. The maximal effect on inositol phosphate production is identical among ATP and analogs such as ADPβS and 2MeSATP during assays of 5 minutes or less. However, with extended incubation where PtdIns(4,5)P_2 may become limiting, the effects of ATP on inositol phosphate production are greater than with the ATP analogs due to the fact that ATP supports substrate production as well as stimulates the P_{2Y}-purinergic receptor.

P_{2Y}-purinergic receptor-mediated stimulation of the turkey erythrocyte phospholipase C is strictly dependent on the presence of guanine nucleotides, with the maximal effect of agonists on enzyme activity in the presence of GTP approximately 15 % of that observed with agonists in the presence of a maximally effective concentration of GTPγS. In the absence of a P_{2Y}-purinergic receptor agonist, GTPγS activates the turkey erythrocyte phospholipase C with a $K_{0.5}$ value of approximately 3 μM. Addition of a P_{2Y}-purinergic receptor agonist such as 2MeSATP produces a concentration dependent increase in the effect of a maximally effective concentration of GTPγS, and the GTPγS activation curve is shifted approximately 50-fold to the left in the presence of a maximally effective concentration of agonist.

Inositol phosphate production in the presence of GTPγS occurs with a time course exhibiting a considerable time lag. This lag is decreased significantly by 2MeSATP. Activation by GTPγS follows first order kinetics and the rate of activation, i.e. the k_{obs} calculated by semilogarithmic transformation of activation time courses, is markedly increased in a concentration dependent and saturable manner by P_{2Y}-purinergic receptor agonists. In contrast to the effects of agonists on the rate of activation of the phospholipase C, the rate of activation of the enzyme in the presence of a fixed concentration of agonist is independent of the concentration of guanine nucleotide (Boyer et al., 1989a).

GDPβS blocks guanine nucleotide stimulation of the turkey erythrocyte phospholipase C in a competitive manner (K_i = 1 μM). By adding high concentrations of GDPβS to turkey erythrocyte membranes previously activated by agonist and GTP, Gpp(NH)p, or GTPγS, the forward reaction in the phospholipase C catalytic cycle can be blocked, and the decay rate of the activated enzymic species can be followed. The off-rate of the GTP-preactivated enzyme is rapid (1.1 min^{-1}), whereas the off-rate of the enzyme preactivated in the presence of the hydrolysis resistant analogs is much slower (0.2 min^{-1}) (Boyer et al., 1989a).

These properties of the turkey erythrocyte phospholipase C are highly reminiscent of the properties of the hormonally regulated adenylate cyclase (Sevilla et al., 1976; Ross et al., 1977; Tolkovsky and Levitzki, 1978; Birnbaumer et al., 1980). Thus, activation of phospholipase C by P_{2Y}-purinergic receptor agonists is dependent upon the presence of guanine nucleotides, and the rate of activation of the enzyme is markedly increased by P_{2Y}-purinergic receptor stimulation but is independent of the concentration of guanine nucleotide. These results suggest that receptor activation promotes guanine nucleotide exchange on the involved G-protein, resulting in transformation of the G-protein into its active GTP- (or GTPγS-) liganded state. As with adenylate cyclase, the life time of the GTP-liganded activated state is apparently determined by the activity of a GTPase. In the case of activation by GTP the consequences of this activity are considerable, whereas with stable GTP analogs this is much less so. Thus, the maximal enzyme activity observed in the presence of GTP and a P_{2Y}-purinergic receptor agonist is much less than that observed with a hydrolysis resistant GTP analog and an agonist (Boyer et al., 1989a).

DIRECT IDENTIFICATION OF THE GUANINE NUCLEOTIDE-REGULATED P_{2Y}-PURINERGIC RECEPTOR

We have adapted [^{35}S]ADPβS as a radioligand to study directly the P_{2Y}-purinergic receptor that regulates phospholipase C in turkey erythrocyte membranes (Cooper et al., 1989). [^{35}S]ADPβS binds to purified turkey erythrocyte membranes with an association rate constant of 1×10^{7} M^{-1} min^{-1} and with a rate constant for dissociation of 4×10^{-2} min^{-1}. Radioligand binding is saturable (B_{max} = 3 pmol/mg protein) and occurs at an apparently homogeneous (K_d = 8 nM) population of binding sites. The binding of [^{35}S]ADPβS is inhibited by ADP and ATP analogs in a concentration-dependent manner and by mass action kinetics, i.e. Hill coefficients are approximately unity. Moreover, the K_i values for a series of seven compounds are consistent with those predicted for a P_{2Y}-purinergic receptor and are essentially identical to the $K_{0.5}$ values of the same compounds for stimulation of turkey erythrocyte phospholipase C. Thus, the binding site identified by [^{35}S]ADPβS is apparently the same as the binding site involved in receptor-mediated activation of the phospholipase C.

An important aspect of the binding site identified by [^{35}S]ADPβS is that it is under regulation by guanine nucleotides. That is, radioligand binding is inhibited by guanine nucleotides with an order of potency (GTPγS > Gpp(NH)p > GTP = GDP > GDPβS >> GMP) consistent with that observed for other G-protein-linked receptors. The effects of guanine nucleotides do not involve competition at the level of the [^{35}S]ADPβS binding site, since in contrast to the competitive inhibition observed with ADP and ATP analogs, inhibition by guanine nucleotides is noncompetitive. In addition, the rate constant for dissociation of radioligand is increased approximately three-fold in the presence of GTPγS (C.L. Cooper, unpublished). In analogy to other more fully studied G-protein-linked receptors, these effects of guanine nucleotides on [^{35}S]ADPβS binding almost certainly are a consequence of P_{2Y}-purinergic receptor agonist-induced association of the receptor and the phospholipase C-linked G-protein. We recently have demonstrated (R.A. Jeffs and C.L. Cooper, unpublished) that the P_{2Y}-purinergic receptor can be solubilized with nonionic detergent from turkey erythrocyte membranes, and directly identified using [^{35}S]ADPβS as a radioligand in an equilibrium binding assay. Furthermore, this soluble [^{35}S]ADPβS binding site is regulated by guanine nucleotides with properties analogous to those observed with the membrane-associated receptor. We have interpreted these results to mean that a P_{2Y}-purinergic receptor-G-protein complex is stable to nonionic detergent solubilization from turkey erythrocyte membranes. The availability of such a species in a soluble form may prove propitious in the eventual identification of the G-protein invovlved in phosphoinositide signalling.

AGONIST-INDUCED DESENSITIZATION OF THE P_{2Y}-RECEPTOR-REGULATED PHOSPHOLIPASE C

Hormone-induced desensitization of the receptor-regulated adenylate cyclase signalling system has been widely studied (Harden, 1983; Sibley and Lefkowitz, 1985; Lefkowitz and Caron, 1988), and steps involved in its occurrence can be described in molecular terms (Harden, 1983; Clark, 1986; Strasser et al., 1986; Sibley et al., 1987; Clark et al., 1988) It is clear from intact cell studies that desensitization of phospholipase C-linked receptors occurs (Nakahata and Harden, 1987; Sugiya et al., 1987; Paris et al., 1988), but due at least in part to the difficulty of directly studying the receptor-regulated phospholipase C in mammalian cell-free preparations, progress has been slow in defining its molecular basis. The availability of the turkey erythrocyte membrane as a cell-free preparation that retains high responsivity to receptor stimulation has provided an opportunity to begin to investigate the molecular basis of agonist-induced desensitization of a phospholipase C-linked receptor.

Preincubation of turkey erythrocytes with ADPβS results in a marked decrease in the capacity of ADPβS plus GTP to stimulate phospholipase C in membranes derived from these cells (Martin and Harden, 1989). The half-time of occurrence of desensitization is 0.5-2.0 minutes and within 10 minutes of incubation of cells with agonist, the responsiveness of the resultant membranes has reached a new quasi-steady state level of 40-50 percent of control. Desensitization is reversible in that transfer of agonist-preincubated erythrocytes to agonist-free medium results in recovery of agonist plus GTP responsiveness of the membrane phospholipase C activity to control levels with a half time of 10-20 minutes. The loss of agonist plus GTP responsiveness occurs as a decrease in maximal response observed with little or no change in the apparent affinity of P_{2Y}-purinergic receptor agonists for stimulation of phospholipase C. Furthermore, induction of desensitization follows a pharmacological profile expected for a P_{2Y}-purinergic receptor-mediated event; 2MeSATP and ADPβS are potent inducers of desensitization whereas App(CH_2)p does not induce desensitization even at high (1 mM) concentrations. Neither the rate of activation nor the final phospholipase C activity attained in the presence of GTPγS alone is altered in membranes from cells preincubated with ADPβS for 15 minutes. Furthermore, AlF_4^--stimulated inositol phosphate formation is not modified in membranes from desensitized cells. In contrast, the capacity of ADPβS to increase the rate of activation of phospholipase C by GTPγS is markedly decreased in membranes from agonist-preincubated cells.

Taken together, these data are highly analogous to those reported in detail for agonist-induced desensitization of the receptor-regulated adenylate cyclase system. They suggest, but do not prove, the following chain of events. Incubation of the turkey erythrocyte with a P_{2Y}-purinergic receptor agonist results in activation of phospholipase C through a process involving acceleration of GTP exchange for GDP on the involved G-protein. With a somewhat slower, but nonetheless relatively rapid time course, a covalent modification of the receptor-G-protein-phospholipase C occurs such that a lesion occurs that is stable to cell lysis and membrane incubation at 30 ^{o}C. The location of this lesion has not yet been defined, but based on the results described above probably does not involve the phospholipase C per se or the protein sequence of the G-protein involved between guanine nucleotide binding and activation of the phospholipase C. In a classical sense, the signalling system is uncoupled at the level of receptor-G-protein coupling. This could be reflective of a modification of the receptor or of that part of the G-protein that is involved in communicating the signal between agonist-occupied receptor and the acceleration of guanine nucleotide exchange at the GTP-binding site. In analogy with the adenylate cyclase system it is attractive to propose that phosphorylation of a component of the phosphoinositide signalling pathway is responsible for desensitization, and indeed, data with cultured mammalian cell lines indirectly suppport such a premise. Along these lines, preincubation of turkey erythrocytes with phorbol ester activators of protein kinase C decrease membrane responsiveness to P_{2Y}-purinergic receptors (M.W. Martin, unpublished observations). Whether these results presage a feed-back regulation of this signalling system by protein kinase C during receptor activation remains to be proven.

SUMMARY AND FUTURE DIRECTIONS

In a number of ways the turkey erythrocyte P_{2Y}-purinergic receptor/phospholipase C signalling system provides an ideal model to study hormone and guanine nucleotide-mediated regulation of phosphoinositide signalling. Responses that are observed are of large magnitude and are long-lived. It is additionally advantagous to work with a homogeneous cell preparation from which plasma membranes can be purifed in large quantities. This is of considerable importance in that neither the G-protein nor the phospholipase C involved in receptor-regulated phosphoinositide signalling have been identified. Attempts to identify the G-protein in turkey erythrocytes by reconstitution of known purified G-proteins or of GTP-binding fractions purified from turkey erythrocyte plasma membranes have not yet been successful. One series of experiments

(Boyer et al, 1989b) do suggest, however, that it is a heterotrimeric G-protein that is involved in P_{2Y}-purinergic receptor-mediated regulation of phospholipase C. That is, reconstitution of G-protein $\beta\gamma$ subunits purified from turkey erythrocyte plasma membranes, bovine brain, or human placenta all have marked effects on turkey erythrocyte phospholipase C activities (AlF_4^--stimulated is inhibited and P_{2Y}-purinergic receptor agonist + GTP-stimulated activity is enhanced) over the same conentration range that AlF_4^--stimulated adenylate cyclase activity is inhibited.

Progress in purification of the turkey erythrocyte phospholipase C enzyme recently has been made. That is, an approximately 150 kDa protein has been purified to near homogeneity (V_{max} = 20 µmol/min mg protein) that expresses high affinity (K_m = 5 µM) for PtdIns(4,5)P_2 and that does not react with antibodies made against other known phospholipase C enzymes (A.J. Morris and G.L. Waldo, unpublished observations). Preliminary data suggest that this enzyme can be reconstituted into turkey erythrocyte ghosts where it can then be regulated by endogenous G-protein and P_{2Y}-purinergic receptors. Our hope is that isolation of the involved phospholipase C will now allow us to unambiguously identify the involved G-protein, and to reconstruct a responsive G-protein-regulated phospholipase C in a model vesicle or membrane system.

ACKNOWLEDGEMENTS

This work on turkey erythrocytes was initiated in Dr. C. P. Downes laboratory, Dept. of Cellular Pharmacology, Smith Kline French Res. Ltd., Welwyn, England, and has continued in part in that laboratory. Work in the North Carolina laboratory was supported by USPHS grants GM 29563 and GM 38213. Dr. Jose L. Boyer is on leave from the Instituto Nacional de Cardiologia "Ignacio Chavez", Mexico.

REFERENCES

Berridge MJ (1987) Inositol trisphosphate and diacylglycerol: Two interacting second messengers. Ann Rev Biochem 56:159-193

Berridge MJ, Irvine RF (1984) Inositol triphosphate, a novel second messenger in cellular signal transduction. Nature 312:315-321

Birnbaumer L, Swartz TL, Abramowitz J, Mintz PM, Iyengar R (1980) Transient and steady stste kinetics of the interaction of guanyl nucleotides with the adenylate cyclase system from rat liver plasma membranes. Interpretation in terms of a simple two-state model. J Biol Chem 255:3542-3551

Boyer JL, Downes CP, Harden TK (1989a) Kinetics of activation of phospholipase C by P_{2Y}-purinergic receptor agonists and guanine nucleotides. J Biol Chem 264:884-890

Boyer JL, Waldo GL, Evans T, Northup JK, Downes CP, Harden TK (1989b) Modification of AlF_4^-- and receptor-stimulated phospholipase C by G-protein $\beta\gamma$ subunits. J Biol Chem, In Press

Burnstock G (1978) A basis for distinguishing two types of purinergic receptors. In: Straub RW, Bolis L (eds) Cell membrane receptors for drugs and hormones: A multidisciplinary approach. Raven Press, New York, p 107

Burnstock G, Kennedy C (1985) Is there a basis for distinguishing two types of P_2 purinoceptors?. Gen Pharmacol 16:433-440

Burnstock G, Kennedy C (1986) A dual function of adenosine 5'-triphosphate in the regulation of vascular tone. Excitatory cotransmitter with noradrenaline from perivascular nerves and locally released inhibitory intravascular agent. Circ Res 58:319-330

Charest C, Blackmore PF, Exton JH (1985) Characterization of responses of isolated rat hepatocytes to ATP and ADP. J Biol Chem 260:15789-15794

Clark RB (1986) Desensitization of hormonal stimuli coupled to regulation of cyclic AMP levels. Adv Cyclic Nucleotide Protein Phosphorylation Res 20:151-209

Clark RB, Kunkel MW, Freidman J, Goka TJ, Johnson RA (1988) Activation of cAMP-dependent protein kinase is required for heterologous desensitization of adenylyl cyclase in

S49 wild-type lymphoma cells. Proc Natl Acad Sci USA 85:1442-1446

Cooper CL, Morris AJ, Harden TK (1989) Guanine nucleotide-sensitive interaction of a radiolabeled agonist with a phospholipase C-linked P_{2Y}-purinergic receptors. J Biol Chem 264:6202-6206

Downes CP, Berrie CP, Hawkins PT, Stephens L, Boyer JL, Harden TK (1988) Receptor and G-protein-dependent regulation of turkey erythrocyte phosphoinositidase C. Phil Trans R Soc Lond B320:267-280

Fleetwood G, Gordon JL (1987) Purinoceptors in the rat heart. Br J Pharmacol 90:219-227

Gordon JL (1986) Extracellular ATP: Effects, sources and fate. Biochem J 233:309-319

Harden TK (1983) Agonist-induced desensitization of the β-adrenergic receptor-linked adenylate cyclase. Pharmacol Rev 35:5-32

Harden TK, Stephens L, Hawkins PT, Downes CP (1987) Turkey erythrocyte membranes as a model for regulation of phospholipase C by guanine nucleotides. J Biol Chem 262:9057-9061

Harden TK, Hawkins PT, Stephens L, Boyer JL, Downes CP (1988) Phosphoinositide hydrolysis by guanosine 5'-[γ-thio]triphosphate-activated phospholipase C of turkey erythrocyte membranes. Biochem J 252:583-593

Harden TK (1989) The role of guanine nucleotide regulatory proteins in receptor-selective direction of inositol lipid signalling. In: Michell RH, Drummond AH, Downes CP (eds) Inositol lipids in cell signalling. Academic Press, London, p 113

Kikkawa U, Nishizuka Y (1986) The role of protein kinase C in transmembrane signalling. Ann Rev Cell Biol 2:149-178

Lefkowitz RJ, Caron MG (1988) Adrenergic receptors. Models for the study of receptors coupled to guanine nucleotide regulatory proteins. J Biol Chem 263:4993-4996

Londos C, Wolff J (1977) Two distinct adenosine-sensitive sites on adenylate cyclase. Proc Natl Acad Sci USA 74:5482-5486

Martin TFJ (1989) Lipid hydrolysis by phosphoinositidase C: Enzymology and regulation by receptors and guanine nucleotides. In: Michell RH, Drummond AH, Downes CP (eds) Inositol lipids in cell signalling. Academic Press, London, p 81

Martin MW, Harden TK (1989) Agonist-induced desensitization of a P_{2Y}-purinergic receptor- and guanine nucleotide-regulated phospholipase C. Submitted for publication.

Nakahata N, Harden TK (1987) Regulation of inositol trisphosphate accumulation by muscarinic cholinergic and H_1-histamine receptors on human astrocytoma cells. Biochem J 241:337-344

Okajima F, Tokumitsu Y, Kondo Y Ui M (1987) P_2-purinergic receptors are coupled to two signal transduction systems leading to inhibition of cAMP generation and to production of inositol phosphates in rat hepatocytes. J Biol Chem 262:13483-13490

Paris S, Magnaldo I, Pouyssegur J (1988) Homologous desensitization of thrombin-induced phosphoinositide breakdown in hamster lung fibroblasts. J Biol Chem 263:13791-13796

Pirotton S, Raspe E, Demolle D, Erneux C, Boeynaems JM (1987) Involvement of inositol 1,4,5 trisphosphate in the action of adenine nucleotides on aortic endothelial cells. J Biol Chem 262:17461-17466

Ross EM, Maguire ME, Sturgill TW, Biltonen RL, Gilman AG (1977) Relationship between the β-adrenergic receptor and adenylate cyclase. Studies of ligand binding and enzyme activity in purified membranes of S49 lymphoma cells. J Biol Chem 252:5761-5775

Sevilla N, Steer ML, Levitzki A (1976) Synergistic activation of adenylate cyclase by guanylyl imidodiphosphate and epinephrine. Biochemistry 15:3493-3499

Sibley DR, and Lefkowitz RJ (1985) Molecular mechanisms of receptor desensitization using the β-adrenergic receptor-coupled adenylate cyclase system as a model. Nature 317:124-129

Strasser RH, Sibley DR, Lefkowitz RJ (1986) A novel catecholamine-activated adenosine cyclic 3',5'-phosphate independent pathway for β-adrenergic receptor phosphorylation in wild type and mutant S49 lymphoma cells: Mechanisms of homologous desensitization of adenylate cyclase. Biochemistry 25:1371-1377

Sibley DR, Benovic JL, Caron MG, Lefkowitz RJ (1987) Regulation of transmembrane signalling by receptor phosphorylation. Cell 48:913-922

Stiles GL (1986) Adenosine receptors: Structure, function and regulation. Trends Pharmacol Sci 7:486-490

Sugiya H, Tennes KA, Putney JW (1987) Homologous desensitization of substance-P-induced inositol polyphosphate formation in rat parotid acinar cells. Biochem J 244:647-653

Tolkovsky AM, Levitzki A (1978) Mode of coupling between the β-adrenergic receptor and adenylate cyclase in turkey erythrocytes. Biochemistry 18:3795-3810

Van Calker D, Muller M, Hamprecht B (1979) Adenosine regulates via two different types of receptors the accumulation of cyclic AMP in cultured brain cells. J Neurochem 33:999-1005

Williams M (1987) Purine receptors in mammalian tissues: Pharmacological and functional significance. Ann Rev Pharmacol Toxicol 27:315-345

DISCUSSION

Dumont: Did you also see the potentiation by βτ if you used instead of GTP, Gpp(NH)p or GTPτS for the assay?

Boyer: Yes we observed it, but it was somewhat more variable than the effects on GTP-stimulated activity. Either with GTP or GTPτS the effects of βτ on phospholipase C activity were not produced by changes in the EC_{50} for the receptor agonist. We did not try Gpp(NH)p.

Iyengar: Did you not try Gpp(NH)p by itself or with agonist, I mean can you get a βτ effect on Gpp(NH)p activation?

Boyer: I have not tried Gpp(NH)p alone or in the presence of agonist.

Houslay: Is your receptor a glycoprotein?

Boyer: I have done one experiment where we treated membranes with neuraminidase and Endo F, we did not see any effect on the mobility of the receptor in that particular experiment.

Caron: A long time ago we purified the β1-adrenergic receptor from turkey erythrocytes, and we had a whole lot of problems to inhibit proteolysis, the β-receptor comes as two bands at about 40,000 kDa and the other about 45,000 kDa, which turns out that they are probably proteolysis products of a larger molecule of about 55,000 kDa, I wonder if you do need protease inhibitors?

Boyer: We use protease inhibitors. In the first experiments we observed the same phenomenon, we observed an additional band of about 45,000 kDa that had exactly the same pharmacology as the receptor. We found that the presence of this band correlated with the increase in the temperature of the sample during the alkylation of the proteins prior the electrophoresis.

Corbin: I have a maybe a very naive question, what is believed to be the natural metabolizable compound that binds to the P_2-purinergic receptor?

Boyer: ATP and probably also ADP.

Corbin: So it is extracellular?

Boyer: Yes this is an extracellular effect of ATP. There are reports on extracellular effects of ATP and its products since about 15 years ago, and now the interest is growing, in a few months there will be a meeting on purine nucleosides and nucleotides in cell signalling in Maryland, and later this year the New York Academy of Sciences is organizing another meeting on extracellular effects of ATP.

Corbin: So in what situations would extracellular ATP increase?

Boyer: It has been shown that ATP can be released by some tissues and from nerve terminals as a cotransmitter with noradrenaline.

Downes: Platelets are full of ATP in the secretory granules as well and it is notable that endothelial cells have a similar P_{2Y}- purinergic receptor and respond in the same way, so there is a pretty good example of a probable interaction, between the aggregating platelets and the endothelium through that receptor.

Cooke: Is your G-protein toxin sensitive? I mean to pertussis toxin.

Boyer: So far we have not been able to demonstrate any effect of pertussis toxin on receptor-stimulated phospholipase C activity. This question is somewhat complicated in our system: Turkey erythrocytes posses very active endogenous ADP-ribosyltransferases and this makes it very difficult to demonstrate that under our conditions (where we do not see effects on phospholipase C), pertussis toxin treatment fully [^{32}P]ADP-ribosylate G_i-like proteins. However, pertussis toxin-treatment of turkey erythrocyte membranes or intact cells under the same conditions in which marked effects are observed in other cell types, have failed to modify phospholipase C activity in turkey erythrocytes.

Cooke: Have you looked using antibodies to find out in what range of the known G-proteins they are?

Boyer: We are looking at that now.

Dumont: Am I wrong or is there a discrepancy between what Ravi Iyengar showed the other day, where with the $\beta\tau$ you would completely knock off the PIP_2 response and here you potentiate it. And if there is a discrepancy is that due to differences in the number of G_i in the two systems or, what is your explanation for that?

Iyengar: I think it makes life interesting, I do not know.

Dumont: But you agree there is a discrepancy?

Iyengar: Yes.

Downes: But all you have to have is that G_o in these cells is in the heterotrimer state. The idea that potentiation can occur is independent of the notion that some of the G-protein G_p is existing as free alpha subunits, so you may only see inhibition if there is a G-protein that is fully in the heterotrimeric form.

Iyengar: Well if you ever do careful purification there is no free $\beta\tau$ floating around, so I am going to be really surprised to find that, I saw that there was may be a twofold potentiation, is that right? (confirmed) so you are talking about a lot of free alpha subunits, and I am willing to think of other explanations but not of that one.

Devreotes: Did you look at the photolabelled band in desensitized cells?

Boyer: After incubation for 30 minutes in the presence of agonist, there is no difference in the labelling.

G-protein activation

MECHANISMS INVOLVED IN G-PROTEIN ACTIVATION BY HORMONE RECEPTORS

K.H. Jakobs, T. Wieland and P. Gierschik
Pharmakologisches Institut
Universität Heidelberg
Im Neuenheimer Feld 366
D-6900 Heidelberg
Federal Republic of Germany

INTRODUCTION

Signal-transducing guanine nucleotide-binding proteins (G-proteins) are heterotrimeric plasma membrane-located proteins. The function of these proteins is to couple receptors for a wide variety of extracellular signalling molecules (neurotransmitters, peptide hormones and auto- and paracrine hormonal factors) to various intracellular signal-forming systems (for reviews see Rodbell, 1980; Stryer, 1986; Stryer and Bourne, 1986; Gilman, 1987; Birnbaumer et al., 1987; Casey and Gilman, 1988; Lochrie and Simon, 1988; Neer and Clapham, 1988; Chabre and Deterre, 1989; Gierschik et al., 1989a). The effector moieties regulated in their activity by G-proteins are either enzymes, e.g., the cyclic AMP-forming adenylate cyclase, the cyclic GMP-hydrolyzing phosphodiesterase in the retina or the polyphosphoinositide-hydrolyzing phospholipase C, or ion channels. Compared to the number of receptors for extracellular signalling hormonal factors, being in the range of one hundred, the number of G-proteins, known so far both from the protein and cDNA level, is rather small, in the range of ten to fifteen. Some of these G-proteins are apparently widely distributed in different cell types of the mammalian organisms, e.g., the G_s- and G_i-proteins, while others exhibit a more restricted distribution, e.g., the G_o-proteins, which are apparently mainly localized in neuronal cells and tissues. Other G-proteins are even confined to one specialized cell type, e.g., the transducins, the G-proteins involved in light signal transduction by the photoexcited receptor rhodopsin.

In spite of the differential distribution and the coupling of some G-proteins to only one specific receptor type, the basic composition and the principal mechanisms of G-protein activation

NATO ASI Series, Vol. H 44
Activation and Desensitization of Transducing Pathways
Edited by T. M. Konijn, M. D. Houslay, P. J. M. Van Haastert

and deactivation are apparently very similar. All the G-proteins known so far to be involved in signal transduction exhibit a heterotrimeric composition, with a larger α-subunit (M_r's on SDS PAGE of 39 to 52 kDa), which is the most distinct subunit in the various G-proteins, a smaller β-subunit (M_r's of 35 and 36 kDa) and a γ-subunit (M_r's between 8 and 10 kDa). The α-subunit is the subunit which binds guanine nucleotides such as GTP, GDP and the various analogs of these endogenously occurring nucleotides and which, in addition, exhibits GTPase activity. Two distinct β-subunits are known so far at a protein and cDNA level, namely β_1 and β_2 with M_r's of 36 and 35 kDa, respectively. These two β-subunits show a high degree of sequence homology. Furthermore, no functional difference has been found between these two distinct β-subunits. The association of the two β-subunits with the various G-protein α-subunits appears to be more cell- than G-protein-specific. While most purified G-proteins contain both subunits (β_1 and β_2), although to a variable relative extent depending on the original tissue or cell type, the transducin α-subunit purified from mammalian rod other segments is apparently associated only with the 36 kDa β_1-subunit. The γ-subunits of G-proteins are apparently more heterogenous than the β-subunits, although so far the primary structure of only one of these subunits has been reported (γ-subunit of transducin). While the α-subunits of G-proteins can be separated from the other two subunits, the β- and γ-subunits are non-denaturing conditions always found associated with each other as βγ-dimers. Considering the possible combinations of one or the other β-subunit with one of the various γ-subunits, a high degree of heterogeneity of the βγ-dimers can in principal be imagined, which may be similarly high as the heterogeneity of the α-subunits. With the exception of transducin, which is apparently only loosely associated with rod outer segment membranes and which can be released from dark-adapted membranes by only reducing the ionic strength of the ambient medium, the other G-proteins are apparently integral membrane proteins. In addition, while transducin, at least its α-subunit, is released from the membranes in the course of its activation by the photoexcited receptor rhodopsin and GTP, the release of other G-protein α-subunits from membranes by agonist-liganded receptors is still a matter of debate.

ACTIVATION - DEACTIVATION OF G-PROTEINS

The principal mechanisms of G-protein activation by agonist-liganded or light-activated receptors are, at least for some G-proteins, e.g., the G_s- and G_t-proteins (transducin), well characterized. Based on numerous functional and structural data, the following sequence of events is assumed to occur (for reviews see e.g. Stryer, 1986; Gilman, 1987; Birnbaumer et al., 1987): In the inactive state, the G-protein α-subunit is associated with the βγ-dimer as a holo-G-protein complex and, in addition, GDP is apparently bound to the G-protein α-subunit. Activation of receptors by agonist binding or by light induces a conformational change of the receptor, now binding to the holo-G-protein. This interaction of the activated receptor with the G-protein induces a conformational change of the G-protein, at least of its α-subunit, which results in a reduced affinity of the G-protein for GDP and a subsequent dissociation of GDP from the α-subunit. The release of GDP, in turn, apparently not only alters the G-protein but, in addition, induces a further conformational change of the receptor. This conformational change of the receptor is measurable as increase in receptor affinity for the bound agonist. When GTP is available, this nucleotide will now bind to the nucleotide-free G-protein α-subunit. The binding of GTP to the α-subunit has apparently three consequences. First, the G-protein dissociates, at least functionally, from the agonist-bound receptor. Second, the affinity of the receptor, now dissociated from the G-protein, for the agonist will be decreased and the agonist can dissociate from the receptor. Third, as best shown with transducin, the G-protein apparently dissociates into its subunits, the GTP-liganded α-subunit and the βγ-dimer. The GTP-liganded α-subunit is apparently the active species of G-proteins responsible for interacting with effector moieties and changing their activities, although for some systems the apparently released βγ-dimers have also been claimed to regulate effector activities. The active state of the G-protein α-subunits is terminated by the GTPase activity of the G-protein, hydrolyzing the bound GTP to GDP and

P_i. In addition, the GDP-liganded α-subunit will reassociate with the βγ-dimer and, thus, form the inactive holo-G-protein complex. This inactive complex can reenter a new activation-deactivation cycle when interacting with an agonist-activated receptor.

In this short review, not all of mechanisms involved in G-protein activation by hormone receptors will be analyzed, instead only two problems, which we focussed on recently in our laboratory, will be discussed, namely first the relationship between agonist-receptor binding and G-protein activation and second the apparent involvement of phosphotransfer reactions in G-protein activation.

RELATIONSHIP BETWEEN G-PROTEIN ACTIVATION AND AGONIST-RECEPTOR BINDING

Based on data mainly obtained in studying interaction of adrenoceptors with G-proteins, it is generally assumed that coupling of agonist-liganded receptors to G-proteins always results in a high affinity agonist-binding state of the receptor and that this high affinity binding is also required for G-protein activation, i.e., the release of G-protein-bound GDP and the subsequent binding of GTP to the G-protein α-subunit (for reviews see Lefkowitz et al., 1983; Gilman, 1987; Gierschik et al., 1988). This assumption is based on the findings that in the physical or functional absence of G-proteins only low affinity agonist-receptor binding is measured. Furthermore, there is a good correlation between the formation of the high affinity agonist-binding state of the receptor and the ability of various receptor agonists (full and partial agonists) to activate G-proteins, measured as regulation of adenylate cyclase activity. The formation of the high affinity agonist-binding state of the receptor is, in most systems, dependent on the presence of divalent cations such as Mg^{2+} and Mn^{2+} at millimolar concentrations. Lastly, activation of G-proteins by GTP or its analogs guanylyl-5'-imidodiphosphate (GppNHp) and guanosine-5'-O-(3-thiotriphosphate) (GTP[γS]) coincides with a reduction in receptor affinity for agonists, while antagonist binding to receptors is usually not altered by guanine nucleotides.

In measuring binding of the formyl peptide receptor agonist, fMet-Leu-Phe (fMLP), to its binding sites in membranes of differentiated human leukemia cells (HL 60), we studied first whether the ternary complex (agonist-receptor-G-protein) always exhibits high affinity for the agonist and second whether the high affinity agonist-binding state of the receptor is required for G-protein activation. In the absence of guanine nucleotides, the receptor exhibited two affinity states for agonist binding, with the appearance of the high affinity agonist-binding state being absolutely dependent on the presence of divalent cations such as Mg^{2+} or Mn^{2+} at millimolar concentrations (Gierschik et al., 1989b). Furthermore, polycations such as neomycin and other aminoglycoside antibiotics were also able to induce the high affinity agonist-binding state (Herrmann et al., 1989). Most important, addition of guanine nucleotides such as GTP, GTP[γS], GDP or guanosine-5'-O-(2-thiodiphosphate) (GDP[βS]) not only reduced the high affinity agonist-receptor binding (by more than 90 %), these nucleotides also decreased agonist-receptor binding in the absence of divalent cations, thus, at a receptor exhibiting low agonist affinity. Furthermore, in the presence of neomycin, which like Mg^{2+} induced high affinity agonist-receptor binding both on formyl peptide receptors in HL 60 cell membranes and on β-adrenoceptors in guinea pig lung membranes, but in contrast to Mg^{2+} did not induce G-protein activation in the presence of GTP or its analogs, agonist-receptor binding was also decreased by guanine nucleotides (Herrmann et al., 1989). These data suggest that interaction of an agonist-liganded receptor with a G-protein does not necessarily induce a high affinity binding state of the receptor for the agonist. Although the relative extent of guanine nucleotide regulation of low affinity agonist-receptor binding was rather high, e.g., guanine nucleotides decreased fMLP receptor binding by more than 60 % under this condition, the absolute extent was rather small, about one tenth, compared to the guanine nucleotide regulation of high affinity agonist-receptor binding. Thus, on adrenoceptors (the model system), where agonist-receptor binding is usually studied as agonist competition of labeled antagonist binding (Lefkowitz et al., 1983), the regulation of low affinity agonist-receptor binding by guanine nucleotides is, because of technical reasons, not detectable.

By measuring agonist-stimulated GTP hydrolysis by G-proteins, we demonstrated that in the low affinity agonist-binding state the receptor not only interacts with G-proteins but is also capable of activating them. At submicromolar concentrations of free Mg^{2+}, which are required but also sufficient for providing agonist-stimulated GTPase activity of G-proteins, which however do not induce high affinity agonist-receptor binding, the agonist-liganded formyl peptide receptor perfectly stimulated a high affinity GTPase in HL 60 membranes (Gierschik et al., 1989b). Increasing the free Mg^{2+} to millimolar concentrations and, thereby, inducing the high affinity agonist-binding state of the receptor caused a marked increase in the potency of the agonist to stimulate the G-protein-dependent GTP hydrolysis. There was a good correlation between the high and low affinity dissociation constants for agonist-receptor binding and the K_{act} values of the agonist to stimulate GTPase activity. In addition, at millimolar concentrations of free Mg^{2+} inclusion of any agent which decreased or reversed the high affinity agonist-binding state of the receptor, such as sodium ions (Gierschik et al., 1989c), GDP or high GTP concentrations, in the GTPase assay decreased the potency of the agonist to stimulate GTPase activity. Finally, when agonist-induced GTPase stimulation was studied under conditions where both high and low affinity agonist-binding states of the receptor were observed, the agonist-induced GTPase stimulation was apparently also by the two affinity states of the receptor (Gierschik and Jakobs, manuscript in preparation).

Thus, in summary, interaction of agonist-liganded receptors with G-proteins does not necessarily induce the high affinity agonist-binding state of the receptor. Even in a low affinity agonist-binding state, the receptor is capable of interacting with and activating G-proteins. It has to be pointed out here that this low affinity agonist-binding state of the receptor requires the G-protein and is, thus, not identical with the low affinity state observed in the absence of G-proteins or when the interaction of receptors with G-proteins is prevented, e.g., after ADP-ribosylation of G_i-proteins by pertussis toxin. Thus, there are apparently at least three states of the receptor with regard to its binding affinity for the agonist, a very low affinity state in the physical or functional absence of G-proteins, a low affinity state, which is usually observed in the absence of divalent cations, and a high affinity agonist-binding state, which usually requires millimolar concentrations of divalent cations or which can also be induced by polycations such as neomycin and other aminoglycoside antibiotics. For the latter two states, G-proteins are required, and in both of these states the agonist-liganded receptor can apparently activate G-proteins. Whether in intact cells with at one hand rather high ambient Mg^{2+} concentrations (sub- to millimolar) and on the other hand rather high levels of the endogenous guanine nucleotides GTP and GDP (up to 100 μM or even more) G-protein activation is induced by the high or low affinity agonist-binding state of the receptor or by both states is more difficult to answer and may even be different for different cell types and receptors.

INVOLVEMENT OF PHOSPHOTRANSFER REACTIONS IN G-PROTEIN ACTIVATION

G-protein activation by agonist-liganded receptors cannot only be induced by GTP added to the incubation medium or present in the intact cell but apparently also by GTP locally formed by the action of nucleoside diphosphokinases (NDP kinase) (Kimura and Shimada, 1983; Seifert et al., 1988; Jakobs and Wieland, 1989). Such enzymes are apparently present in any membrane system (Parks and Agrawal, 1973). As shown in human platelet membranes, inhibition of this kinase with subsequent decreased G-protein activity by high concentrations of UDP, known to lead to an abortive NDP·enzyme complex, can be functionally counteracted by an agonist-activated receptor (Jakobs and Wieland, 1989). These data suggested that the agonist-activated receptor somehow stimulates the formation of GTP by the NDP kinase and/or that the formation of GTP by the NDP kinase and the subsequent binding of GTP to the G-protein are intimately connected reactions and that the newly formed GTP may be a better "substrate" for the G-protein than GTP added to the medium. As shown in human platelet membranes, the NDP kinase can also use adenosine-5'-O-(3-thiotriphosphate) (ATP[γS]) as phosphate donator leading to the formation of GTP[γS] with subsequent G-protein activation (Wieland and Jakobs, 1989a).

In order to see whether newly formed GTP[γS] binds to G-protein with higher apparent affinity than exogenously added GTP[γS], we studied the formation of [^{35}S]GTP[γS] from [^{35}S]ATP[γS] and the subsequent binding of the newly formed labelled GTP[γS] to G-protein in bovine rod outer segment membranes and in membranes of HL 60 cells, both in response to activated receptors and in competition with exogenously added GTP[γS] (Wieland et al., manuscript in preparation). Similarly as observed on a functional basis in human platelet membranes, activated receptors (photoexcited rhodopsin and agonist-activated formyl peptide receptors) stimulated the binding of newly formed [^{35}S]GTP[γS] to G-proteins. While in HL 60 membranes the addition of GDP was required to observe the formation and binding of [^{35}S]GTP[γS], in bovine rod outer segment membranes endogenous GDP or GDP released from transducin in response to interaction with photoexcited rhodopsin was apparently present at sufficiently high concentrations to provide the substrate for formation of [^{35}S]GTP[γS]. Most interestingly, UDP at high concentrations almost completely inhibited the formation and binding of [^{35}S]GTP[γS] under dark conditions, while with illuminated membranes the binding of newly formed [^{35}S]GTP[γS] was only partially reduced by the presence of UDP. These findings suggest that the NDP kinase responsible for formation of [^{35}S]GTP[γS] preferentially uses endogenous GDP, most probably that released from transducin in response to the photoexcited rhodopsin. Finally, competition experiments with unlabelled GTP[γS] on binding of newly formed [^{35}S]GTP[γS] suggested that exogenously added GTP[γS] exhibited an about ten-fold lower apparent affinity for binding to transducin than the newly formed [^{35}S]GTP[γS]. These data suggest that not only the GDP released from the G-protein is the preferred nucleoside diphosphate substrate for the NDP kinase, but, in addition, that GTP or GTP[γS] formed by NDP kinases is the preferred "substrate" for the G-protein. Thus, it appears that in intact membranes a highly selective interaction of G-proteins with NDP kinases takes place, resulting in a concerted conversion of

G-protein-released GDP to GTP and binding of GTP to the G-protein.

An additional phosphate transfer reaction apparently occurring directly on G-proteins was found when we studied phosphorylation of G-proteins by GTP and its analog GTP[γS]. These studies were initiated by the finding that βγ-subunits of transducin purified by two distinct methods had quite distinct effects on G-protein activity. βγ-Subunits prepared from transducin released from illuminated bovine rod outer segment membranes in the presence of GTP had no effect, over a wide range of concentrations, on G-protein activity in HL 60 membranes. In contrast, βγ-complexes purified from transducin released from the membranes with the GTP analog GTP[γS] caused a marked and potent G_i-protein activation. This activation of G_i-proteins by transducin βγ-complexes was heat-sensitive and not due to a contamination by free GTP[γS] (Wieland and Jakobs, 1989b). The "inactive" βγ-subunits purified from transducin eluted with GTP were convertable to "active" βγ-subunits by recombining the βγ-subunits with illuminated, transducin-depleted rod outer segment membranes in the presence of GTP[γS], while the various controls (minus rod outer segment membranes, minus GTP[γS] or rod outer segment membranes plus GTP[γS] alone) had virtually no effect. When we used [^{35}S]GTP[γS] instead of unlabelled GTP[γS], we observed that the β-subunit of transducin was thiophosphorylated under this condition. Finally, the thiophosphate group bound to the β-subunit was transferrable to GDP leading to the formation of GTP[γS] (Wieland and Jakobs, 1989b).

Similar β-subunit (thio)phosphorylations as observed with the β-subunit of transducin were found by studying (thio)phosphorylation of HL 60 membrane proteins by [γ-^{32}P]GTP and [^{35}S]GTP[γS] (Wieland et al., manuscript in preparation). This phosphorylation was specifically inhibited by unlabelled GTP and GTP[γS] but only weakly or not at all by ATP, GppNHp, ATP[γS] and adenylyl-5'-imidodiphosphate. Furthermore, the phosphate group was released from the prelabelled protein by addition of GDP but of its metabolically stable analog GDP[βS]. Finally, in the presence of GDP and the formyl peptide receptor agonist fMLP the phosphate group was not only transferred to GDP leading to the formation of [γ-^{32}P]GTP, but this newly formed GTP was apparently also bound to the G-protein and subsequently hydrolyzed by the G-protein GTPase activity to GDP and $^{32}P_i$.

Thus, in summary, G-protein activation by agonist-liganded receptors is apparently not only induced by dissociation from and binding of guanine nucleotides to G-protein α-subunits, but it appears that, in addition, phosphate group transfer reactions are involved in the activation process of G-proteins by GTP and its analog GTP[γS]. These transfer reactions apparently involve both membrane-associated NDP kinases and the G-protein β-subunits, providing thereby for these subunits also a specific function in G-protein activation. Although it is presently unknown whether or not these reactions are obligatory for G-protein activation by GTP and agonist-activated receptors, it appears that the complexity of the intact signal transduction systems is much higher than previously imagined.

Acknowledgements

The authors' studies reported herein were supported by the Deutsche Forschungsgemeinschaft, the Fritz Thyssen-Stiftung and the Fonds der Chemischen Industrie.

REFERENCES

Birnbaumer L, Codina J, Mattera R, Yatani A, Scherer N, Toro MJ, Brown AM (1987) Signal transduction by G proteins. Kidney Intern 32 (Suppl 23):S14-S37

Casey PJ, Gilman AG (1988) G Protein involvement in receptor-effector coupling. J Biol Chem 263:2577-2580

Chabre M, Deterre P (1989) Molecular mechanism of visual transduction. Eur J Biochem 179:255-266

Gierschik P, McLeish K, Jakobs KH (1988) Regulation of G-protein-mediated signal transfer by ions. J Cardiovasc Pharmacol 12 (Suppl 5):S20-S24

Gierschik P, Sidiropoulos D, Dieterich K, Jakobs KH (1989a) Structure and function of signal-transducing, heterotrimeric GTP-binding proteins. In: Habenicht A (ed) Growth factors, differentiation factors and cytokines. Springer, Berlin Heidelberg New York, in press

Gierschik P, Steißlinger M, Sidiropoulos D, Herrmann E, Jakobs KH (1989b) Dual Mg^{2+} control of formyl peptide receptor-G-protein interaction in HL 60 cells. Evidence that the low agonist affinity receptor interacts with and activates the G-protein. Eur J Biochem in press

Gierschik P, Sidiropoulos D, Steißlinger M, Jakobs KH (1989c) Na^+ regulation of formyl peptide receptor-mediated signal transduction in HL 60 cells. Evidence that the cation prevents unoccupied receptors from activating the G-protein. submitted for publication

Gilman AG (1987) G Proteins: Transducers of receptor-generated signals. Ann Rev Biochem 56:615-649

Herrmann E, Gierschik P, Jakobs KH (1989) Neomycin induces high affinity agonist binding of G-protein-coupled receptors. submitted for publication

Jakobs KH, Wieland T (1989) Evidence for receptor-regulated phosphotransfer reactions involved in activation of the adenylate cyclase inhibitory G-protein in human platelet membranes. Eur J Biochem in press

Kimura N, Shimada N (1983) GDP does not mediate but rather inhibits hormonal signal to adenylate cyclase. J Biol Chem 258:2278-2283

Lefkowitz RJ, Stadel JM, Caron MG (1983) Adenylate cyclase-coupled beta-adrenergic receptors. Structure and mechanisms of activation and desensitization. Ann Rev Biochem 52:159-186

Lochrie MA, Simon MI (1988) G Protein multiplicity in eukariotic signal transduction systems. Biochemistry 27:4957-4965

Neer EJ, Clapham DE (1988) Roles of G protein subunits in transmembrane signalling. Nature 333:129-134

Parks RE, Agrawal RP (1973) Nucleoside diphosphokinases. In: Boyer PD (ed) The Enzymes, vol VIII, Part A, 3rd edn. Academic Press, New York, p 307

Rodbell M (1980) The role of hormone receptors and GTP-regulatory proteins in membrane transduction. Nature 284:17-22

Seifert R, Rosenthal W, Schultz G, Wieland T, Gierschik P, Jakobs KH (1988) The role of nucleoside-diphosphate kinase reactions in G protein activation of NADPH oxidase by guanine and adenine nucleotides. Eur J Biochem 175:51-55

Stryer L (1986) Cyclic GMP cascade of vision. Ann Rev Neurosci 9:87-119

Stryer L, Bourne HR (1986) G Proteins: A family of signal transducers. Ann Rev Cell Biol 2:391-419

Wieland T, Jakobs KH (1989a) Receptor-regulated formation of GTP[γS] with subsequent persistent G_s-protein activation in membranes of human platelets. FEBS Lett 245:189-193

Wieland T, Jakobs KH (1989b) Mechanisms of G-protein activation by GTP[γS]: The β-subunit of transducin serves as a thiophosphorylated intermediate leading to G-protein activation in HL 60 membranes. Naunyn-Schmiedeberg's Arch Pharmacol 339 (Suppl):R33

DISCUSSION

Iyengar: So what fraction of your G-proteins are GDP free, what fraction of the total binding activity can you get? What fraction of the receptor stimulated activity can you see if you bind GTPτS at 5 mM magnesium without fMLP?

Jakobs: When we measured the binding of GTPτS in the absence of any GDP, in the absence of fMLP, we measured total binding of GTPτS to all G-proteins in these membranes, which are not only single transducing, but about 50 % or more are not single transducing small G-proteins.

Iyengar: Can you measure with GDP the blank or the background, or you still have the small molecular range to worry about?

Jakobs: When we add GDP, to see the hormone-receptor action we have to add another nucleotide, then we see the receptor stimulated binding of GTPτS. In the absence of GDP we see only a very small effect. But this is maybe because we have a very large background of GTPτS-binding proteins in these membranes. The number is about 100 pmoles/mg protein.

Simon: The labeled material that you showed is GTPτS covalently bound to β. Do you have any idea what kind of intermediate it looks like?

Jakobs: It is acid-labile, with a labile phosphate group on the β-subunit, it is alkaline stable.

Simon: And the stoichiometry?

Jakobs: The stoichiometry, we don't exactly know.

Downes: On the question of whether high or low affinity forms of receptor are responsible for transducing responses in intact cells. There is quite a substantial amount of evidence of the phospholipase C systems, that low affinity forms were involved and some of the best experiments use more classical pharmacological techniques and muscarinic receptors, they have available propyl benzoyl choline, which is a covalent antagonist. And so you can touch straightout any receptor reserve that exist and therefore measure effectively a dissociation constant for the agonist by measuring the response to that agonist. There are only a few situations where you can actually do that. When you do that for at least some of the muscarinic systems then you find very convincingly that the dissociation constants match those of low affinity receptor forms. You can certainly look at systems, for example the phopholipase C response in rat brain, that is very clearly, I mean there is no receptor reserve for the PLC there, it is very clearly almost exclusively low affinity. There is at least one situation where it seems only to involve low affinity forms of the receptor. I do not think it has been published, but I believe that Nigel Birdsale's group generated data from several systems which all respond to the same type of muscarinic receptor, but show different levels apparently of receptor reserve between different tissues, either in for example rat brain, which seems to have no receptor reserve, parotic gland which has a receptor reserve of maybe up to a factor of 10. And in the lacrimal gland, which is by far more sensitive system; I believe in the lacrimal gland by titrating out the receptor reserve component you still retain apparently a higher affinity dissociation constant compared with prostoglandin of brain, but I could not be absolutely sure about it. So it may actually not be a constant factor, but there are certainly circumstances in the PLC system, where apparently only low affinity forms are involved.

Devreotes: Do you know if there are two states of the receptors or two different proteins or the same proteins?

Jakobs: In the HL60 cells we do not know whether they are different types of fMLP receptors or only one, but we have not cloned or purified the receptor. Our assumption is that there is only one receptor like for example the β-receptor which also shows high and low affinity receptor. It is still possible that there are two fMLP receptors, a high and a low affinity receptor, which are different proteins. But we think it is only one protein, but we have no definite proof for it.

Houslay: Taken that one further which G-protein do you think interacts with this single or more than one receptor?

Jakobs: Gi-proteins; we can measure the coupling between the G-protein and the receptor by measuring the cholora toxin stimulated ADP-ribosylation and this shows the coupling of receptor to the G-protein.

Houslay: Could I just ask you about your β-τ. We and others have shown that β can be phosphorylated for example by the insulin receptors tyrosine kinase. Do you think this might have a functional consequence or not?

Jakobs: β-subunit phosphorylation by the insulin receptor tyrosine phospho- rylation; this is surely not the tyrosine phosphorylation.

Houslay: I was just asking you to speculate whether phosphorylation of β-τ might occur in the function.

Jakobs: We have no data or speculations about this.

*Chabre:*I would like to get back to the low and high affinity because one thing is not clear in my mind. You speak of high and low affinity state of the receptor. I always understood that the high affinity was what you measure when the receptors are coupled for some reason to G-protein and low affinity was the affinity of the receptor itself. The point is to measure the affinity you have, to add agonist and then you induce the binding of the G-protein. Now if you have no nucleotide for some reason the thing would not go on. I mean the G-protein will not dissociate and this reacts on the binding of the agonist and the agonist will release and you measure high affinity. That is classically when you measure affinity without any nucleotide or without anything that allows the GDP-GTP exchange to process.

Jakobs: What I am saying, we have three affinity states; a very low in the absence of G-protein, a low which also is agonist for receptor-G-protein and a high which is millimolar magnesium in addition and agonist for receptor-G-protein. And both the medium low and the high can, as is shown, interact with the G-protein and activate it.

Chabre: They do interact certainly, because if the affinity is medium low or high it is because the G-protein is there. I mean, when you measure high affinity you are measuring not the affinity of the receptor, the affinity of the receptor coupled to the G-protein. It is not the property of the receptor itself.

Jakobs: It is well known that the high affinity receptor is the receptor coupled to the G-protein.

Chabre: Already coupled; is it coupled before you add the agonist or is it the agonist that induce the coupling?
Jakobs: Yes, the agonist induces the coupling.

Visual transduction

REGULATION AND RAPID INACTIVATION OF THE LIGHT INDUCED cGMP PHOSPHODIESTERASE ACTIVITY IN VERTEBRATE RETINAL RODS

M. Chabre, P. Deterre, P. Catty and T.M. Vuong
Laboratoire de Biophysique Moléculaire et Cellulaire
(UA 520 CNRS)
LBIO DRF/CENG - BP 85
F - 38041 Grenoble

INTRODUCTION: THE PROBLEM OF THE FAST TERMINATION OF THE cGMP RESPONSE

The rhodopsin-transducin-cGMP phosphodiesterase cascade triggered by light in visual cells has become an archetype for the now ubiquitous G-protein mediated transduction processes (Stryer 1986, Deterre and Chabre 1989). Within a few hundred milliseconds of the photoisomerization of one rhodopsin in a rod cell, the cascade developps very rapidly and can induce the hydrolysis of hundred thousands of cGMP molecules. These fast kinetics are made possible first by the great mobility of rhodopsins and transducins, respectively in and on the rod disc membrane. The high gain is accounted for by the large number of transducins than can interact in rapid sequence with a single photo-excited rhodopsin. The activated TαGTP subunits of transducins are released in the quasi-two dimensionnal interdiscal space, in which they reach locally a very high concentration (~ 500 μM). They diffuse rapidly in the narrow cytoplasmic cleft and soon interact with the PDE bound onto the two limiting disc membrane surfaces. This interaction results in the derepression of the extremely rapid, cGMP specific, hydrolytic activity of a few hundred PDE molecules, hence inducing a rapid decrease of the cGMP concentration. The drop in cGMP concentration propagates from the interdiscal space to the cytoplasmic region beneath the outer cell membrane, at the level of the illuminated disc. The cGMP dependent cationic channels, present exclusively in the cell membrane,

NATO ASI Series, Vol. H 44
Activation and Desensitization of Transducing Pathways
Edited by T. M. Konijn, M. D. Houslay, P. J. M. Van Haastert

were maintained open in the dark by the binding of fast exchangeable cGMP molecules: they close within milliseconds of the cGMP depletion. The closure of these cationic channels shuts off a significant fraction of a large inward "dark current" carried essentially by entering Na^+ ions, which maintained the rod cell partially depolarised in the dark: the cell hyperpolarizes. Overall, the photoexcitation of one rhodopsin in a disc membrane of a primate rod cell induces within approximately 250 millisecond the closure of a few hundred cationic channels in its outer cell membrane.

This highly amplified response must be regulated and controlled by the cell. Indeed, as soon as it has reached its maximum amplitude, the hyperpolarizing response to a single photon decays dramatically at a rate comparable to that of its rising phase, as if a highly efficient regenerative process was triggered: the physiological response is fully terminated within half a second. At higher flash intensity or under constant illumination other effects, generally referred to as adaptation - equivalent to desensitization in the terminology of hormonal or neuronal transduction - modulate the amplitude of the response. We will not address here these complex, and still very poorly understood adaptation mechanisms, which might involve other G-protein pathways or branched pathways of transducin with other effectors, but we will try to elucidate the mechanisms that underly the fast shut-off of the physiological response to a single photon.

Electrophysiological measurements demonstrate that the channels themselves are not responsible for the fast recovery of the dark-adapted conductance, their response to cGMP being potential independent and non adaptive: in the prolonged absence of cGMP the channels remain permanently closed. In order for the response to be terminated, the cGMP level must therefore be rapidly restored. Two major opposing processes regulate the cGMP level: a synthetic pathway mediated by guanylate cyclase counterbalances the light activated degradative pathway of the PDE. The guanylate cyclase activity is not directly influenced by the light triggered cascade, but has recently been shown to respond indirectly, through a calcium dependance, to the light induced closure of the cationic channels. In the dark, the open channels let enter a significant calcium flux, amounting to a few percents of the sodium influx. The internal homeostasis of calcium is maintained in the micromolar range by the action of sodium/calcium exchangers located in the rod cell membrane. The light induced closure of the cationic channel that shutts off the depolarising Na current also blocks the calcium influx. As the activity of the Na/Ca exchanger is not decreasesd, the internal calcium level drops down. An important observation was that the guanylate cyclase is negatively modulated by calcium: its activity is increased when the calcium level is lowered - probably through the action of a calcium dependent inhibitory protein (Lolley and Racs 1982, Stryer and Koch 1988). The kinetics and amplitude of this secondary guanylate cyclase response are not yet well determined but this process might provide a rapid regeneration of the cGMP level.

HOW TO STOP THE PDE ACTIVITY ?

A delayed increase of guanylate cyclase activity after the light triggered activation of the PDE does not however reset the system back to its dark adapted state: both the synthetic and the degradative pathways are now fully activated, resulting in the regeneration of the cGMP level at the expense of a high consumption of GTP. But the reopening of the cationic channels will finally re-hinder the guanylate cyclase by letting again calcium enter the cell. The temporary increase of the cyclase activity cannot therefore be the full answer to the problem of the termination of the response, if the activation of the PDE is not also soon stopped. A first necessary step to block the cGMP cascade is to block at its first stage the catalytic activity of R*, the photoexcited receptor. Kühn and coworkers (Wilden et al 1986) have shown that the successive actions of rhodopsin kinase, which phosphorylates specifically photoexcited rhodopsin, and of arrestin (or 48K protein or S antigen), which then binds specifically to phosphorylated R*, provide a rapid and efficient quenching mechanism for the photoactivated receptor. Arresting the formation of further active intermediate TαGTP will not however stop the hydrolysis of cGMP as long as the already formed TαGTP continue to interact with the PDE and activate it. This will eventually be terminated only by the decay of TαGTP to inactive TαGTP, through the auto-GTPase activity of Tα. The slow rate of GTPase, as it was usually measured in vitro, poses problems which we have reinvestigated and will discuss. We have further analyzed the mechanism of activation of the PDE by TαGTP, found out that the quaternary structure of the native enzyme was more complex than supposed before, and speculated on a mechanism for a GTPase independent rapid inactivation process that might result from the interaction with TαGTP of this complex multisubunit structure.

ACTIVATION OF PDE BY TαGTP: REMOVAL OF INHIBITORY SUBUNIT(S)

As is it is common for the regulation of a fast hydrolytic enzyme, the activation of the PDE is indeed due to the release of an inhibitor. It had long been shown (Miki et al 1975) that PDE activity could be elicited by a tryptic treatment, suggesting that the enzyme is maintained in its basal state by a proteolyzable inhibitor. Hurley and Stryer (1982) isolated and purified a 13 kDa subunit (I) which could inhibit with high affinity the 85+88 kDa catalytic complex, PDEαβ, of the trypsin activated enzym. The natural activation of the PDE was therefore expected to result from an interaction of TαGTP with I. In the low ionic strength eluate of ROS membranes in which the PDE had been permanently activated by illuminating in presence of GTPγS, we detected a complex of Tα and I. By ion exchange chromatography this Tα-I complex was separated from the activated PDE as well as from the excess inactive PDE (Fig. 1). Gel filtration studies and polypeptide analysis suggested a 1:1 stoichiometry for this Tα-I complex (Deterre et al 1986). It is not clear whether under physiological conditions the complex, that remains membrane bound, separates physically from PDEαβ on the membrane. The fact however that these two components are extracted as separated entities in a low ionic strength medium suggests that they are already separated on the membrane.

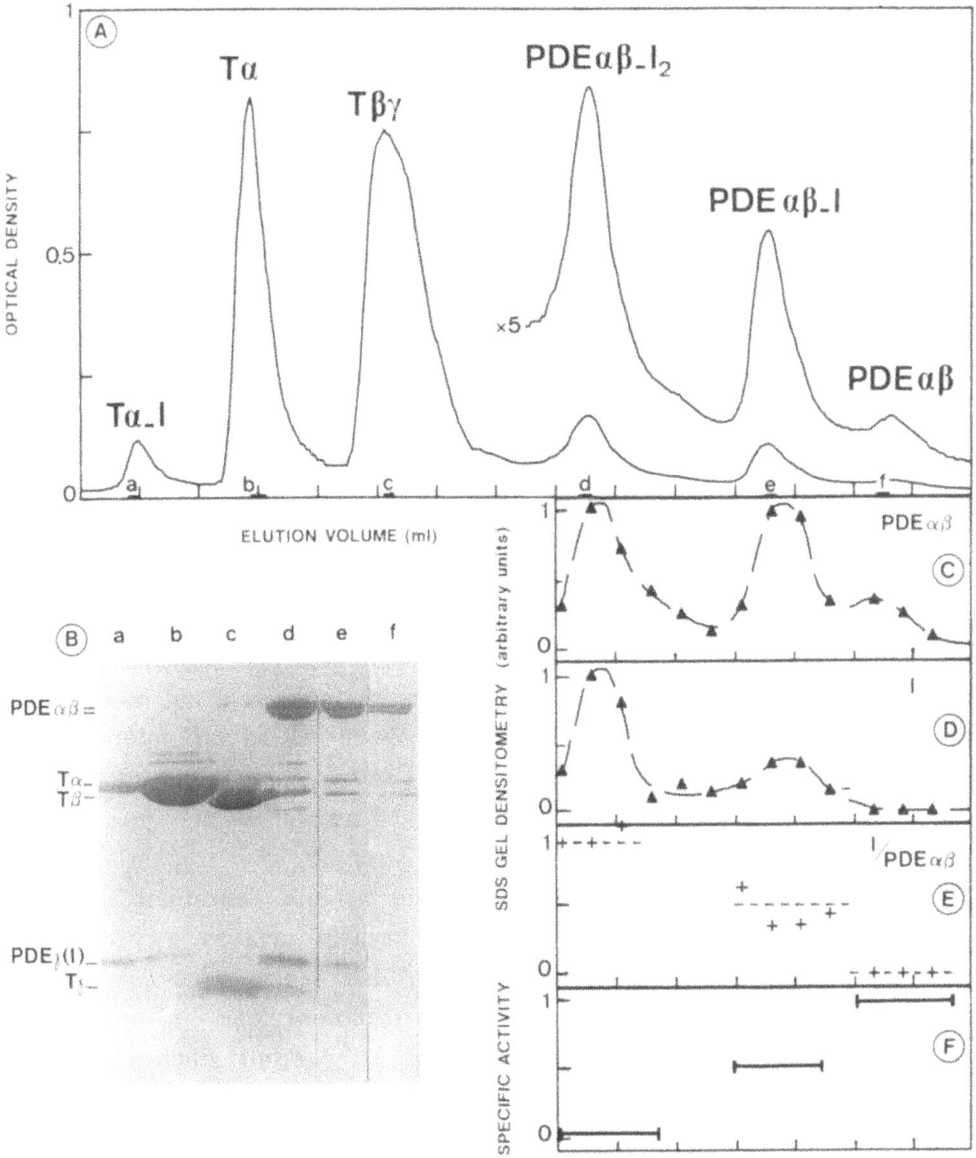

Figure 1: Analysis of the interaction of TαGTP with the PDE complex, demonstrating the formation of Tα-I complexes of two states of activated PDE: I-PDEαβ and PDEαβ - modified from Deterre et al (1988).

A: Complete elution profile of a mixture of transducin and PDE extracted after inducing them to interact by illuminating the ROS in the presence of GTPγS before the extraction (see Deterre et al 1988, for details)

B: Gels of the eluted fractions from the six different peaks

C , D: Densitometry of the stained gel fractions from the three PDE containing peaks. The PDEαβ content (C) and the I content (D) of each fraction were quantitized

E: Quantitation of the I/PDEαβ ratio in each fraction

F: Specific PDE activity of the pooled fractions corresponding to each of the three peaks. The PDE activity is measured at 500 μM cGMP with the same method as in figure 3, and is mormalized at 100% for PDEαβ.

TWO INHIBITORY SUBUNITS PER PDE COMPLEX

In the absence of precise data on the inhibitor abundance, a 1:1:1 stoichiometry had usually been assumed for the α, β and I subunits in the inactive PDE holoenzyme, but our experiments unexpectedly demonstrated that each PDE$\alpha\beta$ catalytic complex is regulated by two inhibitory subunits (Deterre et al 1988). As seen on fig. 1, the chromatography of an activated ROS extract resolved not one but two peaks of activated PDE, beside the peak of excess inactivated PDE. Only the last peak is devoid of inhibitor and the I/PDE$\alpha\beta$ ratio in the first active peak is close to 50% of that found for native PDE. The simplest explanation is that the three PDE peaks, with 2/1/0 relative amounts of inhibitor per catalytic complex, correspond respectively to inactive I_2-PDE$\alpha\beta$ and to two different active states I-PDE$\alpha\beta$ and PDE$\alpha\beta$. The same two states of activated PDE could also be isolated upon progressive mild proteolysis of the native enzyme, which degrades and successively eliminates the two inhibitors. The V_{max} for the I-PDE$\alpha\beta$ peak fraction was about half of that measured on totally stripped PDE$\alpha\beta$. When tested in solution, the activities of the trypsin-activated fractions were comparable to that of the corresponding transducin-activated PDE. However in the presence of membrane and at a physiological ionic strength, the trypsin-activated species no longer bind to the membrane as the transducin-activated ones do: before digesting the inhibitor, trypsin cleaves a very short terminal peptide of the catalytic complex, that seems necessary for its membrane attachment. As membrane attachment can influence the efficiency of the catalytic units, one must be carefull when comparing activities of trypsin-treated species to that of the corresponding transducin-activated ones.

When mixing purified PDE$\alpha\beta$ with native I_2-PDE$\alpha\beta$, exchange of inhibitor is observed (fig. 2), giving rise to the intermediate species I-PDE$\alpha\beta$. The possibility of this exchange, the kinetics of which is not yet known, occurring *in vivo,* precludes an unambiguous estimate of the enzymatic properties (V_{max} and K_m) of the intermediate species. But, as discussed later, this exchange might provide a fast regulation mechanism of the PDE activity, independent of the inactivation of transducin by its GTPase.

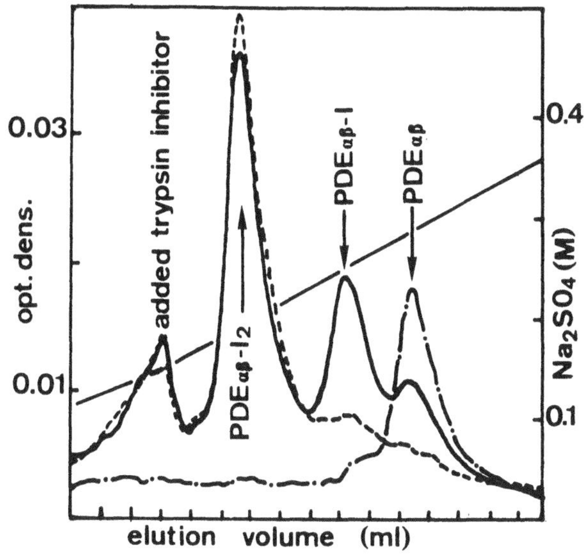

Figure 2: Evidence for the exchange of inhibitor subunits between I_2-PDEαγ and PDEαβ. The elution profiles were obtained by eluting successively on the same ion exchange column: – – – one aliquot of PDEα; – · – · one aliquot of I_2-PDEαβ; —— a mixture of the two preceeding aliquots. One sees in the elution profile of the mixture PDEαβ + I_2-PDEαβ a peak that elutes at the position of I-PDEαβ and is created at the expense of both components of the mixture, most likely by exchange of PDE inhibitors between I_2-PDEαβ and PDEαβ.

WHAT LIFETIME FOR THE ACTIVE STATE TαGTP ? THE GTPase RATE

In any case the hydrolysis of the bound GTP in Tα must definitely terminate the activation of PDE: the γ phosphate of the GTP is released and with a GDP bond, Tα switches back to the conformation which has a low affinity for the PDE inhibitor and a high one for Tβγ. The lifetime of the bound GTP in Tα therefore defines an upper limit to the duration of PDE activation. This lifetime is usually not directly measured but is rather inferred from the GTPase turn-over rate of transducin. As measured *in vitro*, and particularly with model systems reconstituted from purified rhodopsin in vesicles and purified transducin, this turn-over is surprisingly slow: of the order of 3 to 6 GTP hydrolyzed per minute per transducin (Fung 1983). If the hydrolysis of GTP is really the rate limiting step, this would imply a lifetime of 10 to 20 seconds for TαGTP. Such a long lifetime seems hard to reconcile with a turn-off of the PDE in the subsecond range. Even if a secondary increase of cyclase activity helps regenerates temporarly the cGMP level, it looks very uneconomical to have the cell maintaining for many seconds a high cGMP level with the cyclase running at full speed to compensate for a high PDE activity. Furthermore in such a regime the cell would remain unable to react to a second photon, thus implying a long "dead time" after a single photon response.

The lifetime of the active TαGTP state might however be shorter than what is deduced simplistically from GTPase measurements: the rate limiting step in a full cycle of GTPase might not be the hydrolysis of GTP, but the subsequent rebinding to the membrane of TαGDP, which was solubilised in the GTP bound state, and its reassociation to Tβγ which is required for TαGDP to be able to rebind to R* and exchange its GDP for a new GTP. In vitro, due to the solubility of transducin, these steps might be artefactually slowed down by the dilution of the system. Indeed our recent measurements of light dependent GTPases on concentrated native ROS membrane suspensions lead to much higher rates, up to 20 GTP hydrolyzed per minute per transducin (Deterre et al, in preparation). As discussed later, this does not necessarily mean that a soluble accelerating factor of the GAP type (GTPase Activating Factor of ras) is lost in diluted or reconstituted preparations.

We also devised a new way to study the lifetime of TαGTP by measuring by time-resolved microcalorimetry the course of enthalpy release, correlated to the hydrolysis of the γ phosphate of GTP: 8 kcal are released per GTP hydrolyzed, and the time course of the heat release sets an upper limit to that of the conversion of TαGTP to TαGDP. We have designed and built a microcalorimeter based on the use of pyroelectric PVDF film (Vuong and Chabre in preparation). The measurements are done on populations of transducin molecules in which GTP has been loaded by a short pulse of R* created by a low intensity flash and immediately quenched in the presence of high concentration of hydroxylamine. Each sample contains the rod outer segments extracted for 2 to 4 cattle retinas. The sensitivity of the device is on the order of 1 μcal/sec and its impulse response, as measu-red by flashing an absorbing dye, is 0.5 sec. When GTP is added to the ROS suspension, the flash elicits a heat pulse that far exceeds the energy of the flash itself. The duration of the heat pulse depends not only on the lifetime of hydrolysis of the GTP loaded on transducin, but also on the lifetime of R*: as long as the R* are active, they will reload GTP on the transducins that have already hydrolysed their first loaded GTP. To reach the real lifetime of TαGTP, that of R* must be reduced to negligible value. We therefore add in the suspension a high concentration of hydroxylamine which does not act on dark adapted rhodopsin but is known to hydrolyse the Schiff base bond of the retinal as soon as rhodopsin is photoexcited to the active Meta II state, hence inactivating R*. Under this condition, the heat pulse decays with a half life of the order of 5 seconds at 23°C. This confirms the results obtained from light triggered GTPase measurements on concentrated native suspension. No concentration dependence of the heat pulse decay time was observed, for ROS membrane concentration ranging from 1 to 7 mg/ml. This suggests that the concentration dependence sometimes observed in GTPase measurements are due to dilution effects on transducin itself rather than on a putative soluble GTPase activating component. Measurements at 37°C, performed by the GTPase approach, but not feasible by the microcalorimetric technique, suggest a lifetime of 2±1 seconds for TαGTP at physiological temperature. This seems still unsufficient to account for the fast shutt-off of the physiological responses.

MIGHT A DIFFUSION CONTROLLED FAST INACTIVATION OF PDE PRECEDE THE HYDROLYSIS OF GTP IN Tα ?

The exchangeability of inhibitors between inactive I_2-PDEαβ and fully activated PDEαβ raises the possibility of a fast, diffusion controlled decrease of the PDE activity generated by a single photoactivated rhodopsin, that could take place before the hydrolysis of GTP in Tα induces the dissociation of the Tα-I complexes. Around the photoexcited rhodopsin, all the available transducins are rapidly activated and local concentration of TαGTP in the cytoplasmic cleft can raise up to 500 microMolar. This high concentration could be necessary to overcome the high affinity of the last inhibitor molecule for PDEαβ, and quantitatively strip the local pool of PDE of its two inhibitors. But as soon as R* is blocked by phosphorylation and arrestin binding, the limited pool of TαGTP dilutes in the cytoplasm and its concentration might soon become too low to produce any more PDEαβ. The initially formed PDEαβ also diffuse away into membrane area where they encounter an excess of inactivated I_2-PDEαβ, from which they might fast regain an inhibitor by back exchange, producing a double number of I-PDEαβ. If I-PDEαβ were inactive, or had significantly less than half of the activity of PDEαβ, this back exchange process could result in a significant decrease of PDE activity, before the TαGTP decays to TαGDP and release their bound inhibitor.

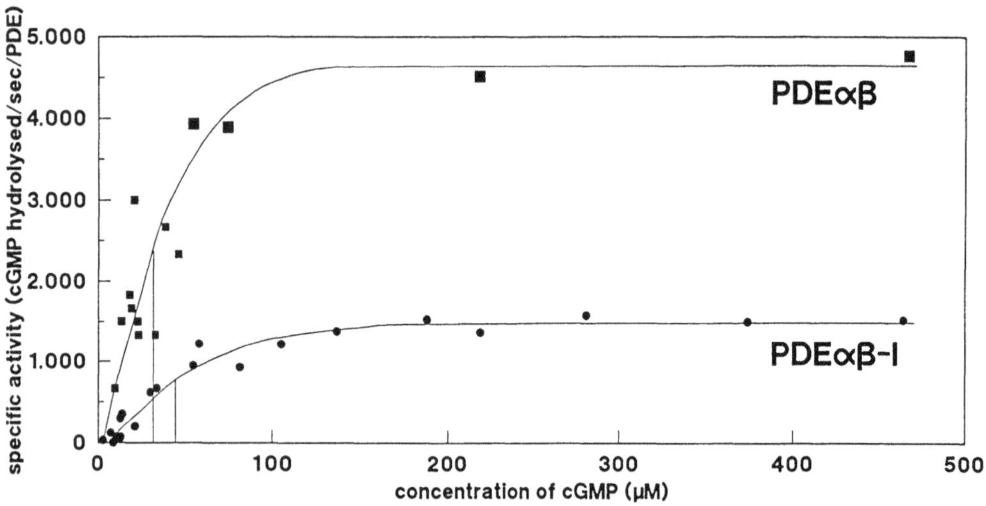

Figure 3: Specific activity of the two active PDE species (PDEαβ and I-PDEαβ) obtained after trypsinization and purification on Polyanion Pharmacia column. The PDE sample is incubated at 25°C in a medium containing: Tris 20 mM pH 7.5, NaCl 120 mM, $MgCl_2$ 5 mM, DTT 1 mM, in the presence of extensively washed ROS membranes containing 2 μM Rh. The PDEαβ is assayed at 0.5 nM. The I-PDEαβ is assayed at 10 nM in the presence of 10 nM PDEαβ-I_2 in order to avoid the formation of PDEαβ by dilution (Wensel & Stryer, 1986). The assay is performed in 200 μl for durations between 6 and 36 sec and at cGMP concentration between 25 and 500 μM. It is terminated by addition of 1 ml of cold PCA 10%. After neutralization and centrifugation the cGMP/5'GMP content of the supernatant is analyzed by chromatography on FPLC Pharmacia Polyanion column.

The model however would require, to be kinetically significant, that the k_{off} of the less bound inhibitor from the holoenzyme: I_2-PDEαβ $\rightarrow$ I + I-PDEαβ, be at least in the range of 10 sec^{-1}, for the back exchange process to be rapid enough. Such a k_{off} value would not be incompatible with rough estimates of binding affinity, but is has not been really determined experimentally. Other difficulties in checking this model lie in the measurement of the cGMP hydrolysis rate of the various states of the PDE under conditions close to the physiological ones, that is at very low free cGMP concentration. The enzyme is extremely rapid, with V_{max} of about 5000 and 2000 cGMP hydrolyzed per second per molecule for respectively PDEαβ and I-PDEαβ. These V_{max} are easily measured with millimolar concentrations of cGMP. The corresponding K_ms are respectively of 35±15 μM for PDEαβ and 70±20 μM for I-PDEαβ (fig 3): assuming Michaelian behaviours, these values suggest indeed that a significant decrease of activity could occur at low cGMP concentration, when one PDEαβ combines to one I_2-PDEαβ to lead to two I-PDEαβ. But a Michaelian extrapolation down to the micromolar range is hazardous particularly when one realises that in situ the substrate concentration is much lower than that of the enzyme. On the other hand the turn over rates of I-PDEαβ is still too fast, when cGMP is down to the micromolar range, to be measurable unless one dilutes the enzymes down to subnanomolar concentration. Then significant dissociation of I from I-PDEαβ might occur, leading to totally stripped PDEαβ.

Therefore, although the existence of two units of inhibitor per catalytic complex must be important for the regulation of the PDE activity, its significance for a fast blocking of the PDE activity is still far from certain. Since the lifetime of the active TαGTP state is much shorter than previously assumed, it may well be that the PDE remains active until the hydrolysis of GTP in the TαGTP induces the release of the inhibitors from the Tα-I complexes. The physiological response could then be shortened simply by the temporary enhancement of the cyclase activity. It remains however to be checked whether the association to I might accelerate the hydrolysis rate of GTP in the bound Tα, further shortening the lifetime of the Tα-I complexes.

REFERENCES

Chabre M, Deterre P (1989) Molecular mechanism of visual transduction. Eur J Biochem 179:255-266

Deterre P, Bigay J, Forquet F, Robert M, Chabre M (1988) cGMP phosphodiesterase of retinal rods is regulated by two inhibitory subunits. Proc Natl Acad Sci 85:2424-2428

Deterre P, Bigay J, Robert M, Pfister C, Kühn H, Chabre M (1986) Activation of retinal rod cGMP phosphodiesterase by transducin: characterization of the complex formed by phosphodiesterase inhibitor and transducin α-subunit. Proteins Struct Funct Genetics 1:188-193

Fung BKK (1983) Characterization of transducin from bovine retinal rod outer segments. I. Separation and reconstitution of the subunits. J Biol Chem 258:10495-10502

Hurley JB, Stryer L (1982) Purification and characterization of the gamma regulatory subunits of the cyclic GMP phosphodiesterase from retinal rod outer segments. J Biol Chem 257:11094-11100

Koch KW, Stryer L (1988) Highly cooperative feedback control of retinal rod guanylate cyclase by calcium ions. Nature (London 313:16-17

Lolley RN, Racz E (1982) Calcium modulation of cyclic GMP synthesis in rat visual cells. Vision Res 12:1481-1486

Miki N, Baraben JM, Keirns JJ, Boyce JJ, Bitenski MW (1975) Purification and properties of the light activated cyclic nucleotide phosphodiesterase of rod outer segments. J Biol Chem 250:6320-6326

Stryer L (1986) Cyclic GMP cascade of vision. Annu Rev Neurosci 9:87-119

Wilden U, Hall SW, Kühn H (1986) Phosphodiesterase activation by photoexcited rhodopsin is quenched when rhodopsin is phosphorylated and binds the intrinsic 48 kDa protein of rod outer segments. Proc Natl Acad Sci 82:4316-4320

DISCUSSION

Iyengar: Is there any negative cooperativity between the two sites you think, when you say it is non-Michaelian?

Chabre: This is open to speculations, it is just an experimental fact, that if you try to get down measurement of V at low cGMP concentrations it looks like it drops down. But these measurements are very difficult when the pool of cGMP is very small, one has to to be very quick. There are also small pools of cGMP that are trapped. We are not sure that the enzyme activity stops or whether there is a pool of cG which is unaccessible to the enzyme and which we extract later. Experimentally it looks like the hydrolysis stops when the cG concentration drops below 20 μM, but I'm not willing to say that's the enzyme itself that stops.

Iyengar: Do you have real sensitive data, because in the add-back experiments for our desensitization, the contrast between cyclase in the native membranes which have some unknown amount of Gs, if you take away the inhibitor I'd come very much towards your thinking along the same line, that there is going to be multiple sites for Gs for regulating catalysts, because that's why and you know we had a fully activated Gs and everybody knows that there is stoichiometrically more Gs than there is catalyst molecules. But yet we can enhance this regulation by adding exogenously Gs.

Chabre: Here we can clearly show that transducin acts by removing an inhibitor. For the cyclase system apparently Gs remains bound on the cyclase, does one know whether one Gs binds or more than one?

Iyengar: No, one doesn't know that, but I'm gonna guess that there is more than one.

Devreotes: Why did you leave free inhibitor out of your pictures and free α and free β. You only had PDE. Why didn't you consider there being a pool of free inhibitor?

Chabre: The affinity of PDEαβ and Tα for inhibitor, are so high that there is not a very significant amount of free inhibitor in the preparation.

Devreotes: So how does the exchange work then? Inhibitor doesn't dissociate and then associate with another. How does it exchange between the PDE's?

Chabre: That we do not know. We don't have the kinetics.

Devreotes: You mean, it does not dissociate from PDEαβ.

Chabre: Yes, the first one might, but the second one doesn't dissociate. There are two, one could be of comparatively low affinity; the second one, the one that you require to remove to get full activity as a high affinity form PDE.

Devreotes: But if the first one will dissociate, why is there no free?

Chabre: There is some free.

Devreotes: Don't you need to consider that in your model?

Chabre: We did, that is discussed in our paper (Deterre et al. 1988). I'm sorry I had misunderstood your question.

Corbin: I was wondering if it's possible that the cGMP binding-sites on the enzyme are activating, so that by lowering the cGMP concentration that would lower the activation of the catalytic units.

Chabre: You are talking about the non catalytic site for binding cGMP on the PDE. That's something we all think about, but we don't have any solid data on. So I'm ready to accept there are non catalytic sites on the PDE, but I don't know anything about them. There are clearly a lot of sites for cGMP in that cell, because the total concentration of cGMP is something like 60 mM and the working concentration seems to be only 5 mM. So in the cell there are binding sites for cGMP. Those could be on PDE, that's possible.

Corbin: I think it was Peter Van Haastert who showed in *Dictyostelium* that in an enzyme like this the binding site was stimulatory, using analogues.

Van Haastert: You mean that the cGMP PDE is stimulated by cGMP. Yes that's true. And there are two different sites, because you can distinguish with analogues. 8 Br-cGMP is an activator, but a very poor substrate and IMP is a substrate but no activator. And the kinetics of that activation are compatible with the function of cGMP degradation. So we measured some non-equilibrium kinetics. Equilibrium at the catalytic site is obtained within a second, but it takes about 10 seconds to activate PDE.

Corbin: And also an enzyme like this in liver and other tissues which is a cAMP PDE, but as a cGMP regulatory site and that is also stimulatory, so may be the same family of enzymes as this, I don't know.

Van Haastert: Well the only difference of the *Dictyostelium* enzyme with the mammalian enzyme is that the *Dictyostelium* enzyme does not hydrolize cAMP.

Clark: In your system, I have not given up on Tβτ yet, but you have not talked about the possibility since Kühn showed that

Yβτ greatly stimulates that GTPase activity. Not Kühn, Fung.

Chabre: What Fung showed was that Tβτ is required to get any GTPase activity. You need some Yβτ. If you have no Tβτ, you have no activity. If there is some Tβτ you get activity, but that doesn't mean you optimize it.

Clark: That's exactly my question, have you ever tried in a system where you get your turnover number of about 3 per minute, adding Tβτ?

Chabre: That doesn't change in diluted systems, the activity saturates with a very low level of Tβτ.

Iyengar: In other G proteins as well, right, Gβτ doesn't affect the K-catalytic of GTPase, it is not a real stimulation.

Chabre: Tβτ is just required to couple back to the receptor.

Clark: What I was wondering if it got diluted, because you have to break the cells, the Tα goes away, and may be it doesn't find Tβτ as well.

Chabre: Yes, the Tβτ stays on the membrane.

Clark: I know it stays, that is exactly my point, it is fixed.

Chabre: Tβτ stays, and what is probably slowing down the process, is the dilution of Tα which has to get back on the membrane. 90% of the Tβτ stays on the membranes, so it is not limiting. Adding more, does not change anything. That's what Fung has shown. The membrane attachment of Tα, and even of Tβτ are very sensitive to the ionic strength of the medium. In a low sodium chloride medium both Tα and Tβτ are solubilized. The GTPase is very very slow.

Clark: What concentration of magnesium do you use when you break rod outer segments?

Chabre: 2 mM.

Devreotes: Have you ever injected PDE_0 into your system, I mean without inhibitors, what happens when you do that?

Chabre: The activity at equilibrium is linear with the amount of PDE_0 added.

Devreotes: Shouldn't that give you the kinetics of it, getting turned off, because you should get the exchange?

Chabre: Yes it should, but we do not see it turning off. We are too slow. But those measurements have only been made at high cGMP concentration, and then our model certainly does no work, since PDE_1 had about half of the activity of PDE_0. As I said, if the exchange model is to work, it will be only at the low, physiological, cGMP concentration: in this micromolar range, and we have not been able to make reliable measurements in this low concentration range.

Auxin-coupled systems

CHARACTERIZATION AND MODULATION OF THE SENSITIVITY OF PLANT PROTOPLASTS TO AUXIN

H. Barbier-Brygoo, G. Ephritikhine, W. H. Shen, A. Delbarre, D. Klämbt* and J. Guern

Laboratoire de Physiologie Cellulaire Végétale
C.N.R.S./I.N.R.A., Bat 15
F-91198 Gif-sur-Yvette Cedex
France

INTRODUCTION

Plant growth and development are regulated by several classes of growth substances, namely auxins, cytokinins, gibberellins, abcissic acid and ethylene. As to the mode of action of these plant hormones, ideas have been influenced by concepts and models relative to animal systems, but the molecular mechanisms by which they exert their regulatory effects are still poorly understood. For auxins, which constitute the most documented system, the few pieces available so far suggest that the signalling pathway for this hormone could involve binding to plasmalemma receptors, modulation of phosphoinositide turnover, liberation of second messengers and regulation of cell activity through the set of phosphorylations/dephosphorylations of membrane and soluble proteins (Guern, 1987; Boss and Morré, 1989).

Although the existence of auxin-binding proteins is well established (see the review of Libbenga *et al*, 1986), very few evidence could be provided to assign to these binding sites a true receptor function in the auxin signalling system (Löbler and Klämbt, 1985b). This has been notably due to the lack of identifiable primary functional responses induced by the binding of the hormone to its receptor. Such a response can be

*Botanisches Institut der Universität Bonn, Meckenheimer Allee 170, D-5300 Bonn 1; Federal Republic of Germany

NATO ASI Series, Vol. H 44
Activation and Desensitization of Transducing Pathways
Edited by T. M. Konijn, M. D. Houslay, P. J. M. Van Haastert

provided by the hormone-induced modifications of the transmembrane potential difference (Em) of plant cells, shown to constitute an early response to auxin application on isolated organs (Senn and Goldsmith, 1988 and references therein). A similar hyperpolarizing effect is induced on plant cells by a fungal phytotoxin, fusicoccin, which mimics the effects of auxin in many respects. The membrane responses of isolated protoplasts to both effectors (auxin and fusicoccin) have been investigated here, with the aim of identifying specific steps of the auxin-signalling pathway and bringing functional evidence for the presence of an auxin receptor at the plasmalemma. This hormone-induced electrical response was then used to characterize biological systems differing in their developmental response to auxins (wild-type and auxin-resistant mutant tobacco genotypes, normal and *Agrobacterium rhizogenes* transformed roots of *Lotus corniculatus*), and likely affected in their sensitivity to the hormone.

RESULTS

Effects of Auxin and Fusicoccin on the Transmembrane Potential Difference of Isolated Protoplasts

Protoplasts isolated from tobacco mesophyll cells, and suspended in the control medium in the absence of effectors, exhibited a small, negative inside, electrical potential difference (Em) as shown in Figure 1A. Upon the addition of either auxin (1-naphtalene acetic acid, NAA) or fusicoccin (FC), the distribution of Em values was homogeneously shifted towards more negative values. The induced Em variations were -8.2 mV and -6.0 mV for 5 μM NAA and 1 μM FC, respectively. From these results, dose-response curves describing the Em variations as functions of the auxin concentrations in the medium were established (Figure 1B). The response to auxin displayed an inverted bell-shaped curve, with an hyperpolarization of protoplasts for NAA doses up to 5 μM and a relative depolarization for higher concentrations. The same

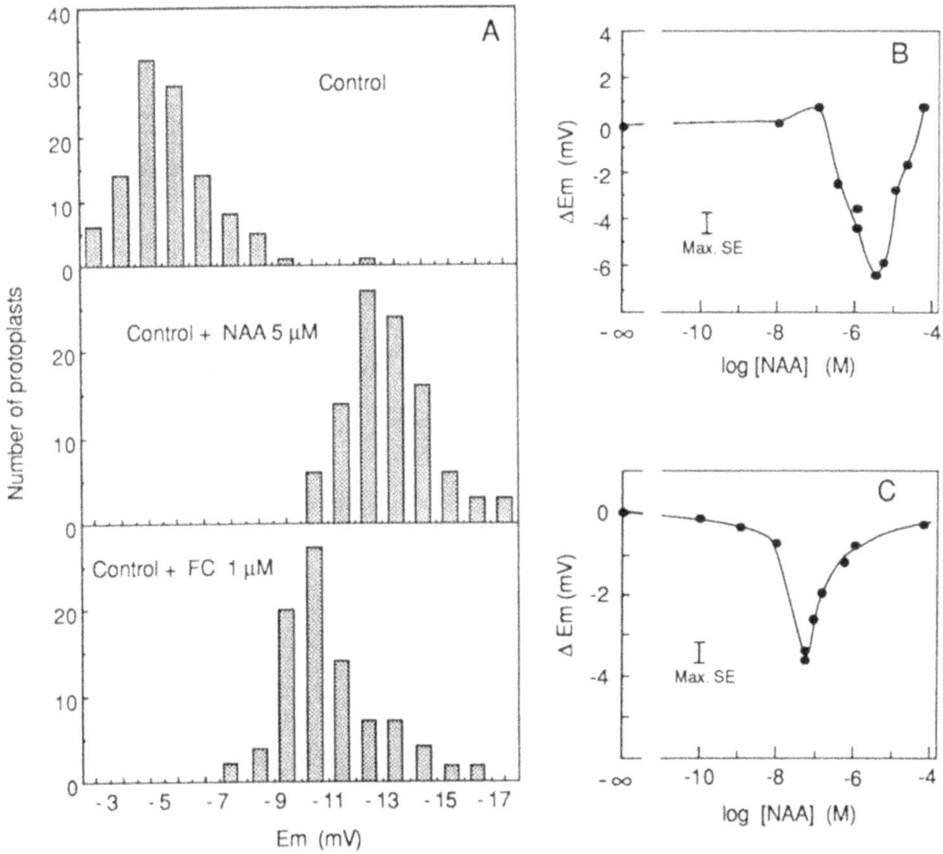

FIGURE 1 . Effect of NAA and FC on the transmembrane potential difference of isolated protoplasts. Mesophyll protoplasts (A and B) were isolated from young tobacco leaves of wild-type plants (*Nicotiana tabacum*, cv Xanthi) according to Caboche (1980). Root protoplasts (C) were isolated from *Lotus corniculatus* root tips as described in Shen *et al* (1988).
(A) Em distributions represent the data pooled from three independant experiments. Em measurements were performed on individual protoplasts by the microelectrode technique, in the complete culture medium in the absence of effector (control) or immediately after the addition of NAA (5µM) or FC (1µM). The mean Em values corresponding to each distribution were calculated : Control : -5.3 ± 0.2 mV, n=84 ; Control+NAA : -13.5 ± 0.2 mV, n=86 ; Control+FC : -11.3 ± 0.2 mV, n=90.
(B and C) Dose-response curves for Em of mesophyll (B) and root (C) protoplasts in the presence of NAA. The NAA-induced Em variations (ΔEm) were calculated from the mean control value mesured in the absence of effector. Each point corresponds to a mean value calculated from 15 to 20 individual measurements. Data from two independant experiments are given, and maximal standard errors are indicated.

type of response was observed for protoplasts isolated from *Lotus corniculatus* root tips (Figure 1C). As to dose-response curves for fusicoccin, a maximal hyperpolarization was reproducibly reached for 1-5 μM and maintained up to the highest FC concentration tested (10-20 μM), but its amplitude varied from one protoplast suspension to another (from -4 to -9 mV).

Analysis of the mechanisms controlling the Em variations induced by auxin and fusicoccin

It has been widely suggested that the auxin- or fusicoccin-induced hyperpolarization on cells involves the stimulation of the plasma membrane-bound H^+-ATPase, which would be one of the physiological targets of both effectors. The hypothesis that auxin hyperpolarizes tobacco protoplasts through this pathway was investigated by using antibodies directed against two plasmalemma proteins possibly involved in the response, namely the auxin-binding protein (ABP) and the H^+-ATPase.

The anti-ATPase IgG fraction, raised to a purified plasma membrane ATPase from yeast (Clément *et al*, 1986), was able to decrease the control Em value, thus indicating that part of the measured Em was dependent on the functioning of the ATPase (Table 1). In the presence of the anti-ATPase IgG, neither NAA nor FC were able to induce Em variation. The anti-ABP IgG, directed against the auxin-binding protein purified from maize coleoptile membranes (Löbler and Klämbt, 1985a; Barbier-Brygoo *et al*, 1989), did not modify the control Em, but completely inhibited the NAA-induced hyperpolarization, without affecting the FC-induced response (Table 1).

Sensitivity of Isolated Protoplasts to Auxin and Fusicoccin

The sensitivity of protoplasts to auxin can be expressed, from the dose-response curve, as the optimal auxin concentration inducing the maximal membrane hyperpolarization. According to this criterium, the sensitivity of tobacco mesophyll protoplasts isolated from the auxin-tolerant

TABLE 1. Effects of anti-ATPase IgG and anti-ABP IgG on the variation of potential difference induced on tobacco protoplasts by auxin or fusicoccin.

Addition to medium	Em (mV)	NAA- or FC-induced Em variation (mV)
None	-6.1 ± 0.3	-
NAA (5μM)	-12.5 ± 0.4	-6.4
FC (1μM)	-10.8 ± 0.4	-4.7
Anti-ATPase IgG (0.4μM)	+0.8 ± 0.6	-
Anti-ATPase IgG + NAA (5μM)	-0.2 ± 0.5	-1.0
Anti-ATPase IgG + FC (1μM)	+0.9 ± 0.6	+0.1
None	-5.6 ± 0.3	-
NAA (5μM)	-12.7 ± 0.3	-7.1
FC (1μM)	-10.2 ± 0.3	-4.6
Anti-ABP IgG (0.4μM)	-5.8 ± 0.2	-
Anti-ABP IgG + NAA (5μM)	-6.0 ± 0.4	-0.2
Anti-ABP IgG + FC (1μM)	-10.3 ± 0.3	-4.5

Em measurements were done in the complete culture medium in the absence or presence of the effectors (NAA and FC). IgG fractions were tested either alone or in combination with NAA or FC. The medium was supplemented with the anti-ATPase IgG (Clément *et al*, 1986) or the anti-ABP IgG (total IgG fraction isolated from antisera of rabbits previously immunized with purified ABP, see Barbier-Brygoo *et al*, 1989) and protoplasts were preincubated for 2.5 h at room temperature before measurements. Em values are presented as mean ± SE calculated from 20 individual measurements for each experimental condition (data from one representative experiment).

mutant (Muller *et al*, 1985) appeared about 10 times lower than that of protoplasts isolated from the wild-type plant (Table 2). This difference is in good agreement with the one observed between the two genotypes for the auxin-induced proliferation of protoplast-derived cells (Caboche *et al*, 1987 ; Ephritikhine *et al*, 1987).

On *Lotus corniculatus* roots, the comparison of protoplasts isolated from normal roots and from roots transformed by *Agrobacterium rhizogenes* showed that the transformation induced

TABLE 2. Sensitivity of isolated protoplasts to auxin and fusicoccin.

Material	Genotype	NAA-induced response		FC-induced response	
		Max ΔEm (mV)	Opt [NAA] (μM)	Max ΔEm (mV)	Opt [FC] (μM)
Nicotiana tabacum[a]	WT	- 6.1	5	- 5.5	1
	M	- 5.1	50	- 6.5	1
Lotus corniculatus[b]	N	- 3.8	0.05	- 5.2	5
	T	- 3.5	0.0001	- 4.8	5
Nicotiana tabacum[c]	N	- 4.2	1	-	-
	T	- 4.1	0.005	-	-

Protoplasts were isolated from tobacco leaves [a,c] ([a]*N. tabacum*, cv Xanthi, WT: wild-type and M : resistant mutant genotypes ; [c]*N. tabacum* cv Maliga Petit Havana, N : normal and T : transformed genotypes) and from *L. corniculatus* roots[b] (N : normal and T : *A. rhizogenes* transformed genotypes). For each genotype, Em variations (ΔEm) were measured as a function of auxin (NAA) or fusicoccin (FC) concentrations in the medium. The maximal ΔEm was deduced from the dose-response curve and the optimal effector concentration inducing this maximal variation was used to characterize protoplast sensitivity to auxin or fusicoccin.

a dramatic increase in protoplast sensitivity to auxin, by a factor of 500 (Table 2). This increase in sensitivity was also expressed in transformed root tips, as revealed by the study of auxin-induced cell elongation or proton excretion (Shen *et al*, 1988). A similar shift towards higher sensitivity to auxin was also observed on mesophyll protoplasts isolated from *A. rhizogenes* transformed tobacco plants in comparison to their normal counterparts (Table 2). Interestingly, protoplasts from wild-type and mutant tobacco genotypes on one hand, and from normal and transformed roots on the other hand, do not differ significantly in their sensitivity to fusicoccin (Table 2 and Figure 2). This suggests that what has been modified in the mutant or transformed protoplasts resides at the reception or early transduction steps of the auxin signal.

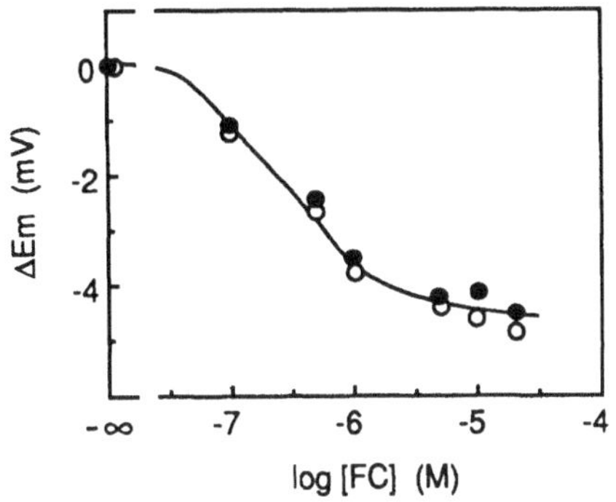

FIGURE 2.Dose-response curve for Em of root protoplasts in the presence of FC. Protoplasts were isolated from normal (open symbols) and *Agrobacterium rhizogenes* transformed (black symbols) roots of *Lotus corniculatus*, according to Shen *et al* (1988). Em variations were measured as described in the legend of Figure 1.

Modulation of Protoplast Sensitivity to Auxin by the Number of Receptors at the Plasmalemma

The hypothesis that protoplasts from normal, mutant and transformed tobacco genotypes differed in the number of auxin receptors exposed at the plasmalemma was investigated. The results of immunotitrations by a monospecific anti-ABP antibody, using the auxin-induced ΔEm as a functional response of hormone binding to its receptor, are presented in Figure 3. This figure illustrates the progressive inhibition of the NAA-induced ΔEm on wild-type protoplasts by increasing concentrations of the anti-ABP IgG. From this curve, it is possible to determine the concentration of anti-ABP IgG inducing 50% inhibition of the NAA-induced response, i.e. 0.3 nM IgG. The same experiments, performed on tobacco protoplasts isolated from the mutant and transformed plants gave immunotitration curves parallel to the one obtained for wild-type protoplasts (data not shown). However, it appeared that the 50%-inhibiting IgG concentration, determined above on wild-type protoplasts, induced approximately 90% inhibition of NAA-induced response on mutant protoplasts, but only 20% inhibition on transformed protoplasts. To obtain 50% inhibition of the maximal ΔEm, about 10 times less IgG are needed to block the response of mutant protoplasts, whereas 10 times more IgG are necessary on transformed protoplasts. If these differences

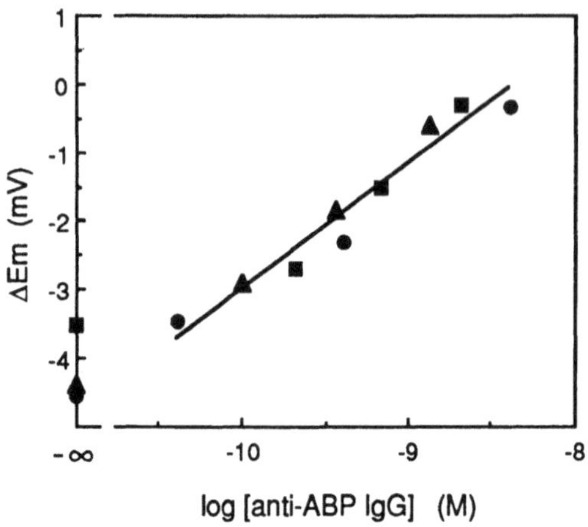

FIGURE 3. Inhibition of the auxin-induced Em variations on tobacco protoplasts by a monospecific polyclonal antibody directed against an auxin-binding protein (ABP) from corn coleoptile membranes.
Protoplasts isolated from wild-type plants (*Nicotiana tabacum*, cv Xanthi) were incubated ($5x10^4$ protoplasts.ml^{-1}) with various concentrations of anti-ABP IgG (monospecific anti-ABP IgG, purified from the total IgG fraction by affinity chromatography on an ABP-Sepharose column, see Barbier-Brygoo *et al*, 1989) during 1.5 h at room temperature. NAA was then added in the suspension, at the concentration inducing the maximal variation in the absence of IgG (1 µM). The NAA-induced Em variations (ΔEm) were measured as described in the legend of Figure 1. Standard errors did not exceed 0.7 mV. Different symbols represent independant experiments.

between genotypes in efficient IgG concentrations are interpreted as reflecting differences in the amounts of functional receptors at the membrane surface, they are in good accordance with the observed variations in protoplast sensitivity to auxin. The mutant protoplasts, with less receptors than the wild-type ones, display a decreased sensitivity to the hormone, whereas transformed protoplasts exhibit an enhanced responsiveness linked to an increase in the number of receptors.

In order to confirm the assumption that receptor density could influence the sensitivity of protoplasts to auxin, we tried to modify the responsiveness of wild-type protoplasts by changing their number of functional binding sites at the plasmalemma. Figure 4 shows that large opposite shifts of the dose-response curves to NAA were obtained by blocking part of the receptor population with the anti-ABP antibody, or by supplementing protoplasts with purified ABP. It was thus possible, starting from the wild-type genotype, to reproduce mutant or transformed phenotypes, in terms of sensitivity to auxin, only by modulating the number of receptors at the protoplast surface. The immunotitration of wild-type protoplasts complemented with 10^{-10} M exogenous corn ABP revealed that the number of titratable receptors was increased by a factor of ten, in comparison with non-complemented protoplasts (data not shown).

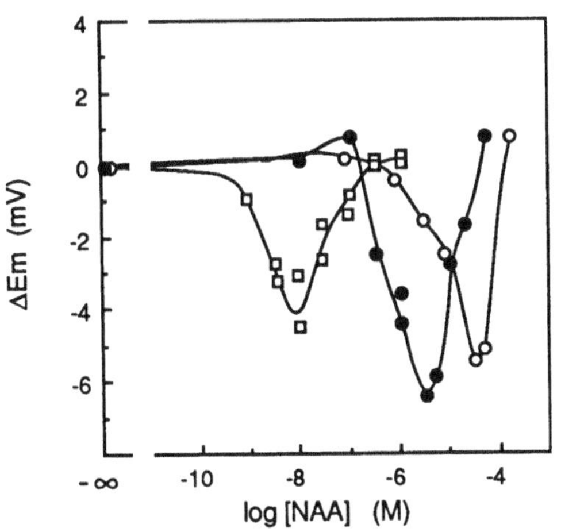

FIGURE 4. Dose-response curves for Em of tobacco protoplasts in the presence of NAA. Mesophyll protoplasts were isolated from the wild-type genotype and Em variations (ΔEm) were measured as a function of auxin concentrations in the medium (black symbols). The protoplasts ($5x10^4$ protoplasts.ml^{-1}) were then incubated either with the monospecific anti-ABP IgG ($4x10^{-9}$ M IgG, 1.5 h at room temperature), or with a purified ABP preparation (10^{-10} M ABP, 15 min at room temperature) (see Barbier-Brygoo *et al*, 1989, for the preparation of the anti-ABP IgG and the purified ABP fractions). Dose-response curves for NAA were then established in the presence of the anti-ABP IgG (O) or the purified ABP (□).

DISCUSSION

The results presented here provide direct evidence for the presence of a membrane receptor for auxin at the plasmalemma, the binding of the hormone to this receptor leading to the activation of the proton-pumping ATPase. They also show that at least the primary steps of activation by NAA are distinct from those of the FC-induced response, in good agreement with the fact that auxin and fusicoccin binding sites, although both located at the plasma membrane, are different protein entities (Klämbt, 1987, references therein).

Large differences in sensitivity to auxin are induced, either by mutagenesis or by transformation, and can be measured in the short term by the auxin-induced ΔEm of isolated protoplasts. These differences in sensitivity to the hormone were shown to be correlated with modifications in the number of functional receptors at the plasmalemma, with regards to the situation of wild-type protoplasts. A decrease in the amount of receptors, as seen on protoplasts from the NAA-tolerant mutant or on wild-type protoplasts treated with the anti-ABP IgG, was associated to a reduced sensitivity. At the opposite, an enhanced sensitivity was observed on protoplasts from *A. rhizogenes* transformed plants, or on wild-type protoplasts supplemented with purified ABP, in correlation with their increased receptor pool. However, in the case of transformed or ABP-complemented protoplasts, no proportionality was observed between the increases in sensitivity (100 to 200 times) and in the receptor number (about ten times).

A good description of these experimental data could be achieved by using a model in which variations in the number of receptors at the plasmalemma are able to displace the dose-response curves along the auxin concentration range without modifying the amplitude of the response (Figure 5). This model is able to account for transformed or complemented situations, where the sensitivity varies as a function of the square power of the total number of receptors, provided the active form of the receptor is a dimer. This assumption of an oligomeric nature of the auxin binding site is in good

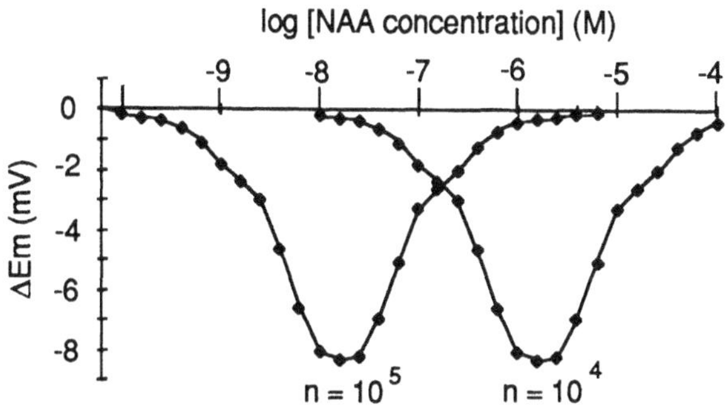

FIGURE 5. Dose response curves calculated for two protoplast types differing by a factor of 10 in their number of plasmalemma receptors and assuming that the active form of the receptor is a dimer.
The model used assumes that (1) two receptor monomers associate with an affinity constant K = 10^{-10} M. The affinity of the dimer for NAA is KS = 10^{-7} M (estimated from binding experiments) and two values (n = 10^4 and 10^5) are taken for the number of monomers present at the surface of one protoplast (rough estimation from NAA binding to isolated plasmalemma vesicles) (2) a second messenger A is produced as a direct function of the number of occupied dimers, A = k. RRH with k= 500 (3) the response to A is inhibited by an excess of A according to the classical equation E = Emax (1 + KA/A + A/K'A) with Emax = -20 mV, KA = 5.10^{-11}, K'A = 10^{-10}.

agreement with biochemical data showing that the native form of the auxin receptor in corn coleoptiles is a dimer of 20-21 kDa monomers (Löbler and Klämbt, 1985a ; Shimomura *et al*, 1986). To describe with this model the behaviour of the mutant genotype (sensitivity and number of receptors decreased by 10) one has to assume that the mutation, besides its effect on the receptor number, affected also the properties of the receptor monomer (ability to dimerize, affinity for the hormone or efficiency of coupling with the transducing system).

Evidence is provided here for the first time that the sensitivity of plant cells to hormones can be controlled by the density of receptors at the plasmalemma. The most puzzling results concern the complementation of tobacco protoplasts by heterologous corn ABP. This raises many questions as to the mode of interaction of ABP with the membrane, to the transmission of the auxin signal to the cell interior and to the real specific role of ABP itself in the signalling pathway.

ACKNOWLEDGEMENTS

The authors are grateful to M. Caboche (Biologie Cellulaire, I.N.R.A, Versailles, France) for providing tobacco plants form the auxin-tolerant mutant, and to P. Costantino (Dipartimiento di Genetica e Biologia Molecolare, Università di Roma, Italy) for supplying tobacco plants transformed by *A. rhizogenes*.

REFERENCES

Barbier-Brygoo H, Ephritikhine G, Klämbt D, Ghislain M, Guern J (1989) Functional evidence for an auxin receptor at the plasmalemma of tobacco mesophyll protoplasts. Proc Natl Acad Sci USA 86:891-896

Boss WF, Morré DJ (1989) Second Messengers in Plant Growth and Development, Plant Biology, vol 6, Alan R Liss, New York

Caboche M(1980) Nutritional requirements of protoplast-derived, haploid tobacco cells grown at low cell densities in liquid medium. Planta 149:7-18

Caboche M, Muller JF, Chanut F, Aranda G, Cirakoglu S (1987) Comparison of the growth promoting activities and toxicities of various auxin analogs on cells derived from wild-type and a nonrooting mutant of tobacco. Plant Physiol 83:795-800

Clément JD, Ghislain M, Dufour JP, Scalla R (1986) Immunodetection of a 90000-Mr polypeptide related to yeast plasma membrane ATPase in plasma membrane from maize shoots. Plant Sci 5:43-50

Ephritikhine G, Barbier-Brygoo H, Muller JF, Guern J (1987) Auxin effect on the transmembrane potential difference of wild-type and mutant tobacco protoplasts exhibiting a differential sensitivity to auxin. Plant Physiol 83:801-804

Guern J (1987) Regulation from within : the hormone dilemma. Ann Bot 60:75-102

Klämbt D (1987) Plant Hormone Receptors, NATO ASI Series, Series H : Cell Biology, Vol 10, Springer-Verlag, Berlin Heidelberg New-York London Paris Tokyo

Libbenga KR, Maan AC,van der Linde PCG, Mennes AM (1986) Auxin receptors.In : Chadwick CM, Garrod DR (eds) Hormones, Receptors and Cellular Interactions in Plants. Cambridge Univ. Press, Cambridge, p1

Löbler M, Klämbt D (1985a) Auxin-binding protein from

coleoptile membranes of corn (*Zea mays* L.). I Purification by immunological methods. J Biol Chem 260:9848-9853
Löbler M, Klämbt D (1985b) Auxin-binding protein from coleoptile membranes of corn (*Zea mays* L.).II Localization of a putative auxin receptor. J Biol Chem 260:9854-9859
Muller JF, Goujaud J, Caboche M (1985) Isolation *in vitro* of naphtalene-acetic acid tolerant mutants of *Nicotiana tabacum* which are impaired in root morphogenesis. Mol Gen Genet 199:194-200
Senn AP, Goldsmith MHM (1988) Regulation of electrogenic proton pumping by auxin and fusicoccin as related to the growth of *Avena* coleoptiles. Plant Physiol 88:131-138
Shen WH, Petit A, Guern J, Tempé J (1988) Hairy roots are more sensitive to auxin than normal roots. Proc Natl Acad Sci USA 85:3417-3421
Shimomura S, Sotobayashi T, Futai M, Fukui T (1986) Purification and properties of an auxin-binding protein from maize shoot membranes. J Biochem 99:1513-1524

DISCUSSION

Mulle: Are the receptors for hormones in plants cloned and sequenced?
Barbier-Brygoo: The response will come probably in the next talk. In fact up to now the only receptor which has been isolated and sequenced is the auxin binding protein from maize coleoptile membranes.
Mulle: Are there any channels; you talked about transporters.
Barbier-Brygoo: There are channels in the plasma membrane of isolated protoplasts but we do not know if these channels are involved in the response to auxin.
Mulle: Are they modulated by anything?
Barbier-Brygoo: Yes, they are voltage dependent. Some of them are modulated by calcium, others are stretched-activated. Some evidence, obtained not directly by the patch-clamp technique, but only by using some inhibitors suggest that calcium channels are present at the plasmalemma and could be involved in the response to developmental signals. In fact in our laboratory we are interested in developing the patch-clamp technique to get more information about the molecular mechanism involved in the electrical response induced by auxin.
Dumont: I have a very naive question. Why do these plants spend ATP to extrude protons. Is that pump going to the other side too. Can the system be used to synthesize ATP.
Barbier-Brygoo: No, not in the plasma membrane. This ATPase, extruding protons outside, participates in the regulation of the cytoplasmic pH. Furthermore, it creates and maintains an electrochemical proton gradient which is used thereafter for the cotransport of ions, glucose or amino acids.
Downes: Do you know how much the pH in the protoplast changes?
Barbier-Brygoo: It is very difficult to evaluate. We tried to do pH measurements using the NMR technique. The problem in the plant cells is that almost the whole volume of the cell is occupied by the vacuole and the pH changes we are interested in are in the cytoplasm. We could only see a very small acidification of the cytoplasmic pH of 0.1 -0.2 pH units for

high NAA concentrations (>20μM NAA). This acidification could perhaps be related to the supraoptimal effect of the dose response curve. We could not find any significant change in the cytoplasmic pH using lower NAA concentrations, which were expected to induce cytoplasmic alkalinisation, if their primary effect is to activate the proton pump.

Clementi: Do you know where these antibodies bind to the receptor?

Barbier-Brygoo: We do not know, the antibody we used was a polyclonal antibody.

Clementi: Do you know if these antibodies induce down-regulation of the receptor as in animal cells?

Barbier-Brygoo: We know nothing about down-regulation or desensitization of the auxin receptor. The only evidence we have concerning the Em change induced by auxin is that it is maximal within one minute and stable for the time needed for a series of measurements i.e. 40-50 min.

Devreotes: When you change the number of the receptors up or down under different conditions, it shifts the dose response and not the magnitude. Does it tell you anything about the ratio of the receptors to the ATPase?

Barbier-Brygoo: It is difficult to say; we tried to establish some models in order to describe this response. Such shifts of the dose-response curves to auxin, without any modification of the amplitude, can be accounted for by models where only the number of receptors is changed without any other modification of the reception-transduction pathway, for example in the receptor affinity for the hormone or in the efficiency of the transduction steps leading to the activation of the ATPase.

Clementi: I think you can justify this with just changing the affinity of the receptor without changing the number of receptors.

Barbier-Brygoo: Sure you get exactly the same shift in the dose response curve if you change either the affinity, or the number, or even the efficiency of the transduction. But when establishing the models we tried to account for the immunotitration results which indicated that we had modifications in the number of receptors.

Devreotes: If the cells which were under different conditions had more or less response did the ATPase antibodies block at the same dose of antibody.

Barbier-Brygoo: These results of modified sensitivity on mesophyll protoplasts are quite new so we have not done it yet. Concerning the in vitro cultured roots, the protoplasts isolated from *A.rhizogenes* transformed roots exhibited a dramatic increase in their sensitivity to auxin, but do not differ from their normal counter parts in their sensitivity to fusicoccin. As we know that the plasma membrane ATPase is involved in both responses (to auxin and fusicoccin), it can be inferred that this ATPase is not modified by the transformation event leading to an increased sensititvy to auxin.

THE AUXIN-BINDING-PROTEIN FROM MAIZE COLEOPTILES

Gero Viola, Ulrich Tillmann, Thomas Hesse [x], Beate Knauth, Klaus Palme [x], Marian Löbler and Dieter Klämbt

Botanisches Institut
Universität Bonn
Meckenheimer Allee 170
D-5300 Bonn 1
FRG

[x] MPI für
Züchtungsforschung
D-5000 Köln
FRG

There is good evidence that auxin-dependent cell elongation (Löbler, Klämbt (1985b) and H^+ secretion (Barbier-Brygoo et al., 1989) is mediated by an auxin-binding-protein (ABP). The ABP is isolated from maize coleoptiles, purified and characterized by Löbler and Klämbt (1985 a,b) and Shimomura et al. (1986). Recently cDNA clones of ABP were isolated, sequenced and further characterized (Hesse et al., 1989; Tillmann et al., 1989; Inohara et al., 1989). These reports confirmed earlier results. The pre-ABP contains a signal peptide and one glycosylation site.

During protein synthesis at the membrane associated ribosoms, the polypeptide chain crosses the ER membrane and the signal peptide is splitted immediately. There is one N-glycosylation site - Asn-Thr-Thr - and a carboxy terminal sequence - Lys -Asp-Glu-Leu - which is characteristic for ER-residental Proteins (Pelham, 1988). The localization of the reactive ABP is still uncertain. Early reports (Dohrmann et al., 1978) located the binding moiety - Site I - within the ER vesicles. Later solubilization of the binding site led to the ABP which showed binding characteristics very similar to Site I (Venis 1977; Tappeser et al., 1981;, Löbler, Klämbt 1985a; Shimomura et al., 1986, 1988).

Two important results were obtained to test whether ABP is involved in auxin-regulated cell elongation of maize coleoptiles: First, ABP is mostly localized in the outer epidermal cells of maize coleoptiles, which are the auxin-dependent, actively elongating cells. Second, highly puri-

NATO ASI Series, Vol. H 44
Activation and Desensitization of Transducing Pathways
Edited by T. M. Konijn, M. D. Houslay, P. J. M. Van Haastert

fied anti-ABP antibodies interfere with the auxin action upon cell elongation (Löbler, Klämbt 1985b). The latter result can be interpreted only by the assumption that the ABP mediating the auxin effect upon cell elongation is localized at the outer surface of the plasmalemma. Barbier-Brygoo et al. (1989) came to the same conclusion for the ABP mediated auxin effect upon H^+ secretion in tobacco protoplasts. They used a heterologous system employing antibodies against maize ABP in these membrane potehtial measurements on tobacco protoplasts. Very recent experiments demonstrated that maize ABP applied to tobacco protoplasts at different concentrations altered the auxin sensitivities for optimal hyperpolarization (Barbier-Brygoo et al., 1989 in preparation). These data consequently predict that, although auxin molecules easily diffuse into plant cells, auxin has to be recognized by ABP or a homologon at the outer surface of the plasmalemma of the reacting cells.

All available data of ABP-cDNA sequences (Hesse et al., 1989, Tillmann et al., 1989, Inohara et al., 1989) show the carboxy-terminal KDEL signal. It is expected that KDEL at plant proteins functions in a very similar manner as in KDEL containing animal proteins. Either there are genetically instructed ABP isoforms lacking the KDEL sequence or during ABP lifetime carboxy-peptidases or endopeptidases will shorten the ABP molecule from the carboxy end. Both alternatives would result in a secretable ABP.

We used two different affinity matrices (kindly provided by Steve Fuller) containing the carboxy-terminal octa-peptide from rat PDI (Edman et al., 1985) or carboxy-terminal undeca-peptide from rat BIP (Bole et al., 1986) and we selected antibodies specific to the KDEL terminus from crude IgGs from antisera raised against SDS-denatured ABP. These IgG preparations labelled only one of the regularly obtained ABP double band on PAGE by immunostaining of Western blots. This result indicates that besides one or several isoforms of ABP with the carboxy terminus of KDEL, there exist one or several isoforms missing the complete

KDEL-signal. These isoforms are insufficient to bind the KDEL-specific antibodies and to be retained within the ER. There are no indications so far of the origin of the latter ABP fraction. It could be derived from a special mRNA lacking the information of the complete KDEL C-terminus, or just by protein digestions within the ER. If the KDEL is the ER-sorting signal as we expect, it has to be exposed on the ABP molecule and is therefore susceptible to carboxy-peptidases as well.

Two questions arise at this point:
1. What are the functions of the ABP molecules within the ER? In vitro auxin can be bound by ER-residental ABP. In vivo binding of auxin to ABP within the ER may be questionable. There is no plausible function for such an auxin-binding, except for being connected with ABP and auxin secretion.
2. Which pathway does auxin take to the outside to be finally recognized by ABP?
Auxin biogenesis seems to be connected with merismatic cell growth. Shoot tips and cambial regions are the main auxin sources. In general a basipetal auxin transport is reported. The vectorial movement is explained by an active auxin secretion mainly at the base of the polar orientated cells (Rubery, Sheldrake 1974, Cande, Ray 1976, Goldsmith 1977). Auxins as weak aromatic acids easily diffuse from the apoplastic space at pH around 5.5 into these cells. Symplastic auxin transport through plasmodesmata was not detectable because specific inhibition of auxin secretion by NPA or TIBA results in no polar auxin movement. We tested by the use of isolated epidermal strips of maize coleoptiles whether auxin secretion is an essential step for auxin regulated cell elongation.
10 mm long epidermal strips from young maize coleoptiles do not elongate if incubated in phosphate buffer. Strips elongate in phosphate buffer containing various amounts of auxin. Strips incubated for 2 h in $2x10^{-4}$M IAA, washed for 30 min and incubated for 6 h onto phosphate buffer containing 2% sucrose and 0,7% agar elongate as

well. Strips treated similarly and incubated on phosphate buffer containing 2% sucrose, 0,7% agar and 10^{-4}M NPA did not elongate at all. If during the 6 h incubation onto NPA-agar IAA in various concentrations was added, cell elongation was restored (Knauth, Klämbt 1989).These data lead to the conclusion that IAA taken up into the reactive cells stimulates cell elongation as long as auxin secretion is not blocked. NPA inhibits auxin secretion but does not inhibit cell elongation when auxin is present in the incubation medium. Therefore auxin secretion seems to be an essential step in cell elongation.

CONCLUSION

The reported results give evidence that cells, able to grow actively by elongation and cells able to secrete protons both in response to auxin, regulate these functions by two secretion processes: Secretion of ABP and of auxin. Auxin binds to ABP at the outer surface of the plasmalemma. The resulting ABP-auxin complex is the essential structure for regulating cell elongation and H^+ secretion.

The auxin reactivity of protoplasts is more sensitive if the ABP content at the membrane surface increases as shown by hyperpolarization in tobacco protoplasts (Barbier-Brygoo et al., 1989 , in preparation). These data are in agreement with the suggestion that for signalling optimal cell activity by the ABP-auxin-complexes a certain amount of this complex is necessary. These observations lead to the suggestion that ABP may be a proteo-hormone as the essential part of a proteo-hormone instead of being a hormone receptor. (Guern, Klämbt 1988 unpublished). This model is in agreement with the experimental data that ABP is a non membrane integrated protein but just membrane associated. The functional auxin molecules have to be recognized by ABP at the outer surface of the plasmalemma. These complexes have to bind to signal transducing, membrane integrated proteins which are functioning as ABP-auxin-receptors (Figure 1).

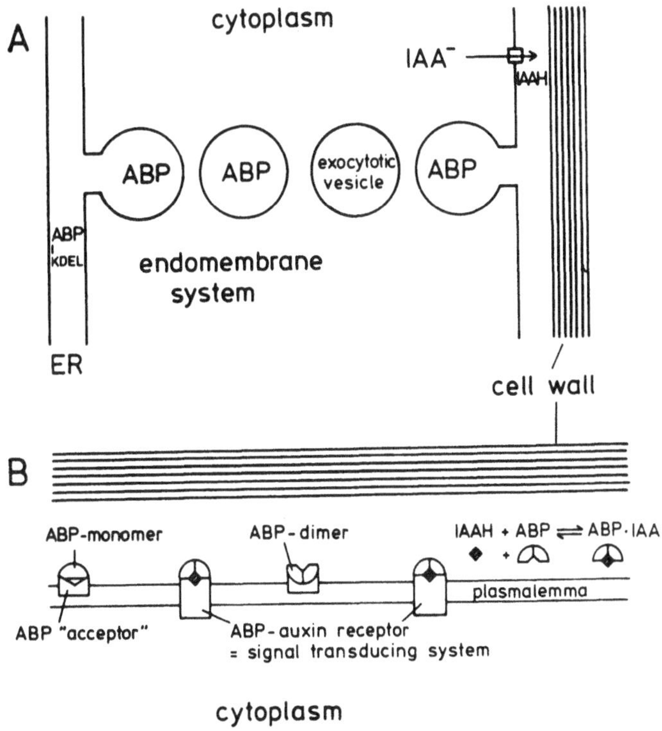

Figure 1: Model of auxin regulation for cell elongation
A: ABP and auxin secretion
B: IAA binding to ABP and ABP·IAA to the signal transducing system.

This reported situation resembles the function of epidermal growth factors (EGFs) in animals.

The data on hyperpolarization of tobacco protoplasts predict that auxin reactive cells contain a certain limited number of the assumed ABP-auxin-receptors within their plasmalemma. Then auxin sensitivity can be modulated by different amounts of ABP at the outer surface of the plasmalemma. This modulation can be achieved by different ABP secretion or different protease activities at the outer surface. Auxin secretion of the same cells seems to be a further essential stepin the regulation of auxin depending functions such as elongation and H^+-secretion.

ACKNOWLEDGEMENTS

This work was supported by grants of the Deutsche Forschungsgemeinschaft and of the EEC (BAB 0207-1).

REFERENCES

Barbier-Brygoo H, Ephritikhine G, Klämbt D, Ghislain M and Guern J (1989) Proc. Natl. Acad. Sci. USA 86: 891-895

Bole DG, Hendershot LM and Kearny JF (1986) J. Cell biol. 102: 1558-1566

Cande WZ and Ray PM (1976) Planta 129: 43-52

Dohrmann U, Hertel R and Kowalik H (1978) Planta 150: 371-379

Edman JC, Ellis L, Blacher RW, Roth RA and Rutter WA (1985) Nature 317: 267-270

Goldsmith MHM (1977) Ann. Rev. Plant. Physiol. 28: 439-478

Hesse T, Feldwisch J, Balshüsemann D, Bauw G, Puype M, Vandekerckhove J, Löbler M, Klämbt D, Schell J and Palme K (1989) EMBO J. submitted

Inohara N, Shimomura S, Fukui T and Futai M (1989) Proc. Natl. Acad. Sci. USA in press

Knauth B and Klämbt D (1989) Botan. Acta submitted

Löbler M and Klämbt D (1985a) J. Biol. Chem. 260: 9848-9853

Löbler M and Klämbt D (1985b) J. Biol. Chem. 260: 9854-9859

Pelham HRB (1988) EMBO J. 7: 913-918

Rubery PH and Sheldrake AR (1974) Planta 118: 101-121

Shimomura S, Sotobayashi T, Futai M and Fukui T (1986) J. Biochem., 99: 1513-1524

Shimomura S, Inohara N, Fukui T and Futai M (1988) Planta 175: 558-566

Tappeser B, Wellnitz, D and Klämbt D (1981) Z. Pflanzenphysiol. 101: 295-302

Tillmann U, Viola G, Kaysser B, Siemeister G, Hesse T, Palme K, Löbler M and Klämbt D (1989) EMBO J. submitted.

Venis MA (1977) Nature 266: 268-269

DISCUSSION

Dumont: How can you exclude the fact that ABP-auxin acts directly on the protein ATP-ase?

Klämbt: The ABP is only an associated protein. You have to have some association structures. From my point of view the ABP has no real receptor function, otherwise we would not understand that an increase of ABP at the protoplasts can lower the auxin concentration for optimal response.

Dumont: Yes, I agree, but I don't see why you have to postulate a transducing system and not as a first hypothesis the direct action of your complex ABP-auxin on the proton ATP-ase?

Klämbt: Perhaps I should add some explanations. Even though we have an acid growth theory explaining auxin action by auxin dependent proton extrusion, there are many facts against this theory. We believe in two different auxin effects: the stimulation of proton secretion analysed in the laboratory in Gif-sur-Yvette and the stimulation of cell elongation analysed in our laboratory. If you agree that both auxin effects are not regulated through only one regulative mechanism, direct action on the proton ATPase cannot describe both effects of ABP-auxin. Nevertheless such an interaction would specify a special form of transducing system.

Devreotes: Why don't you speak of the hydrophobic stretch on the N-terminal as possible membrane domain? Why do you eliminate this? You showed a very hydrophobic region on the N-terminal of ABP and now you are saying that you don't think it crosses the membrane.

Klämbt: We looked very intensely for the signal peptide. We could not get any information that the signal peptide, which is 38 amino acids long, is still present in the extractable ABP. Extracting ABP by transferring our plant material in boiling SDS buffer, we were not able to find it. Therefore an anchor function of the signal peptide is not realistic.

Chabre: Did you have in the purification procedure detergent in all your columns or none at all?

Klämbt: We only used detergent for SDS gel electrophoresis. ABP demonstrated by gel filtration seems to be a dimer. SDS PAGE results in separation of monomers of about 20 kDa.

Chabre: I think that solves the question whether the hydrophobic spot was there, because if it was the protein would not be solvable.

Houslay: Is it possible to look for the membrane site your ABP binds using the chemical cross-linking and then trying to immuno-precipitate the complex?

Klämbt: We search for it, but at the moment without success.

Houslay: Can I ask a second question? You said, you cloned the receptor. I presume, you have been able to express it and to prove it?

Klämbt: There are some experiments to transform cells with this vector but until now without any results.

Zbell: I will only give a comment on your new model of the transducing system which is very strange in comparison to animal systems. You have no amplification system because you say your transducing apparatus is limited. You find normally in animal cells a very low number of receptors but a high number of transducing elements. You get during signal processing an amplification of the signal which is percepted by the membrane receptors. I believe it may be another strategy which is used by the plant cells.

Klämbt: You should realize that the situation in transformed cells is not normal. Nevertheless normal cells can also contain less transducing complexes than ABP molecules at the outside.

Clementi: You said, the auxin is secreted from the cell to the outside. Do you have evidence that this auxin could be bound to these proteins in the same vesicles and could be secreted together?

Klämbt: Actually, it doesn't make any sense. Interpreting the fact that ABP resides primarily within the ER, we found such an idea attractive. Experiments, blocking auxin secretion by NPA, demonstrate that auxin free ABP is localized at the outer cell surface. Therefore the function of the ER residental ABP has to be different from mediating auxin secretion.

Dumont: This is more a question about the statement that was made here. In general you would have many more transducing systems in mammalian cells than receptors. I agree that would be the case in the retina with tranducin and rhodopsin. Is it also true in cells responding to the β-adrenergic receptor? Are the numbers of β-adrenergic receptors much lower than the numbers of G_s in the membrane? What is the order of magnitude?

Iyengar: There is probably 10 Gs to 1 receptor, but I have to guess there is half a cyclase for each receptor.

Downes: I think we are getting a little bit out of context. I thought the way you explained the transducing elements, was not equivalent to our G-protein. It is equivalent to the membrane spanning domain of the receptor. The question whether it is in excess, is incorrect. You have yet to find the equivalent to the G-protein in the system. What I am saying is, as a matter of semantics, the transducing element Dr. Klämbt is talking about is equivalent to the membrane spanning domain of the receptor. It is still compatible with animal cell notions of signal transduction so far. Maybe we will hear more about it tomorrow.

Clark: Returning to your statement you should not relate each signal transducing system to the rhodopsin-transducin system, the total amplification is 10^{+5} of one flash of light.
This is an enormous amplification. There are two steps, each of them amplifies the original signal approximately one thousand to five hundred fold.

Premont: The discussion is focussing on the possibility that the auxin is binding on the auxin binding protein. Then you focus right on and say, well, this complex is obviously binding to a transducing element. What I am thinking of, in terms how animal cells work, the only system I can recall that seems anything like this one, would be the system importing something into the cells. It occurs to me that maybe you have structures similar to a steroid binding protein which is binding to some kind of specific receptor on the cell and that are taking it into the cell. The auxin may be working like a steroid hormone. There is no reason to expect that a system has to transduce a signal from the cell surface.

Klämbt: There is no difficulty to get auxin into the cells by pH dropping. Auxin, a weak acid, is undissociated at the outside pH of 5.5 and easily lipid soluble. It dissociates at the inside and so every cell is catching auxin. Plant cells have no possibility to evaluate how much auxin is coming in. Therefore auxin action seems to be essentially connected to auxin secretion. It appears strange even to botanists but evidences are convincing.

Iyengar: If there is auxin outside of the cell what fraction gets into the cell and what binds to the cell surface? Is there a side of regulation?

Klämbt: Auxin is not any more functional if the ABP is blocked by any type of IgG and without auxin. As we heard from Hélène Barbier-Brygoo, the higher the concentration of ABP at the outside, the lower the concentration of auxin necessary to get optimal reactions. These results contradict the interpretation of ABP as a receptor, and lead to the assumption that ABP may be a proteohormone.

Is a GTP-Binding Protein Involved in the Auxin-Mediated Phosphoinositide Response on Plant Cell Membranes ?

B. Zbell, H. Hohenadel, I. Schwendemann and C. Walter-Back

Botanical Institute
Rupprecht-Karls-University
Im Neuenheimer Feld 360
D- 6900 Heidelberg
Federal Republic of Germany

INTRODUCTION

Three main types of transduction mechanisms for extracellular signals are found on plasma membranes of animal cells: ligand-gated ion channels (Hucho,1986), transmembrane receptors with intrinsic tyrosine kinase activity (Yarden and Ullrich,1988), and G-protein coupled receptor-stimulated effector systems (Dohlman et al.,1987). For the latter mechanism the stimulus generated by the perception of an extracellular signal on its receptor is transduced across the plasma membrane and thereby amplified by specific G-proteins to the adenylate cyclase (Levitzki,1987) or phosphoinositidase C (Cockcroft and Gomperts,1985) as intracellular effector systems. During the process of transmembrane signalling the heterotrimeric G-protein exhibits a cycle of complex association and dissociation reactions of its α-subunit with the $\beta\gamma$-subunit complex which includes a GDP/GTP exchange reaction for its activation, an intrinsic GTPase activity for its inactivation, as well as concomitant coupling and uncoupling reactions to the receptor and effector, respectively (Gilman,1987; Casey and Gilman,1988). As a consequence, G-proteins are recognized as important functional elements of the signal transduction pathway on animal plasma membranes. In contrast to the highly sophisticated knowledge about animal cells our recent understanding of signal transduction processes on plant plasma membranes can be evaluated as a dilemma (Guern,1987). In some recent reports a scheme of transmembrane signalling in plant cells has been proposed, but its weakness stems from its strict homology with the animal model and an acute lack of plant-specific characteristics. However, the plant plasma membrane contains the compounds and several of the functional elements which are known to be involved in the signal-dependent PI [+)] response on the plasma membranes of animal cells.

First, a recent chemical analysis which included deacylation of phosphoinositides and subsequent conversion of the polar head groups to specific polyols by periodate oxidation, reduction, and dephosphorylation, has revealed that plant polyphosphoinositides contain (1,4)IP_2 and (1,4,5)IP_3 (Irvine et al.,1989). The latter compounds are known to be released as products by the phosphoinositidase C from (4)PIP and (4,5)PIP_2 in animal cells (Berridge,

[+)] GDPβS, guanosine 5'-[β-thio]diphosphate; GTPγS, guanosine 5'-[γ-thio]triphosphate; IAA, indoleacetic acid; (1,4)IP_2, inositol (1,4)bisphosphate; (1,4,5)IP_3, inositol(1,4,5)trisphosphate; PI, phosphoinositide; (4)PIP, phosphatidylinositol(4)monophosphate; (4,5)PIP_2, phosphatidylinositol(4,5)bisphosphate.

NATO ASI Series, Vol. H 44
Activation and Desensitization of Transducing Pathways
Edited by T. M. Konijn, M. D. Houslay, P. J. M. Van Haastert

1987). (4)PIP and (4,5)PIP_2 were previously found to be present as minor phospholipids in plasma membranes of carrot cells (Wheeler and Boss,1987). PI and PIP kinases for the formation (Sommarin and Sandelius,1988) and the phosphoinositidase C for the hydrolysis (Melin et al.,1987) of the polyphosphoinositides were detected as enzymatic activities in wheat plasma membranes. One report that GTP or GTPγS promote the release of inositol polyphosphates from plant microsomes (Dillenschneider et al.,1986) contrasts with other findings that the G-nucleotides have no stimulative effect on the phosphoinositidase C activity in plant cell extracts (Melin et al.,1987; McMurray and Irvine,1988). There are also preliminary reports of the occurrence of high-affinity GTP-binding sites in higher plants, detected by: binding of the nonhydrolyzable GTP-analogue [^{35}S]GTPγS to membrane preparations (Hasunuma et al.,1987; Zbell et al.,1989); detection of GTP-binding proteins by [α-^{32}P]GTP blotting (Drøbak et al.,1988); or by immunoblots using very specific antibodies against the G-nucleotide binding domain of animal G-proteins (Blum et al.,1988; Jacobs et al.,1988). Since functions of GTP-binding proteins in the process of transmembrane signalling in plants have not been demonstrated, it can be only speculated that they might be involved in the signal-dependent PI response in plant cells.

Light (Morse et al.,1987,1989) as well as auxin (Ettlinger and Lehle,1988; Zbell and Walter,1987; Zbell and Walter-Back,1988) were found to function as signals to promote the PI response in plant cells. We have introduced an *in vitro* system for the analysis of the functional elements which might be involved in the auxin-mediated PI response in plant membranes. This experimental approach is based on our previous findings that microsomal membranes prepared from *Daucus carota* cell suspension cultures are able to utilize [γ-^{32}P]ATP as a substrate for rapid lipid phosphorylation reactions *in vitro* and that in the presence of IAA, the [^{32}P]-label of the phosholipids is significantly reduced and a concomitant release of [^{32}P]-labeled inositolphosphates is seen (Zbell and Walter,1987; Zbell and Walter-Back,1988). We have used this experimental system further in order to identify putative functional elements such as the auxin receptor and GTP-binding proteins (Zbell et al,1989). In this report, we present our recent findings on the ability of G-nucleotides to promote or inhibit the PI response. We also present data on the detection of high-affinity binding for G-nucleotides and its modulation by IAA.

MATERIAL AND METHODS

Chemicals. Nucleotides were purchased from Boehringer (Mannheim, FRG), and all other chemicals and biochemicals with analytical grade were obtained from Merck (Darmstadt,FRG).

Plant material. Cell cultures of *Daucus carota* were maintained and used for membrane preparation as previously described (Zbell and Walter-Back, 1988).

Preparation of microsomes. In comparison with the previously described method (Zbell and Walter-Back, 1988) microsomal membranes were prepared by a shortened procedure. Immediately after homogenization of carrot suspension cells in buffer (50 mM Tris·HCl, pH 8.0, 250 mM sucrose, 1 mM EDTA, 0.1 mM $MgCl_2$), the extract was centrifuged twice at 2,000 x g for 10 min to remove cell debris. The 2,000 x g supernatant was concentrated by centrifugation at 50,000 x g for 60 min, and after resuspension in buffer (25 mM Tris·HCl, pH 7.5, 250 mM sucrose), a membrane suspension (2mg protein ml^{-1}) was frozen in liquid nitrogen and stored at -80°C until use.

PI phosphorylation assay. The phosphorylation of microsomal membrane lipids was performed *in vitro* as described (Zbell and Walter-Back, 1988). The standard assay contained 93 kBq [γ-^{32}P]ATP (0.55 or 1.07 TBq mmol^{-1} from the Radiochemical Centre Amersham, Buckinghamshire, Great Britain), 100 µM Na_2·ATP, 10 mM $MgSO_4$, 25 mM LiCl, 125 µg membrane proteins in a final volume of 500 µl buffer at pH 7.5. The reaction was started by the addition of the energy substrate, incubated at room temperature, and terminated by the addition of 1 ml ice-cold stop solution containing 2-propanol/conc. HCl (100/1;v/v).

Lipid extraction and fractionation of inositol phosphates. The lipids were extracted from the acidified propanolic solutions with a solvent system using n-hexane as organic solvent (Zbell and Walter-Back, 1988) with the addition of 50µl 3 mM neutral red for the visualization of the phase boundary. The inositol phosphates of the aqueous phase were separated by anion exchange chromatography using 1 ml columns filled with Dowex AG 1-X2 resin (200-400 mesh, Biorad, Munich, FRG) and step gradient of ammonium formate in formic acid as eluent (Zbell and Walter-Back, 1988). [^{3}H]labeled inositol phosphates (NEN Research Products, Dreieich, FRG) were used as reference standards. [^{32}P]-label of the extracts was measured directly via the Cerenkov-radiation, whereas [^{3}H]-label after addition of a scintillant in a liquid scintillation counter.

Binding assay for G-nucleotides. Binding of either [^{3}H]GDP (307 GBq mmol^{-1}, NEN Research Products, Dreieich, FRG) or [^{35}S]GTPγS (48.8 TBq mmol^{-1}, NEN Research Products) was assayed in a final volume of 40 µl using 4 ml-polypropylene tubes. The microsomes were incubated at +25°C in 25 mM Tris·Cl (pH 7.5) with 250 mM sucrose, 1 mM $MgSO_4$, 1mM EDTA, and 100 mM KCl in the presence of 1.5-2.0 kBq [^{35}S]GTPγS and various concentrations of unlabeled nucleotides, as indicated. The binding reaction was started by addition of membranes equivalent to 10 µg protein, and stopped by dilution with an excess amount of ice-cold buffer and subsequent rapid vacuum filtration of the assay through nitrocellulose filters (0.45 µm, Schleicher & Schüll, Dassel, FRG). The dried filters were solubilized by ethylene glycol monomethyl ether and the radioactivity was determined after addition of a scintillant mixture by liquid scintillation counting.

RESULTS AND DISCUSSION

Effects of G-nucleotides and NaF on the PI response. In previous investigations on the auxin-mediated PI response in carrot microsomes we could observe the hormone effect even in the absence of exogenously applied GTP; this effect was stabilized by micromolar amounts of GTP (Zbell and Walter, 1987; Zbell and Walter-Back, 1988). According to these results it was unclear, whether GTP is really a necessary cofactor for the auxin-mediated PI response, or whether G-nucleotides can stimulate the PI response in the absence of auxin. In this connection it should be noted, that the commercially available ATP which is used for the lipid phosphorylation *in vitro* is contaminated by traces of GTP. Also, it cannot be excluded that membrane-bound GDP is phosphorylated to GTP by an intrinsic nucleoside diphosphate kinase utilizing ATP as energy substrate. In order to look for G-nucleotide effects on the release of inositolphosphates as an indication of a possible role of a GTP-binding protein on the phosphoinositidase C activity we have performed experiments with GTP and GTPγS as stimulating and GDP and GDPβS as inhibitory G-nucleotides.

In animal cells, micromolar concentrations of GTP as well as of GTPγS were found to promote the release of inositolphosphates from plasma membranes (Cockcroft and Taylor, 1987; Harden et al. 1988). In the assays with carrot microsomes we found that GTPγS stimulates the release of inositolphosphates in a typical dose-dependent manner and that its effect is stronger for the release of IP_3 than of IP_2 (Fig. 1). The stimulative effect of GTP was evident only of concentrations lower than 50μM, since higher concentrations may have reduced the extent of lipid phosphorylation as consequence of the competition of GTP with ATP as energy substrates for the lipid kinases (results not shown). However, the alternative mechanism which cannot be excluded is that GTP is dephosphorylated by membrane-bound phosphatases to GDP which is known to inhibit the action of a GTP-binding protein (Gilman, 1987).

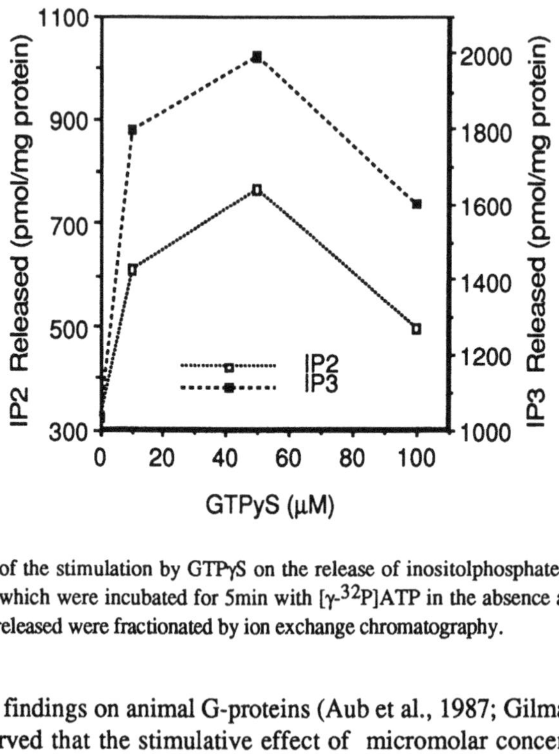

Fig. 1. Dose response of the stimulation by GTPγS on the release of inositolphosphates. After lipid extraction of carrot microsomes, which were incubated for 5min with [γ-^{32}P]ATP in the absence and presence of GTPγS, the inositolphosphates released were fractionated by ion exchange chromatography.

In accordance with findings on animal G-proteins (Aub et al., 1987; Gilman, 1987; Ohtsuki et al., 1987) we observed that the stimulative effect of micromolar concentrations of GTP or GTPγS on the PI response can be progressively blocked by increasing amounts of GDP or GDPβS (results not shown). Moreover, we observed an inhibitory action by GDP (Fig.2) or GDPβS (results not shown) on the release of inositolphosphates even in the absence of exogenously applied GTP or GTPγS. These results confirm our previous suggestion that traces of GTP which are introduced in the assays with the addition of ATP as energy substrate for lipid phosphorylation are sufficient to stimulate the release of inositolphosphates. The effect of GDP is rather complex, since low concentrations <10μM of the G-nucleotide can slightly stimulate the PI response (Fig.2). This phenomenon can be explained by the possible phosphorylation of low amounts of GDP to GTP by a membrane-bound nucleoside diphosphate kinase activity as it is known to occur also in animal cells (Aub et al., 1987; Ohtsuki et al., 1987).

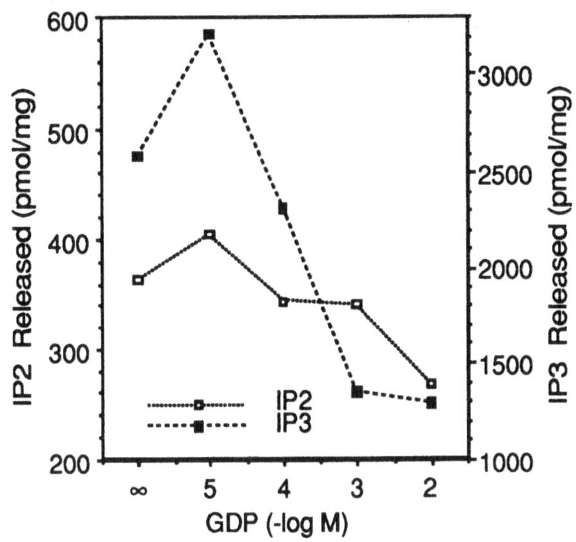

Fig. 2. Dose response effect of GDP on the release of inositolphosphates. Carrot microsomes were incubated for 5min with [γ-^{32}P]ATP in the absence and presence of GDP. After lipid extraction the inositolphosphates released were fractionated by ion exchange chromatography.

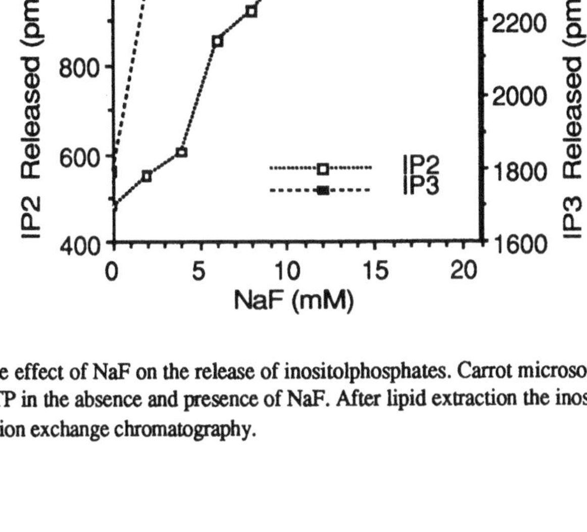

Fig. 3. Dose response effect of NaF on the release of inositolphosphates. Carrot microsomes were incubated for 5min with [γ-^{32}P]ATP in the absence and presence of NaF. After lipid extraction the inositolphosphates released were fractionated by ion exchange chromatography.

Another argument favouring the idea concerning an involvement of a GTP-binding protein in the PI response on carrot microsomes is based on our finding that sodium fluoride stimulates the release of inositolphosphates in a dose-dependent manner (Fig.3). As was the case for GTP and GTPγS (Fig.1) we found a stronger effect by NaF on the release of IP_3 than on IP_2. This observation corresponds well to similar findings on the activation of the phosphoinositidase C by NaF in animal cell membranes (Cockcroft and Taylor, 1987).

Interaction of G-nucleotides and IAA in the PI response. Results from preliminary experiments with carrot microsomes indicate an interaction of IAA and G-nucleotide effects on the PI response. The release of inositolphosphates stimulated by G-nucleotides could be amplified by the addition of 1μM IAA, and the hormone effect was more prominent at low concentrations of G-nucleotides (<10μM). Because GTPγS exhibits a stronger potency than GTP, the corresponding hormone effect was slightly reduced. The inhibitory effect of GDP or GDPβS on inositolphosphate release was reduced by 1μM IAA. The reciprocal experiment demonstrated that the dose-dependent release of IP_3 by IAA was enhanced in the presence of 50μM GTPγS, resulting in a shift of the dose response of IAA. These G-nucleotide effects on the IAA-mediated release of inositolphosphates as well as auxin effects on the G-nucleotide-effected release, support the hypothesis that a GTP-binding protein might be involved in the auxin-dependent PI response as proposed previously (Zbell et al., 1989).

High-affinity binding of [^{3}H]GDP and [^{35}S]GTPγS and its modulation by IAA. One of the criteria which must be fulfilled for a clear identification of a GTP-binding protein involvement in a signal transduction process is the demonstration of high-affinity binding of G-nucleotides and its signal-dependent modulation (Gilman, 1987). Using [^{3}H]GDP or [^{35}S]GTPγS as labeled ligands, we were able to detect such high-affinity sites for either G-nucleotide on carrot microsomes. The binding of [^{35}S]GTPγS was reversible, exhibited a temperature dependency with an optimum at +30°C, and was stimulated by the divalent cations Mg^{2+} or Mn^{2+} in the range below 1mM. The binding is specific for G-nucleotides, since the binding affinities of other nucleotides are three or four orders below that of GTPγS (Tab. 1).

Table 1. IC_{50} Values of Various Nucleotides for the Binding of [^{35}S]GTPγS.

Nucleotides	Inhibition Constants (M)
GTPγS	$1.3 \cdot 10^{-8}$
GTP	$3.4 \cdot 10^{-7}$
GDP	$2.8 \cdot 10^{-7}$
GMP	$> 10^{-4}$
ATP	$> 10^{-3}$
ADP	$>> 10^{-3}$
AMP	$>> 10^{-3}$
CTP	$1.2 \cdot 10^{-5}$
CDP	$>> 10^{-3}$
CMP	$>> 10^{-3}$

The binding of [^{3}H]GDP on carrot microsomes showed a similar nucleotide specificity (results not shown). Scatchard analysis of binding experiments with [^{3}H]GDP revealed a dissociation constant, $K_d \approx 1\mu M$ and a capacity of n≈ 100pmol mg^{-1} membrane protein (result not shown). The corresponding analysis for the binding of [^{35}S]GTPγS gave a lower $K_d \approx 722nM$, but a similar capacity of n≈ 122pmol mg^{-1} membrane protein (Fig. 4), indicating the same magnitude of binding sites for either G-nucleotide.

The detection of high-affinity binding sites for G-nucleotides is, *per se*, no clear indication that these sites are involved in the signal transduction process. It is also necessary to demonstrate signal-dependent stimulation of either the high-affinity GTP binding, the GDP/GTP exchange reaction or the intrinsic GTPase reaction (Gilman, 1987). Consequently, we have performed experiments to determine whether IAA can stimulate the binding of [^{35}S]GTPγS to carrot microsomes. The Scatchard analysis of the binding data indicate a hormone-mediated increase of the binding capacity without any significant change of the affinity for the ligand (Fig. 4). The magnitude of the IAA-mediated stimulation is comparable with results found for the hormone-stimulated GTP-binding on animal cell membranes (Avissar et al.,1988).

We found that, without preincubation of membranes with G-nucleotides, the auxin stimulation was difficult to reproduce. However, if the microsomes were first preincubated with GDP in submicromolar amounts, the IAA effect on the high-affinity binding was stabilized. Kinetic analysis revealed that the hormone-mediated stimulation of the high-affinity binding of [^{35}S]GTPγS is strongest at incubation periods of less than 1min, and it decreases progressively with time. The kinetics of the hormone effect on high-affinity binding of [^{35}S]GTPγS corresponds well with that of the auxin effect on the PI response which was previously found to be most pronounced at less than 1min (Zbell and Walter,1987; Zbell and Walter-Back,1988).

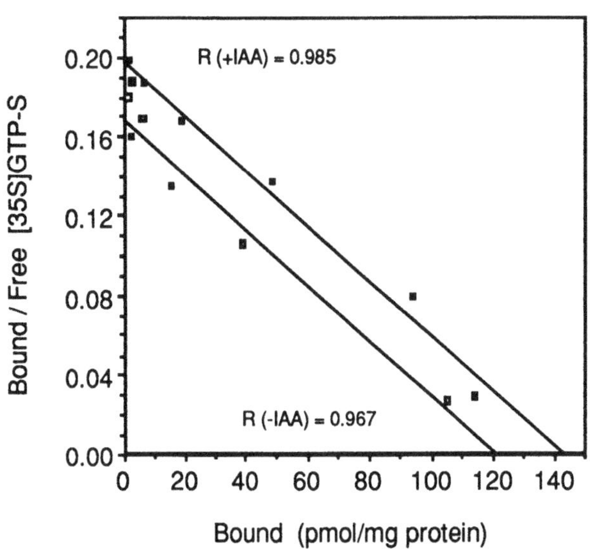

Fig. 4. Scatchard plot of the high-affinity binding of [^{35}S]GTPγS on carrot microsomes in the absence and presence of IAA. The assays were incubated for 20min at +25°C with increasing concentrations of the unlabeled ligand using the standard conditions.

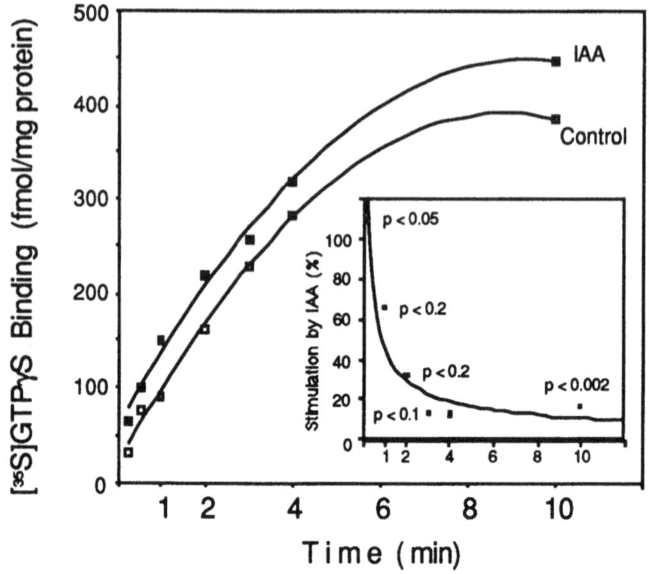

Fig. 5. Kinetics of [^{35}S]GTPγS-binding stimulated by IAA. Carrot microsomes were preincubated for 30min with 10 nM GDP, before the binding of [^{35}S]GTPγS in the absence or presence of 1 μM IAA was started for the incubation periods indicated. The relative magnitude of the auxin effect is plotted against the incubation time (insert).

CONCLUSIONS

The results presented here favour our previous hypothesis that a GTP-binding protein might be involved in the auxin-mediated PI response of carrot microsomal membranes (Zbell et al.,1989). It is also evident that microsomal membranes from carrot suspension cells are a suitable experimental system for a detailed analysis of the processes involved in the transmembrane signalling of auxin. For such experiments, one important advantage is that these membranes retain functional coupling between the putative auxin receptor, the GTP-binding protein, and the phosphoinositidase C even after membrane isolation. Future research will focus on the subcellular localization of the transduction system as well as on the detection of the GTP-binding moiety by the use of specific antibodies which are raised against the common GTP-binding domain of animal G proteins.

ACKNOWLEDGEMENT

The financial support of the Deutsche Forschungsgemeinschaft (DFG) is appreciated. We are also thankful to Prof. Dr. K.H. Jacobs and Dr. P. Gierschik (Institute of Pharmacology, University of Heidelberg) for their advice. Moreover, we express our thanks to Dr. D.O. Perdue (Boyce Thompson Institute for Plant Research at Cornell University, Ithaca) for reading and styling the manuscript.

LITERATURE

Aub DL, Gosse ME, Cote TE (1987) Regulation of thyrotropin-releasing hormone receptor binding and phospholipase C activation by a single GTP-binding protein. J Biol Chem 262: 9521-9528

Avissar S, Schreiber G, Danon A, Belmaker RH (1988) Lithium inhibits adrenergic and cholinergic increases in GTP binding in rat cortex. Nature 331: 440-442

Berridge MJ (1987) Inositol trisphosphate and diacylglycerol: two interacting second messengers. Ann Rev Biochem 56: 159-193

Blum W, Hinsch K-D, Schultz G, Weiler EW (1988) Identification of GTP-binding proteins in the plasma membrane of higher plants. Biochem Biophys Res Commun 156: 954-959

Casey PJ, Gilman AG (1988) G protein involvement in receptor-effector coupling. J Biol Chem 263: 2577-2580

Cockcroft S, Taylor JA (1987) Fluoroaluminates mimic guanosine 5'-[γ-thio]triphosphate in activating the polyphosphoinositide phosphodiesterase of hepatocyte membranes. Role for the guanine nucleotide regulatory protein G_p in signal transduction. Biochem J 241: 409-414

Cockcroft S, Howell TW, Gomperts BD (1987) Two G-proteins act in series to control stimulus-secretion coupling in mast cells: use of neomycin to distinguish between G-proteins controlling polyphosphoinositide phosphodiesterase and exocytosis. J Cell Biol 105: 2745-2750

Dillenschneider M, Hetherington A, Graziana A, Alibert G, Berta P, Haiech J, Ranjeva R (1986) The formation of inositol phosphate derivatives by isolated membranes from *Acer pseudoplatanus* is stimulated by guanine nucleotides. FEBS Lett 208: 413-417

Dohlman HG, Caron MC, Lefkowitz RJ (1987) A family of receptors coupled to guanine nucleotide regulatory proteins. Biochemistry 26: 2657-2664

Drøbak BK, Allan EF, Comerford JG, Roberts K, Dawson AP (1988) Presence of a guanine nucleotide-binding protein in a plant hypocotyl microsomal fraction. Biochem Biophys Res Commun 150: 899-903

Ettlinger C, Lehle L (1988) Auxin induces rapid changes in phosphatidylinositol metabolites. Nature 331: 176-178

Gilman AG(1987) G proteins: transducers of receptor-generated signals. Ann Rev Biochem 56: 615-649

Guern J (1987) Regulation from within: the hormone dilemma. Ann Bot 60 (Suppl 4): 75-102

Harden TK, Hawkins PT, Stephens L, Boyer JL, Downes CP (1988) Phosphoinositide hydrolysis by guanosine 5'-[γ-thio]triphosphate-activated phospholipase C of turkey erythrocyte membranes. Biochem J 252: 583-593

Hasunuma K, Furukawa K, Tomita K, Mukai C, Nakamura T (1987) GTP-binding proteins in etiolated epicotyls of *Pisum sativum* (Alaska) seedlings. Biochem Biophys Res Commun 148: 133-139

Hucho F (1986) The nicotinic acetylcholine receptor and its ion channel. Eur J Biochem 158: 211-226

Irvine RF, Letcher AJ, Lander DJ, Drøbak BK, Dawson AP, Musgrave A (1989) Phosphatidylinositol (4,5)bisphosphate and phosphatidylinositol (4)monophosphate in plant tissues . Plant Physiol 89: 888-892

Jacobs M, Thelen MP, Farndale RW, Astle MC, Rubery PH (1988) Specific guanine nucleotide binding by membranes from *Cucurbita pepo* seedlings. Biochim Biophys Res Commun 155: 1478-1484

Levitzki A (1987) Regulation of adenylate cyclase by hormones and G-proteins. FEBS Lett 211: 113-118

McMurray WC, Irvine RF (1988) Phosphatidylinositol 4,5-bisphosphate phosphodiesterase in higher plants. Biochem J 249: 877-881

Melin P-M, Sommarin M, Sandelius AS, Jergil B (1987) Identification of Ca^{2+}-stimulated polyphosphoinositide phospholipase C in isolated plant plasma membranes. FEBS Lett 223: 87-91

Morse MJ, Crain RC, Satter RL (1987) Light-stimulated inositolphospholipid turnover in *Samanea saman* leaf pulvini. Proc Natl Acad Sci USA 84: 7075-7078

Morse MJ, Crain RC, Coté GG, Satter RL (1989) Light-stimulated inositol phospholipid turnover in *Samanea saman* pulvini. Increased levels of diacylglcerol. Plant Physiol 89: 724-727

Ohtsuki K, Yokoyama M, Uesaka H (1987) Physiological correlation between nucleoside-diphosphate kinases and the 21-kDa guanine-nucleotide binding proteins copurified with the enzymes from the cell membrane fractions of Ehrlich ascites tumor cells. Biochim Biophys Acta 929: 231-238

Sommarin M, Sandelius AS (1988) Phosphatidylinositol and phosphatidylinositolphosphate kinases in plant plasma membranes. Biochim Biophys Acta 958: 268-278

Wheeler J J, Boss WF (1987) Polyphosphoinositides are present in plasma membranes isolated from fusogenic carrot cells. Plant Physiol 85: 389-392

Yarden Y, Ullrich A (1988) Molecular analysis of signal transduction by growth factors. Biochemistry 27: 3113-3119

Zbell B, Walter C (1987) About the search for the molecular action of high-affinity auxin binding sites on membrane-localized rapid phosphoinositide metabolism in plant cells. In: NATO ASI Series, Vol. H10, Plant Hormone Receptors (Klämbt D ed), Springer Berlin Heidelberg New York, pp 141-153

Zbell B, Walter-Back C (1988) Signal transduction of auxin on isolated plant cell membranes: Indications for a rapid polyphosphoinositide response stimulated by indoleacetic acid. J Plant Physiol 133: 353-360

Zbell B, Schwendemann I, Bopp M (1989) High-affinity GTP-binding on microsomal membranes prepared from moss protonema of *Funaria hygrometrica*. J Plant Physiol (in press)

Zbell B, Walter-Back C, Hohenadel H, Schwendemann I (1989) Polyphosphoinositide turnover and signal transduction of auxin on isolated membranes of *Daucus carota* L In: Plant Growth Substances 1988. Proceedings of the 13th International Conference on Plant Growth Substances (Pharis RP, Rood SB eds) Springer Verlag Berlin Heidelberg New York (in press)

DISCUSSION

Downes: I could start with asking you about the experiments where you show auxin stimulation of inositol phosphate generation and also GTPτS effects. Did you do the other experiments which is to look for auxin potentiation of the response to GTPτS.

Zbell: We have obtained data on dose response curves which demonstrate that you get a shift in the auxin dose response to the left. So you need a lower hormone concentration in the presence of GTP. But we haven't demonstrated that for a full range of concentration for auxin. So we get a higher potency at lower concentrations but what is missing is the right part of the dose response curve. So I haven't included the data in my presentation.

Downes: Did you look at the kinetics. That would be the most appropriate thing to look at in this kind of experiments.

Zbell: Time dependence?

Downes: Time dependence of accumulation of inositol phosphates in the presence and absence of hormones.

Zbell: If you have no hormones then you get a very low level.

Downes: Yes, but I mean by looking for an effect of auxin on the response to GTPτS.

Zbell: No, we have only used 30 seconds as a time period.

Corbin: I like to ask is there evidence for the cAMP cascade being important in plants?

Zbell: No, I don't think so. As yet there is no evidence that there is a cAMP cascade in plants. What you can find is cAMP as a metabolite in plant cells. But as yet there is no evidence that there is signal dependence accumulation of cAMP or signal dependent stimulation of adenylate cyclase. There is no evidence for a cAMP dependent protein kinase. So there are some things missing. This problem has not been investigated for the last years, because 10 years ago there was a review in Annual Review of Plant Physiology denying a physiological role of cAMP in plant cells.
Caron: Can I ask you what the structure of 2.4D looks like and a follow up of that is wouldn't it be a better ligand to look at your receptors because as you showed from the plant response it is several orders of magnitude more potent than the compound you were using.
Zbell: 2.4 dichlorophenoxyacetic acid is a compound which is similar to IAA This is a synthetic compound which is used as a standard herbicide. We have started our investigation with indoleacetic acid as the natural compound first. What we know at the moment is that the binding of 2.4D to the auxin binding site is also one order of affinity higher than IAA. The advantage of 2.4D to IAA is that it can not be degraded by enzymes. So it is a much stronger hormone than IAA. What would be very interesting for receptor identification will be to get a photoaffinity agent of 2.4D, but I don't know if it yet exists.
Devreotes: You showed the binding to carrot microsomes. Can you say what is in there? Is it plasma membrane, is it ER?
Zbell: The microsomes contain all membrane types except for mitochondrial and chloroplast membranes because the membranes are prepared from non-photosynthetic cells which have no chlorophyll and only proplastids. We perform the preparation by a first 2000 g centrifugation to remove mitochondria and cell nuclei a.s.o. Then the supernatant was centrifuged at 150000 g to yield the microsomes. In earlier times we have always used purification steps using a sucrose gradient system or Renografin gradient system. But we have found that we loose some activity and so we have decided to use this short preparation method. Now we have started also with phase partitioning and found GTP binding at purified plasma membranes. I know from other laboratories that GTP binding is in the plasma membranes as well as endomembranes. And it must be looked in the future what is the difference between GTP binding sites in the plasma membrane and endomembranes, and which have any function in the signal transduction pathway.
Mulle: I would just like to know if IP3 is involved in calcium signaling in plant cells.
Zbell: Yes, there exist several reports. We have also tested our system by loading the microsomes with ^{45}Ca; you can observe a release of calcium by IP3.
Mulle: It is released from what organelle?
Zbell: Microsomal vesicles. Other people have used isolated vacuoles or tonoplast vesicles; they have clearly demonstrated that commercially available IP3 can release ^{45}Ca. But what is not done at the moment is the characterization of the IP3 binding sites, as it was performed in animal cells. At the moment the data indicate that not the endoplasmatic reticulum

but the tonoplast, i.e.the vacuolar membrane, is the site of IP3 action.

Houslay: Do you know whether there are any proteins in your membranes which interact with specific antipeptide antibodies for G proteins, mammalian G proteins.

Zbell: We have looked in cooperation with Peter Gierschik,and have used antibodies against α subunits of Gs, Gi and transducin but we got no specific cross reactivity. What is perhaps a useful tool is to look for cross reaction with the α common antibody. This strategy was successfully used by 2 groups, and GTP binding proteins were detected in several plant species and their molecular weights are always in the same range, it looks like a 45-50 kD protein.

Snaar-Jagalska: Is GTPτS binding magnesium dependent?

Zbell: We have investigated the cation requirement and found a very low requirement of magnesium. It is below 5-1 mM.

Ion conductance

REGULATION OF THE DESENSITIZATION OF THE ACETYLCHOLINE NICOTINIC RECEPTOR

C.MULLE, P.BENOIT, F.REVAH AND J.P.CHANGEUX

URA CNRS 0210, Laboratoire de Neurobiologie Moléculaire, Département des biotechnologies, Institut Pasteur, 75015 Paris France

Among the several mechanisms which may potentially control synapse efficacy, several prominent ones have been assigned to the functional properties of receptors for neurotransmitters. Up to now, two types of receptors for neurotransmitters have been characterized : the ligand-gated channels are involved in fast synaptic transmission and the receptors coupled to a G-protein which indirectly regulate the opening ionic channels are involved in slow synaptic transmission.

The best characterized example of ligand-gated channels is the acetylcholine nicotinic receptor (AchR) of fish electric organ and vertebrate skeletal muscle (see Changeux et al 1984 for review). The AchR is a glycoprotein composed of four different polypeptide chains assembled into an heterologous transmembrane pentamer. Reconstitution experiments have shown that the receptor molecule contains the ionic channel and all the structural elements responsible for the regulation of its opening.

The AchR undergoes several distinct categories of allosteric transitions. The fast opening and closing of the ion channel leads to a large increase in the permeability of post-synaptic membranes to cations which is responsible for the transmission of the signal at the level of the synapse. In addition, the neurotransmitter acetylcholine and a variety of allosteric effectors (non competitive blockers, fatty acids, Ca++, ...) elicit slower reversible transitions towards states in which the AchR becomes

NATO ASI Series, Vol. H 44
Activation and Desensitization of Transducing Pathways
Edited by T. M. Konijn, M. D. Houslay, P. J. M. Van Haastert

refractory to activation by agonist and are refered to as "desensitized states". Desensitization has been resolved into at least two processes, both by rapid kinetic analysis (Heidmann and Changeux, 1979) and by electrophysiological methods (Sakmann et al., 1980 ; Feltz and Trautmann, 1982) : a "fast phase" at a rate in the range of 2-75 sec-1 and a "slow phase" at a rate in the range of 0.01-0.1 sec-1. The different conformations differ by their affinities for the ligands that bind reversibly to the various categories of sites carried by the receptor molecule (Heidmann et al., 1983). Finally the allosteric properties of the AchR can be modified by covalent modifications such as phosphorylation (Huganir et al., 1986, Hopfield et al., 1988).

The AchR can thus be taken as an example of a transmembrane allosteric protein capable of integrating categories of pharmacological signals. Based on these properties, Changeux and Heidmann (1987) have proposed a simple model that could account for short-term regulation of synapse efficacy. The aim of our experiments was thus to analyse physiological situations in which the efficacy of AchR response could be regulated in an allosteric manner by endogenous effectors of peptidic nature.

Regulation of the AchR desensitization by thymopoietin

In a search for endogenous substances that may regulate AchR desensitization, we studied the effects of the thymic hormone thymopoïetin (TP). TP is a polypeptide consisting of 49 amino acid residues which has been isolated from the bovine thymus (Goldstein, 1974). Apart from its role in the maturation and the function of the immune system, Tp produces definite effects on neuromuscular transmission. Indeed, thymic extracts contain a factor responsible for the impairment of neuromuscular transmission, with increased fatigability at nerve stimulation of 50Hz. This assay led to the isolation of thymopoietin as a factor responsible for the effect (Goldstein, 1974).

With *Torpedo Californica* AchR-rich membrane fragments, Tp at high concentration (>100nM) behaves as a competitive antagonist in a Ca independent manner. It interacts with the α-toxin binding site (Venkatasubramanian, 1986). We have demonstrated that a second type of effect occurs at significantly lower concentrations of Tp (Revah et al., 1987).

Tp alters the kinetics of Ach-activated single channel currents from mouse C2 myotubes :

Single channel recording with C2 myotubes in the cell-attached mode reveals that at low concentration (fig 1), Tp (2 to 20nM) causes the appearance of long closed periods with a mean duration of several hundreds of ms as illustrated by the presence of an additional slow component in closed time distribution. In adition TP progressively reduces

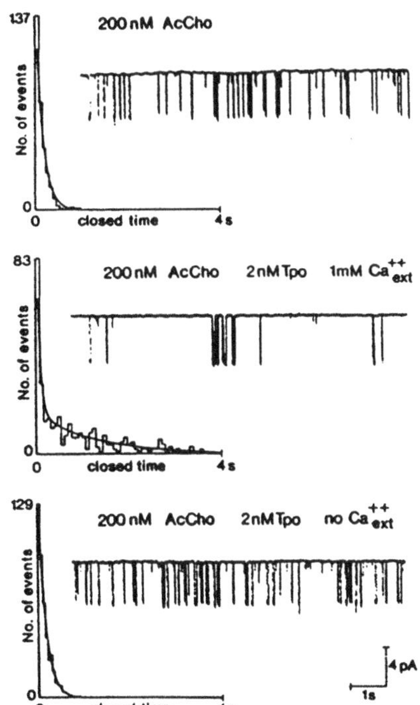

Fig1: Cell-attached recordings of single channel-currents activated by 200nM Ach with and without TP (pipette potential : +100mV) and corresponding distributions of channel closed times. These experiments were performed on mouse C2 myotubes in culture (taken from Revah et al, 1987).

the frequency of channel opening by more than 90%, with no change in mean open time nor in elementary conductance. Such long closed times are reminiscent of the pattern of distribution of channel opening in the presence of desensitizing concentrations of Ach (>5μM) (Patlak et al 1981). A possible interpretation of this effect is thus that Tp favors the desensitization of the AchR.

Cell-attached recordings performed in a Ca^{++}-free extracellular medium failed to demonstrate any effect of low Tp concentrations on AchR desensitization, as illustrated in fig. 1A3 where the distribution of closed times can be fitted by a single exponential component, and only moderate effects were observed for higher concentrations of Tp (fig1B).

The AchR contains a non selective cationic channel which is in particular permeant to Ca^{++}. The role of Ca^{++} was further addressed in outside-out experiments. Rapid chelation of Ca^{++} entry by high concentrations of EGTA (10mM) prevented the occurence of long closed times. However, Tp was still effecive if Ca^{++} that entered through the opening of the AchR ionic channel was allowed to temporarly increase by the use of low concentrations of EGTA (0.2mM). It thus appears that the conjunction of Tp and Ca^{++} on opposite faces of the membrane is required for the effect of Tp on AchR desensitization.

Biochemical studies with *Torpedo marmorata* AchR-rich membrane fragments confirm that Tp increases desensitization of the AchR.

Experiments performed at equilibrium with AchR from *T. marmorata* have shown that in the presence of Ca^{++}, Tp at low concentration enhances the binding of (^{3}H)acetylcholine and causes a decrease in the dissociation constant for the non competitive blocker (^{3}H)phencyclidine (fig.2). Both ligands have a higher affinity for the AchR in its desensitized conformations than in its resting state. Tp was also shown to alter the rapid kinetics of the interaction of the fluorescent agonist Dns-C6-Choline with AchR-rich membranes, by increasing the relative amplitude of the rapid fluorescent signal (Revah et al., 1987). From these two series of data it was concluded that Tp displaces the conformational equilibrium towards a high affinity desensitized state and increased the transition rate towards the same state.

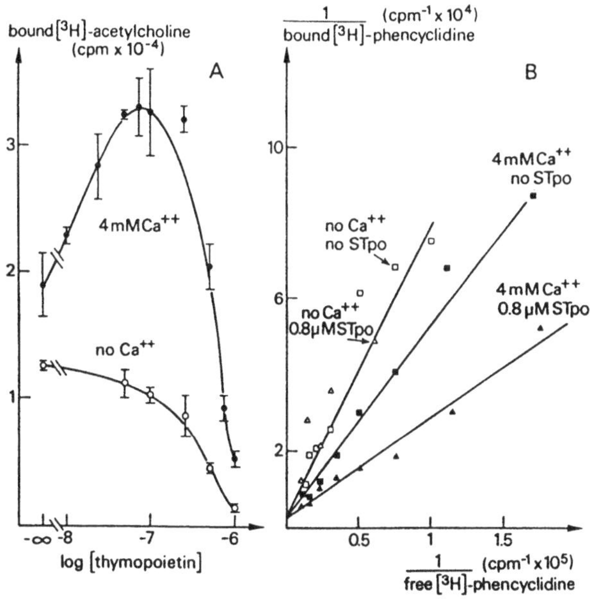

Fig2 : Effect of native thymopoïetin on acetylcholine and (3H) phencyclidine binding to AchR on a preparation of AchR-rich membrane fragments purified from Torpedo Marmorata electric organ (taken from Revah et al, 1987).

The nature of the site(s) at which Tp enhances desensitization is still unknown. These experiments still do not prove that Tp interacts directly with the AchR as an allosteric effector. Yet, at least some of them were performed under conditions (outside-out recordings, membrane preparations) where no second messenger mediated mechanism or ATP-dependent covalent modifications could be expected.

A role for Tp in myasthenia gravis ?

The present results may account for the physiological data suggesting that a hormone secreted by the thymus affects cholinergic transmission at the neuromuscular junction. The pathogenesis of neuromuscular disorders in myasthenia gravis yet is not clear. The presence of autoimmune antibodies directed against the AchR is thought to be the major cause of muscular dysfunction. However the lack of systematic correlation between the level of antibodies and the severity of neuromuscular impairment suggest that these autoantibodies do no represent the only pathogenic mechanism. For instance myasthenia gravis is often associated with an autoimmune thymitis that could result in the elevation of circulating Tp. The dysfunction of cholinergic neuromuscular transmission in patients with myasthenia gravis could thus be better explained by the conjunction of two processes, (1) the loss of available AchR by autoimmune mechanism which will cause a decrease in well-known "safety factor" at the neuromuscular junction and, (2) the increased concentration of Tp which will increase the desensitization of AchR.

Regulation of AchR desensitization by calcitonin gene related peptide (CGRP)

Apart from direct reversible effects of allosteric effectors on the equilibrium and kinetics of interconversion between various conformations of the AchR, the question can be addressed as to whether covalent modifications of the AchR may affect its allosteric properties and in particular its desensitization.

The role of phosphorylation is rather well established for the regulation of voltage dependent channels (Levitan, 1988), although the structural and biochemical basis for this regulation is just starting to be unraveled. The nicotinic receptor from Torpedo has been known for some time to be phosphorylated by various endogenous protein kinases (cAMP dependent protein kinase, protein kinase C, and tyrosine kinases) (see Miles and Huganir 1988 for review). More recently, the AchR from rat and mouse muscle was also shown to be phosphorylated in vivo by agents that elevate cAMP and activate pKA (Miles et al, 1987 ; Smith et al, 1987) .

However the physiological significance of this phosphorylation is poorly understood. Huganir and his collegues have compared the properties of phosphorylated versus non-phosphorylated Torpedo AchR reconstituted into vesicles and have shown in ion flux experiments (Huganir et al., 1986) or with single-channel recording techniques that both AchR phosphorylated in vitro by pKA or tyrosine kinase (Hopfield et al., 1988) displayed an enhanced rate of desensitization. Similarly, the rate of desensitization of the muscle AchR was shown to be increased by agents which raise the intracellular level of cAMP, such as forskolin (Middleton et al., 1988). The later effect is however debated because of a possible direct blocking action of forskolin (at concentrations higher than 10µM) on the AchR (Wagonner and Palotta, 1988). Nevertheless, it was of interest to examine the consequences of activation of the adenylate cyclase pathway in muscle cells by an endogenous signal, calcitonin-gene related peptide (CGRP).

CGRP has been identified in the spinal cord of several vertebrate species (Hökfelt et al., 1986) and in the motor nerve endings, especially at the rodent neuromuscular junction. This peptide was shown to stimulate the biosynthesis of the AchR (Fontaine et al., 1986 ; New and Mudge, 1986). Furthermore, CGRP increases cAMP synthesis in primary cultures of chick myotubes (Laufer et al., 1987). We thus examined the effects of CGRP on

AchR desensitization in a mouse cell line derived from mouse soleus muscle (Mulle et al, 1988).

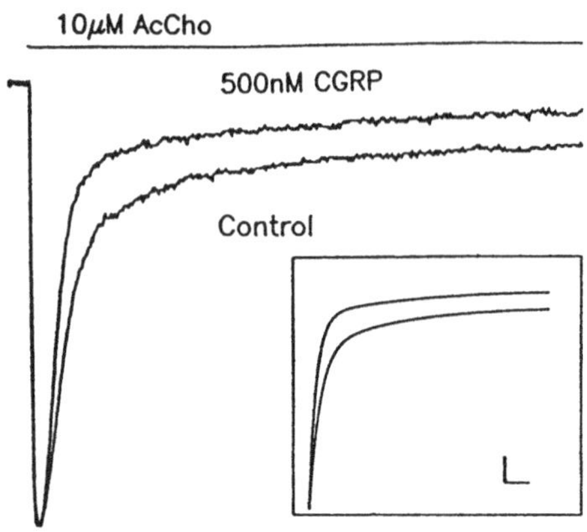

Fig3 : Enhancement of desensitization rate of AchR by CGRP. Macroscopic currents were recorded from muscle cells voltage-clamped at -60mV. Current decay is described by the sum of two exponential curves (shown in inset). CGRP was incubated for 5 minutes (taken from Mulle et al, 1988).

Recording of macroscopic currents using the whole-cell variation of the patch-clamp technique on small cells from the muscle cell line Sol8, showed that CGRP produced a progressive and reversible enhancement of the rapid decay phase of desensitization of the AchR (fig. 3). At a concentration of 500nM, the time constant for the fast phase decreased by 28%, and the relative amplitude of the fast componant increased by 72%. Single channel data further indicated that this effect was not accompanied by any decrease of mean open time nor of unitary current amplitude. However, CGRP affected the frequency of channel openings in cell-attached experiments, under conditions where the patch pipette physically isolated the recorded AchR from the CGRP-containing extracellular medium. This result indicates that CGRP increases desensitization of the AchR by an indirect second-messenger mediated mechanism.

Even though evidence is still incomplete our results support the conclusion that the second-messenger involved is cAMP (fig.4) : (1) CGRP stimulates accumulation of cAMP in mouse Sol8 muscle cells, (2) the effects of CGRP were mimicked, in whole-cell recordings, by addition of cAMP (1mM) in the patch-pipette, (3) this latter procedure prevented any additional effect of CGRP on the desensitization of the AchR.

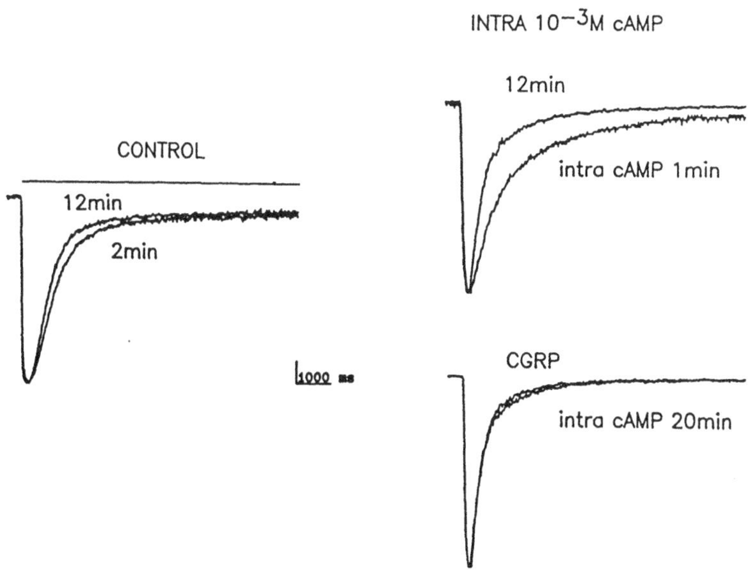

Fig4 : Possible involvment of cAMP in the effect of CGRP on the desensitization of the AchR. In A, the intracellular pipette medium contained ATP (1mM). In B, it contained ATP (1mM) and cAMP (1mM).

The contribution of other indirect mechanism cannot be excluded at this stage. For instance activation of protein kinase C might regulate AchR function (see Miles and Huganir 1988), but this effect is expected to depend on the level of calcium. We checked that the effect of CGRP was not dependent on the levels of internal Ca^{++}. Recently , Miles and collegues (1989) have demonstrated that CGRP stimulates phosphorylation of the AchR in rat primary myotubes in a manner comparable to that caused by forskolin. This finding strongly supports the idea that phosphorylation of the AchR is responsible for the effects of CGRP on the desensitization of the AchR. In addition CGRP, via a cAMP pathway, may play a long-term regulatory role in the development of cholinergic synapses CGRP, as a neuropeptide coexisting with Ach in motor nerve endings, may thus contribute to both short- and long-term regulation of synapse properties.

CONCLUSIONS

Desensitization of the nicotinic AchR is an example of a very general feature of receptors for neurotransmitters. The role of desensitization of the muscle AchR is not clear. There is consensus that in the normal adult muscle, desensitization does not play a major role under physiological conditions of nerve activity, due to the rapid hydrolysis of Ach in the synaptic cleft by the enzyme acetylcholinesrease (Magleby and Palotta, 1981). Neverthless, there are increasing reasons to think that desensitization could be operative at other synapses involving ligand-gated channel receptors (see for instance Tang et al., 1989, in the case of glutamatergic synapses). The nicotinic AchR is a representative member of the superfamilly of these ligand-gated ion channel receptors and serves as an instructive model for understanding the behavior of these other receptors. The data presented here show that desensitization of the AchR can be modulated either by direct interaction with the AchR or by indirect mechanisms involving a covalent modification of the AchR. Both mechanisms have been identified with ligand-gated channels that operate at central nervous system synapses. It will be important to test their physiological significance in the context of short-term synaptic plasticity.

REFERENCES

Changeux J-P, Heidmann T (1987) Allosteric receptors and molecular models of learning In : Synaptic Function (Edelman G M , Gall W E, Cowan W M Eds.) John Wiley and Sons, New York, 549-601

Changeux J-P, Devillers-Thiery A, Chemouilli P (1984) Science 225: 1335-1345

Goldstein G (1974) Isolation of bovine thymin : a polypeptide hormone of the thymus Nature 247 : 11-14

Feltz A, Trautmann A. (1982) Desensitization at the frog neuromuscular junction : a biphasic process J. Physiol. 322 : 257-272

Fontaine B., Klarsfeld A., Hokfelt T., Changeux J-P A peptide present in the spinal cord motorneurons increases the number of acetylcholine receptors in primary cultures of chick embryo myotubes Neurosci. Lett. 71 : 59-65

Heidmann T, Changeux J-P (1979) Fast kinetic studies on the interactions of a fluorescent agonist with the membrane-bound acetylcholine receptor from Torpedo Marmorata Eur. J. Biochem. 94: 255-279

Heidmann T, Oswald RE, Changeux J-P (1983) Multiple sites of action for noncompetitive blockers revealed by (3H) phencyclidine binding to AchR-rich membrane fragments from Torpedo marmorata Biochemistry 22: 3112-3127

Laufer R Changeux J-P (1987) Calcitonin related gene related peptide increases cAMP synthesis in chick embryo myotubes EMBO J 6 : 901-906

Levitan IB (1988) Modulation of ion channels in neurons and other cells Ann Rev Neurosci 11 : 119-136

Magleby KL, Palotta BS (1981) A study of desensitization of acetylcholine receptors using nerve-released transmitters in the frog J Physiol 316: 225-250

Middleton P, Rubin LL, Schuetze SM (1988) Modulation of acetylcholin recptor desensitization in rat myotubes J Neurosci 8 : 3405-3412

Miles K, Anthony DT, Rubin LL, Greengard P, Huganir RL (1987) Regulation of nicotinic acetylcholine receptor phosphorylation in rat myotubes by forskolin and cAMP Proc Natl Acad Sci USA 84: 6591-6595

Miles K, Huganir RL (1988) Regulation of nicotinic acetylcholine receptor by protein phosphorylation Molecular Neurobiology 2: 91-124

Miles K, Greengard P, Huganir RL (1989) (to be published) Calcitonin gene-related peptide regulates phosphorylation of the nicotinic acetylcholine receptor in rat myotubes Neuron

Mulle C, Benoit P, Pinset C, Roa M, Changeux J-P (1988) Calcitonin gene-related peptide enhances the rate of desensitization of the nicotinic acetylcholine receptor in cultured mouse muscle cells Proc

Natl Acad Sci USA 85: 5728-5732

New H V and Mudge A W Calcitonin gene related peptide regulates muscle acetylcholine recptor synthesis Nature 323 : 809-811

Revah F, Mulle C, Pinset C, Audhya T, Goldstein G, Changeux J-P (1987) Calcium-dependent effect of the thymic polypeptide thymopoïetin on the desensitization of the nicotinic acetylcholine receptor Proc Natl Acad Sci USA 84: 3477-3481

Sakmann B, Patlak J, Neher E (1980) Single acetylcholine activated channels show burst kinetics in presence of desensitizing concentrations of agonist Nature 286: 71-73

Smith MM, Merlie JP, Lawrence JC (1987) Regulation of phosphorylation of nicotinic acetylcholine receptors in mouse BC3H1 myocytes Proc Natl Acad Sci USA 84 6601-6605

Takami K, Kawai Y, Shiosaka S, Lee Y, Girgis S, Hillyard CJ, MacIntyre I, Emson PC, Toyama M Immunohistochemical evidence for the coexistence of calcitonin gene related peptide and choline acetyltransferase -like immunoreactivity in neurons of the rat hypoglossal Brain Res. 328 : 386-389

Tang C-M, Dichter M, Morad M. (1989) Quisqualate activates a rapidly inactivating high conductance ionic channel in hippocampal neurons Science 243: 1474-1477

Venkatasubramanian K, Audhya T, Goldstein G (1986) Binding of thymopoïetin to the acetylcholine receptor Proc Natl Acad Sci USA 83 3171-3174

Wagoner PK, Pallotta BS (1988) Modulation of acetylcholine receptor desensitization by forskolin is independent of cAMP Science: 240: 1655-1657

DISCUSSION

Clark: Are these cAMP kinase sites on the subunits different?

Mulle: About the phosphorylation of the α-subunit, it has been said in Huganir's and Greengard's laboratory, that the phosphorylation of the α-subunit could be by an indirect mechanism, which would not involve phosphorylation of the receptor directly by protein kinase A.

Cooke: In the experiments where you describe the addition of EGTA to the inside of the cell through your patch clamp, you are adding 5 mM EGTA in order to inhibit your response. Is that necessary to add so much, what is the intracellular concentration of calcium, is it not in the nanomolar range?

Mulle: When we perform whole cell-experiments, we dialyze all intracellular medium, so we wash out other things, and especially you change the concentration of calcium. Normally you work with 10 mM EGTA so as to reach a concentration of calcium around 10^{-8}M, but Ca^{++} is buffered at this level throughout the experiment. If you want to see any calcium dependent effect, for instance calcium dependent potassium channels, you must not add so much EGTA, because in this case you do not see anything. If you only add 0.2 or 0.5 mM EGTA, then you can see the activation of this calcium dependent potassium channel, which is a good proof that there is a transient increase in the calcium level inside the cell.

Cooke: Wouldn't that 0.2 mM still chelate all your calcium?

Mulle: Yes it does chelate. The question is the rapidity with which it is chelated, the rate of chelation.

Corbin: I have a question on a related point. I do not understand completely your patch and how it works, but you are adding millimolar cAMP and that seems like a very high concentration.

Mulle: Possibly yes. The problem is that you have problems with phosphodiesterases, if you put less than that. This is a usual concentration, well 500 µmolar is a usual concentration for the study of cAMP-dependent effects when added inside a cell. I could use IBMX for instance, but the problem is that IBMX has a direct effect on the receptor.

Dumont: Maybe I missed it, but didn't you say that CGRP had a trophic effect on the ACh-receptor?

Mulle: Yes

Dumont: You increase the number of receptors?

Mulle: Yes.

Dumont: Is protein synthesis involved?

Mulle: Yes.

Dumont: Is it just the receptor, or a lot of proteins are induced?

Mulle: I think that just the nicotinic receptor has been looked at.

Dumont: So it is not a general trophic effect, it is just a specific effect on this receptor?

Mulle: Well, I do not know if it is so specific, but the only protein the people have looked at is the nicotinic receptor, which is a good indication of the maturation of the muscle cell. But I suppose that in muscle you have other effects. At least activation of adenylate cyclase has. In fact it has also been shown by Rolf Laufer that CGRP could also increase IP3 through a cAMP dependent mechanism. He suggested that an increase of cAMP inside the cell will lead to phosphorylation of calcium channels and a subsequent entry of calcium through the calcium channels which are active when phosphorylated, and this increase could activate phospholipase C.

Dumont: Is the nicotinic receptor gene controlled by cAMP, is the gene known and the promotor of the gene?

Mulle: The cAMP regulatory sequence is not known.

Dumont: The nicotinic receptor gene is known and it's promotor. And in this promotor, is there any cAMP responsive element?

Mulle: If there is a responsive element, it has not been identified.

Cellular systems

THE CONTROL OF HUMAN THYROID CELL FUNCTION, PROLIFERATION AND DIFFERENTIATION

S. Reuse, C. Maenhaut, A. Lefort, F. Libert, M. Parmentier, E. Raspé, P. Roger, B. Corvilain, E. Laurent, J. Mockel, F. Lamy, J. Van Sande, G. Vassart and J.E. Dumont.

Institute of Interdisciplinary Research (IRIBHN),
Free University of Brussels, School of Medicine,
Campus Erasme,
B - 1070 Brussels,
Belgium.

INTRODUCTION

The study of thyroid regulation at the cellular level is the main interest of our laboratory since many years. The complex picture emerging from these studies leads to conclusions of general relevance. The regulation of the thyroid cell was once a classical example of the concept one hormone - one cell type - one intracellular secondary messenger with its pleiotypic effects. It should now rather be considered as a network of crosslinked regulatory steps where the extracellular and intracellular signal-molecules act on their receptors as bits of information in an electronic circuit, i.e., express on/off regulations with no definite general physiological meaning per se. Such networks differ from one cell type to another and for a given cell type from one species to another. In the case of the thyroid, many apparent discrepancies in the literature are explained if this is taken into account. In this presentation, we wish to draw mainly on the results of our group to illustrate this point with regard to the regulation of function, proliferation and differentiation of the thyroid cell.

I. THE CASCADES INVOLVED IN THYROID CELL REGULATION

Our main interest is the regulation of the human thyrocytes.

NATO ASI Series, Vol. H 44
Activation and Desensitization of Transducing Pathways
Edited by T. M. Konijn, M. D. Houslay, P. J. M. Van Haastert

However as dog tissue is still easier to obtain than human tissue, our main experimental object is the dog thyrocyte. We therefore develop our methods and concepts on the dog thyroid, to adapt them to the human tissue thereafter. In fact, apart from a few differences that I shall mention, the two tissues are similar. In the dog thyroid, the main and best known circuit involves thyrotropin (TSH) stimulation of plasma membrane adenylate cyclase (Dumont et al., 1981; Lamy et al., 1987). Cyclic AMP produced by the cyclase is the intracellular signal molecule, which, by activating cAMP-dependent protein kinases, will enhance the main functions of the gland: the iodination of thyroglobulin and iodothyronine formation, i.e., the synthesis of thyroid hormones, and the uptake of thyroglobulin and its hydrolysis, i.e., the secretion of thyroid hormones and the synthesis of thyroglobulin at the level of gene transcription. We have now shown that effects on iodination in fact reflect the control of H_2O_2 generation, i.e. the substrate supply of thyroperoxidase (Björkman et al., 1988; Corvilain et al., 1988). The high generation of H_2O_2 in the stimulated thyroid raises the question of its toxicity. This might be involved in the pathogenesis of endemic cretinism, as selenium deficiency and consequently GSH peroxidase deficiency has been observed in Central Africa, i.e. in the geographical zone of endemic cretinism. In the dog thyroid, the main fundamental effects of TSH are reproduced by cyclic AMP analogs and agents such as forskolin and cholera toxin which increase cyclic AMP accumulation in many tissues by activating adenylate cyclase and its GTP binding stimulating transducing protein Ns. They are thus mediated by cyclic AMP. By analogy with the β -receptor mediated action of norepinephrine they have been called B-effects.

Other extracellular signal molecules, prostaglandins of the E type and to a lesser extent norepinephrine through β receptors, activate thyroid adenylate cyclase and mimic the TSH effects. The abnormal thyroid stimulating immunoglobulins (TSI), which appear in the serum of patients with

Graves'disease, also activate adenylate cyclase, presumably by binding to the TSH receptors. In thyroid as in other systems, adenylate cyclase is negatively regulated by receptor activated inhibitory GTP binding transducing protein Ni. In the dog thyrocyte, norepinephrine through α_2-receptors exerts this control. Negative feedback is also exerted on the TSH stimulatory pathway by the substrate of thyroid specialized metabolism : iodide (through a not yet defined oxidized derivative XI). Negative feedbacks by the thyroid hormones themselves and even thyroglobulin has been suggested, but their physiological relevance is still unknown.

The second major cell signalling system, the Ca^{++}-phosphatidylinositol cascade has also been demonstrated in the dog thyroid cells. Acetylcholine through a muscarinic receptor enhances free calcium intracellular concentration (as shown by Quin 2 fluorescence), $^{45}Ca^{++}$ translocation and the generation of Ins P1, Ins(1,4)P2 and Ins(1,4,5)P3 (Sheela et al., 1985; Raspé et al., 1986; Graff et al., 1987). The first phase of intracellular free Ca^{++} rise is independent of extracellular Ca^{++} and thus presumably originates from intracellular stores. The second phase is dependent of extracellular Ca^{++} which suggests that it is caused by an influx of this Ca^{++}. By analogy with other cell models, it is therefore inferred that acetylcholine activates, through its muscarinic receptor and a GTP binding transducing protein, a membrane phospholipase C. This enzyme hydrolyses phosphatidyl-inositol 4,5 phosphate (PtdIns(4,5)P2) and thus generates two intracellular signal molecules : myo-inositol 1,4,5 phosphate (Ins(1,4,5)P3) and diacylglycerol (DAG). IP3 would then cause the release by endoplasmic reticulum of stored Ca^{++} and be responsible for the first phase of Ca^{++} rise. DAG would activate thyroid protein kinase C. The role of the two branches of this cascade can be evaluated using as probes Ca^{++} ionophore A23187, and phorbol esters. The ionophore A23187 allows the influx of extracellular Ca^{++} and thus the activation of Ca^{++} dependent systems; phorbol esters are specific long acting analogs of DAG. In the dog thyroid, all the effects of acetylcholine appear to be mediated by

Ca^{++}: they are reproduced by the ionophore A23187 in the presence of extracellular Ca^{++}, or by high concentrations of extracellular Ca^{++}; they are inhibited in Ca^{++} depleted cells or by Ca^{++}-channel blockers such as Co^{++} or Mn^{++}. These effects are: the activation of protein iodination and glucose oxidation, the enhancement of cyclic GMP accumulation, the synthesis of prostaglandin and the inhibition of cyclic AMP accumulation and thyroid hormone secretion (Decoster et al., 1980). The inhibition of cyclic AMP accumulation is caused by an activation by Ca^{++} of Ca^{++} calmodulin-dependent cyclic nucleotide phosphodiesterase. The inhibition of thyroid hormone secretion is caused both by this inhibition of cyclic AMP accumulation and by a direct effect on the secretory mechanism (Unger et al., 1984). The effects of the other intracellular signal generated by phospholipase C, diacylglycerol, can be inferred from the action of phorbol esters. These tumor promoters, as Ca^{++}, enhance protein iodination and inhibit thyroid hormone secretion. On the other hand, they appear to inhibit the first steps of the Ca^{++}-phosphatidylinositol cascade (IP_3 generation and Ca^{++} influx) and the consequent effect of the free intracellular Ca^{++} rise : the enhancement of cyclic GMP accumulation. DAG could therefore exert a negative feedback on the cascade (Mockel et al., 1987). PGF_2 *, TRH and NaF reproduce some of the effects of acetylcholine in the dog thyrocyte. In dog thyroid cells, TSH enhances ^{32}P phosphate and ^{3}H inositol incorporation into phosphoinositides. It also stimulates $^{45}Ca^{++}$ efflux from prelabelled dog thyroid cells. These have been called A effects. This may suggest that TSH also activates the Ca^{++}-phosphatidylinostol cascade. However, TSH fails to enhance IP_3 generation in dog thyroid slices. Increased incorporation of ^{3}H inositol in IP_3, observed in some experiments, merely reflects increased incorporation in the precursor PiP_2. The meaning of the TSH A effects on dog thyrocyte Ca^{++} and phosphatidylinositol metabolism remains therefore obscure.

Figure 1 is a scheme summarizing the regulation of the human thyrocyte as we know it now (Van Sande et al., 1980; Van Sande et al., 1988). Thyrotropin (TSH), through cyclic AMP, activates all specialized functions of the tissue: the transport of iodide, its oxidation, thyroid hormone secretion and growth. Some of these stimulations are acute (within minutes) and require no prior protein synthesis (thyroid hormone synthesis and secretion) while others (iodide transport and growth) require just such a step. In the human thyrocytes, TSH and some neurotransmitters (ATP, bradykinin, etc) also activate the phosphatidylinositol 4-5 phosphate (PiP_2) cascade releasing Ins(1,4,5)P_3 and DAG in the cell

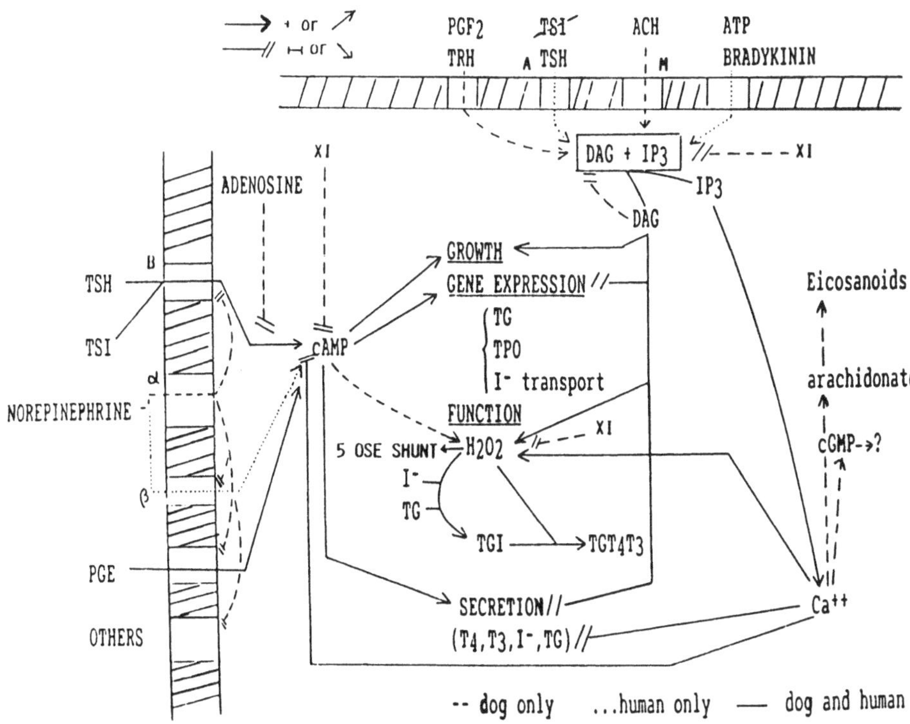

Fig. 1. Controls in the thyrocyte.
Abbreviations : Ach : acetylcholine; DIACG : diacylglycerol; IP_3 : myo-inositol 1,4,5 phosphate; PG : prostaglandin; TG : thyroglobulin; TGI : iodinated thyroglobulin; TPI : phosphatidylinositol 5-biphosphate; TSI : thyroid stimulating immunoglobulins; XI : unknown iodinated inhibitor. Lines indicate controls : straight lines : proved controls; interrupted lines : postulated: controls; + : positive control = stimulation; - : negative control = inhibition.

(Laurent et al., 1987). IP_3 elicits the release of calcium from endoplasmic reticulum and raises intracellular free calcium levels (Ca++i). DAG and Ca++ act in parallel on function, inhibiting secretion and activating iodination and thyroid hormone synthesis. The dual action of thyrotropin may involve two different receptors as drawn on Fig. 1, or one receptor activating the GTP binding transducing proteins(G proteins) controlling the two cascades. The effect of TSH on the PtdIns(4,5)P2 cascade is less sensitive to the hormone and slower than the action on cyclic AMP accumulation.

This general scheme applies to other species we have studied, except for a few important differences :

1) the action of TSH on the Pi cascade does not occur at physiological hormone levels in some species, among which dog
2) the neurotransmitters acting on the two cascades differ from one species to another. For example the Pi cascade is activated by acetylcholine in dog but not in human thyrocytes. β adrenergic cyclic AMP response also varies much between species (small in dog, higher in human)
3) iodination and thyroid hormone synthesis catalyzed by the same thyroperoxidase H2O2 generating system is controlled by H2O2 generation. This step is activated by cyclic AMP in dog thyrocytes, but only barely if at all by cyclic AMP in human thyrocytes. Ca++ and DAG activate H2O2 generation and iodination in both species.

As others, we are trying to clone the human TSH receptors and until now have failed. However, using an approach based on homology with known receptors controlling G proteins, we have managed to clone 7 such receptors : two were known : the β2 adrenergic receptor and the 5HT1 serotonine receptor; one is the dog correspondent of the α1 adrenergic receptor cloned at the same time; the four others RDC1, RDC4, RDC7 and RDC8 are unknown. RDC4 is clearly related to the known serotonin receptors; by Northern blot analysis, its expression has not been detected in any dog tissue !. RDC1 has some homology to substance K receptor; it is expressed in heart, kidney and thyroid. RDC7 and RDC8 belong to a new family of receptors with little extracellular NH2 terminal peptide and no glycosylation site; they are expressed in brain and for RDC7 also in thyroid. We are now busy trying to identify and characterize these receptors. But, as none of them is expressed exclusively in the thyroid, we doubt that the TSH receptor is one of them (Libert et al., 1989).

II. CONTROL OF THYROID CELL GROWTH

Many extracellular signal molecules, such as hormones, neuro-transmitters, and growth factors acting as autocrine or paracrine stimulants, elicit the mitogenic response by activating their membrane receptors and the intracellular biochemical cascade that they control. The study of conventional model systems of established fibroblastic cell lines has allowed a partial characterization of two such pathways (Fig. 2) : (i) Receptors for some growth factors (e.g. epidermal growth factor, EGF) possess an intrinsic protein tyrosine kinase activity as is the case for one class of transforming proteins encoded by oncogenes. The identity and function of many substrates for tyrosine kinase remain elusive and there is no absolute proof that this activity is sufficient in normal cells to induce mitogenesis; (ii) other membrane receptors are coupled via a GTP-binding protein to a phospholipase C that cleaves phosphatidylinositol 4,5 bisphosphate into diacylglycerol and inositol 1,4,5 triphosphate (IP3). The analogs of diacylglycerol, the phorbol esters tumor promoters, activate, in some cell types, mitogenesis. Intracellular Ca^{++} also triggers cell proliferation in specific cells by little known mechanisms. Although recently shown to be clearly distinct in their initial part (receptor, transducer, first intracellular signal) both pathways rapidly converge on several events such as activation of Na+/H+ exchange and of several transporters, phosphorylation on tyrosine residues of 42K proteins and increase in c-fos and c-myc protooncogene mRNA. These early events are assumed to be necessary for growth stimulation, but causative relationship with late commitment to DNA replication remains unclear.

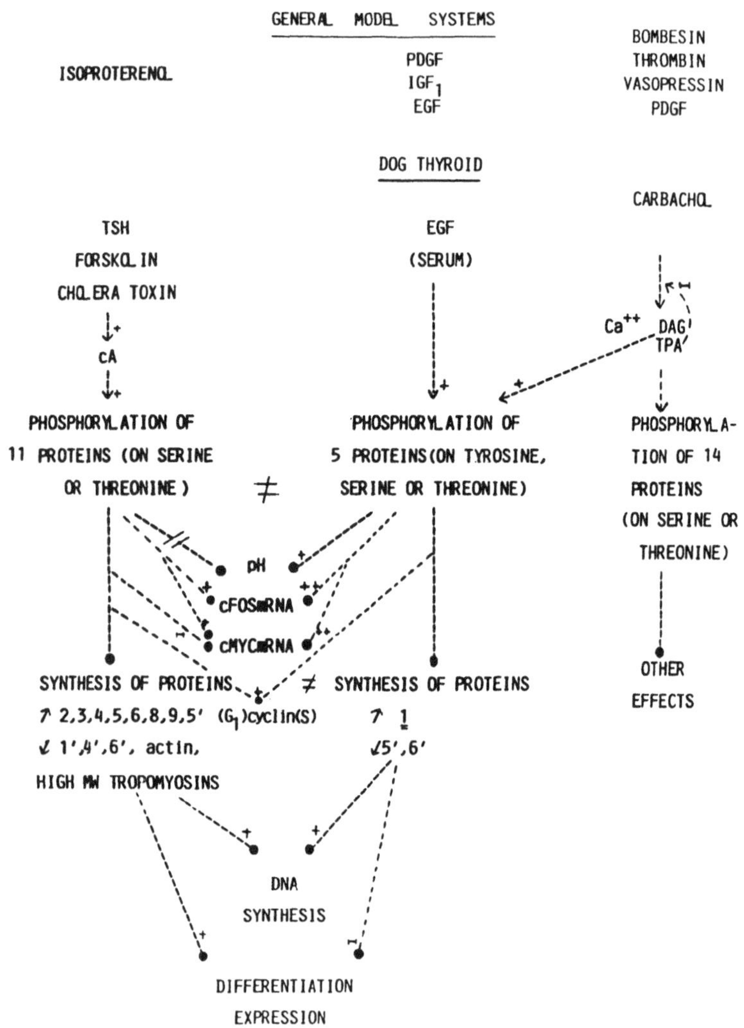

Fig. 2. Mitogenic pathways operating in model systems and in the thyroid.
Abbreviations : EGF : epidermal growth factor; IGF : insulin like growth factors; PDGF : platelet derived growth factor. --> causal relationships; ---O sequential, possibly causal relationships; + positive i.e. stimulations; - negative i.e. inhibitions 1, 2, 3 ...protein signals on 2D gel autoradiograms.

A third major regulatory cascade involved in the regulation of cell proliferation is the cyclic AMP system (Dumont et al., 1989). Despite some early evidence to the contrary, in the seventies, there was an overwhelming acceptance of the concept that cAMP was a negative regulator of cell proliferation. This was based on the evidence, mostly obtained with established cell lines of fibroblastic of tumoral origins, that cAMP derivatives or cAMP elevating agents inhibited growth, and that cAMP levels, in such cells in culture, inversely correlated with proliferation. Although, in some systems, these observations are correct and confirmed by genetic analysis of resistant mutants, many other similar studies are questionable because they are based on the use of high concentrations of cAMP analogs with no suitable controls. Nevertheless, the concept of cAMP as a negative growth signal persists to this day and any evidence against it, is often regarded as an exception to the rule. Perhaps because of these discrepancies, the role of cAMP in the control of growth is often ignored in general reviews on this subject (Baserga, 1985).

In sharp contrast with this background, there is a series of cells in which cAMP enhances or initiates proliferation. Moreover, the demonstrated role of cAMP as an intracellular signal for replication in yeast Saccharomyces Cerevisiae is now giving respectability to such a control in upper eucaryotic cells.

Although there is no doubt that TSH in vivo stimulates the proliferation of thyroid cells, there was in the 1970s no evidence that this was a direct effect. Indeed, the ACTH trophic effect on the adrenal appears to be indirect. The first results obtained in cultures were quite contradictory. Indeed, even now, while in dog thyroid cells in primary culture, in rat thyroid follicles in suspension, in ovine cell lines (OVNI), and in rat cell line (FRTL) thyrotropin has been demonstrated to enhance or induce cell proliferation, to our knowledge, no such effect had been obtained in porcine, calf or ovine thyroid cells in primary culture. Whether this is due to inaccessibility of the TSH receptor(s), lack of an essential element in the culture medium, alteration of cell program in culture or true unresponsiveness to direct TSH action, is not known. In dog thyroid cells in primary culture, TSH enhances proliferation in the presence of serum and induces it in its absence. This has been demonstrated using several methods. Thus in this species at least, TSH directly induces thyroid cell proliferation. In FRTL cell line, TSH is also required for proliferation, however the fact that such cells may die in the absence of TSH and serum might not allow to distinguish between a general trophic effect of the hormone or a definite proliferation signal. It should be noted that, in dog thyroid cells, TSH stimulates proliferation while maintaining the expression of differentiation. Differentiation expression, as evaluated by iodide transport, or thyroperoxidase and thyroglobulin mRNA content, or nuclear transcription, is induced by TSH, forskolin, cholera toxin and cyclic AMP analogs, in dog thyroid cells. Similar results, albeit partial, have been obtained in human and calf cells. These effects are obtained in all the cells of a culture, as shown by in situ hybridization experiments. They are reversible; they can be obtained after the arrest of proliferation. Some of these effects, thyroglobulin but not thyroperoxidase gene expression, require the presence of insulin or IGF1. Epidermal growth factor (EGF) also induces proliferation of dog thyroid cells. This effect is often weaker than the effect of TSH. However, the action of EGF is accompanied by a general and reversible loss of differentiation expression as assessed above. The effects of EGF on differentiation can be dissociated from their proliferative action. Indeed, they are obtained in cells that do not proliferate in the absence of insulin. EGF also stimulates the growth of thyroid cells from other species in culture (eg. porcine, ovine, bovine and human but not of the FRTL cell line which lacks EGF receptors). In dog thyroid cells, insulin is necessary for growth in the presence of EGF and in half the cultures in the presence of TSH. A requirement for insulin has also been observed in rat cells in primary culture and for the FRTL5 rat thyroid cell line. Serum and fibroblast growth factor also induce growth in dog and calf thyroid cells. Finally, phorbols esters, the pharmacological probes of the protein kinase C system, also enhance the proliferation and the dedifferentiation of dog thyroids cells. These effects are transient owing to desensitization of the system by protein kinase C inactivation (Lamy et al., 1987). It must be emphasized that, in the absence of serum, our cells in primary culture respond to stimulation by one or at the maximum 2 rounds of proliferation, while with one percent serum they might achieve up to 4 rounds.

Thyrotropin induces within minutes a striking morphological change in dog thyroid cells in culture : a rounding up following the disruption of the actin network. All the cells are affected. TSH also enhances the accumulation of cyclic AMP in these cells within less than five minutes. Cyclic AMP remains elevated for 48 hours in the continuous presence of TSH. The question arises of the role of cyclic AMP in all these TSH effects. In the dog thyroid cells, analogs of cyclic AMP as well as general cyclase activators (forskolin, cholera toxin) reproduce all the effects of TSH : acute morphological changes, proliferation, expression of differentiation. Moreover, combinations of cyclic AMP analogs which are synergistic on the two cyclic AMP dependent kinases isoenzymes are also synergistic on these effects. Cyclic AMP is therefore a general intracellular signal for function, proliferation and differentiation in the dog thyroid cells. For proliferation, similar results have been obtained with rat thyroid cells in culture and, despite a first contradictory report (Valente et al., 1983), in FRTL5 cells (Dere et al., 1986).

There are indications that TSH may stimulate the Ca++ phosphatidylinositol cascade in thyroid cells. In our dog thyroid cells, TSH, contrary to acetylcholine, does not enhance the generation of inositol phosphates. There is little evidence that the A pathway of TSH action may be involved in the growth effect of TSH in this system.

The effects of EGF on dog thyroid cell (proliferation, inhibition of differentiation expression) are mimicked by phorbol esters tumor promoters. However, these compounds also inhibit EGF action : combined with EGF, they lower the proliferation level to the level induced by them. In several cell types, EGF, not only activates a tyrosine specific protein kinase, but also induces a rapid rise in cytoplasmic free Ca++ concentration. This rise in Ca++ concentration following EGF stimulation has been linked to an activation of the phosphatidylinositol Ca++ cascade although it has been suggested recently that it might result from an entry of extracellular Ca++ through the plasma membrane. It would therefore be conceivable that EGF action in the thyroid cell might result from an increase in Ca^{++} entry or from an activation of the phosphatidylinositol Ca^{++} cascade with generation of diacylglycerol, the action of which is mimicked by phorbol esters. Indeed EGF induces a rise in intracellular Ca^{++} in porcine thyroid cells, but it definitively fails to do so in dog cells. On the other hand, neither EGF nor phorbol esters enhance cyclic AMP accumulation in these cells. It is therefore likely that EGF acts through the phosphorylation of key proteins on tyrosyl residues. We have studied the phenomenology of EGF and TSH proliferative action on quiescent cells with the aim to identify common steps in this action. Three biochemical aspects of the proliferative response occurring at different times of the prereplicative phase have been studied. The pattern of protein phosphorylation induced within minutes by TSH is reproduced by cyclic AMP analogs (Contor et al., 1988). The phosphorylation of at least 11 proteins is increased or induced. NaOH treatment of the gels does not reveal any remaining phosphorylation on these proteins suggestive of tyrosine phosphorylation. In EGF stimulated cells, the phosphorylation of 5 proteins is stimulated, two of which become phosphorylated on tyrosines

(42K). These two proteins are similar (isoelectric points, approximate molecular weight, composition in phosphorylated amino acids) to the two 42K proteins described in other systems and which have been implicated in the mitogenic response to diverse agents. Phorbol esters induce the phosphorylation of 19 proteins, including the tyrosine phosphorylated proteins mentioned above. There is no overlap in the patterns of protein phosphorylation induced by TSH and cyclic AMP enhancers on the one hand, and by EGF and phorbol esters on the other hand.

The expression of c-myc and c-fos mRNA has been studied by Northern analysis of RNA extracts. As in other types of cells, EGF and TPA enhance first c-fos, then c-myc mRNA concentrations. On the other hand, TSH or forskolin enhance strongly but shortly c-myc mRNA concentration and with the same kinetics as for EGF/TPA c-fos mRNA concentration. In fact, cyclic AMP first enhances, then decreases c-myc mRNA accumulation. This second phenomenon is akin to what has been observed in fibroblasts in which cyclic AMP negatively regulates growth. In TSH and cyclic AMP action on the thyrocyte, the first stimulation might be necessary to trigger the proliferation cascade, while the later shut off might be necessary for the retention of the differentiated phenotype. The rapid downregulation of c-myc also found in forskolin treated cells suggests that this inhibition is mediated by cyclic AMP. The decrease in c-myc expression does not occur if protein synthesis is blocked by cycloheximide, suggesting that a labile protein (maybe neosynthetic implicating an autoregulatory mechanism) is involved in the inhibition at the transcriptional level or at the stabilization of the mRNA. Preliminary experiments using Actinomycin D transcription blockade indicate that enhancement of the transcription rate of the c-myc gene is mostly responsible for the enhancement of the c-myc mRNA level detected after one hour of action.

The pattern of proteins synthesized in response to the various proliferation stimuli has been studied (Lamy et al., 1986).

Again two patterns emerge. TSH and forskolin induce the synthesis of at least 8 proteins and decrease the synthesis of 5 proteins. Epidermal growth factor, phorbol ester and serum induce the synthesis of at least 1 protein and decrease the synthesis of 2 proteins. The only overlap between the two patterns concerns the decrease in the synthesis of a protein (18K) which is also reduced by EGF after proliferation has stopped. Only one protein has been shown to be synthesized in response to the three pathways : cyclin, but the kinetics of this synthesis are very different, with an early synthesis in the cyclic AMP cascade, (consistent with a role of signal) and a late, S phase synthesis in the other cascades. Thus, obviously two different phenomenologies are involved in the proliferation response to TSH through cyclic AMP on the one hand, and epidermal growth factor and phorbol ester, presumably through protein tyrosine phosphorylation, on the other hand. Although this conclusion needs to be further substantiated, it certainly suggests that the proliferation of dog thyroid cells is controlled by at least two largely independent pathways.

While the TSH cyclic AMP pathway stimulates function and promotes both the expression of differentiation and proliferation, the converging EGF tyrosine protein kinase and phorbol esters - protein kinase C pathways induce proliferation but inhibit differentiation expression. A priori, it would therefore be tempting to relate hyperthyroidism caused by thyroid stimulating immunoglobulins or by an hyperfunctioning adenoma to an enhancement of the cAMP cascade and dedifferentiating tumors to the activation of the growth factor and phorbol esters cascades. In this regard, it is interesting that TSI, at the concentration existing in pathological sera, in human thyrocytes only activates the cyclic AMP system and not, contrary to TSH, the phospholipase C pathway. Similarly, in preliminary experiments, TSH only stimulates the cyclic AMP system, but not phospholipase C, in autonomous nodules !

Previous studies of human thyroid cells in culture (mostly from pathological tissues) failed to demonstrate a mitogenic effect of thyrotropin (TSH), leading to the proposal that the growth effect of TSH in vivo might be indirect. In order to reexamine the influence of TSH on DNA synthesis and cell proliferation, we established primary cultures of normal thyroid tissue from 9 subjects (Roger et al., 1988). When seeded in a 1%-serum-supplemented medium, thyroid follicles released by collagenase/dispase digestion developed as a cell monolayer that responded to TSH by rounding up and by cytoplasmic retraction. When seeded in serum-free medium, the cells remained associated in dense aggregates surrounded by few slowly spreading cells. In the latter condition, the cells responded to TSH and to other stimulators of cyclic AMP production such as cholera toxin and forskolin by displaying very high iodide trapping levels. Exposure to serum irreversibly abolished this differentiated function. TSH stimulated the proliferation (as shown by DNA content per culture dish) of 1% serum cultured cells (doubling times were reduced from 106h to 76h) and increased by 100% the ^{3}H-thymidine labelling indices. In serum-free cultured cells (dense aggregates or cell monolayers after seeding with serum), control levels of DNA synthesis were lower and up to 8-fold stimulation of DNA synthesis occurred in response to 100 uU/ml TSH (stimulation was consistently detected with 20 uU/ml), based on measurements of ^{3}H-thymidine incorporation into acid-precipitable material and counts of labelled nuclei on autoradiographs (up to 40% labelled nuclei within 24h). The mitogenic effect of TSH required a high insulin concentration (8.3×10^{-7}mol/L) or a low insulin-like growth factor-1 concentration. The mitogenic effects of TSH were mimicked by cholera toxin, forskolin and dibutyryl cAMP. Thus, there is no doubt that TSH, through cyclic AMP, is mitogenic on human thyroid cells. Epidermal growth factor and phorbol myristate ester also stimulated thyroid cell proliferation and DNA synthesis, but they potently inhibited TSH-stimulated iodide transport.

Thus, in human thyrocytes, the regulation of growth and differentiation expression seems to operate through the same pathways as in the dog thyrocyte. However, our knowledge of the human cell is much less advanced. For instance, we know nothing of the patterns of protein phosphorylation or synthesis, or on protooncogene expression in this model. Moreover our experiments suggest that TSH action on proliferation may involve more than cyclic AMP, perhaps the Ca^{++}-phosphatidylinositol cascade. Before analyzing the alterations of the control systems which may be involved in the pathogenesis of goiter or thyroid tumors, a lot more work on the human thyrocyte is necessary (Roger et al., 1987).

III. CALCYPHOSIN

Another distinction between the two types of mitogenic pathways in dog thyroid cells is the synthesis of a 23K protein, which is enhanced by the cAMP pathway, and depressed by EGF, serum and TPA, even after cessation of growth. This protein thus appears as a possible marker of differentiation. Interestingly, this protein is acutely phosphorylated in response to TSH or cyclic AMP enhancers, which suggests a role in functional activation (Fig. 3).

With antibodies raised against protein 23K, we have screened a λgt11 dog thyroid cDNA expression library and obtained corresponding clones. The longest cDNA insert has been cloned and sequenced . The sequence reveals a striking analogy with calmodulin with 4 EF hands; the first contains a putative cAMP dependent protein kinase phosphorylation site, the fourth one is interrupted by a non related aminoacid sequence. This suggested that the 23K protein might be a calcium binding protein. Indeed, in Western blot, this protein, identified by immunoblotting, was shown to bind Ca^{++}. The 23K protein, a marker of differentiation of dog thyrocytes, therefore binds calcium, and is phosphorylated in response to cyclic AMP; it has been named : calcyphosin. Calcyphosin has also been identified in the human thyroid and in other dog organs such as brain and salivary gland. It is the only protein known which is phosphorylated in response to cyclic AMP on the same

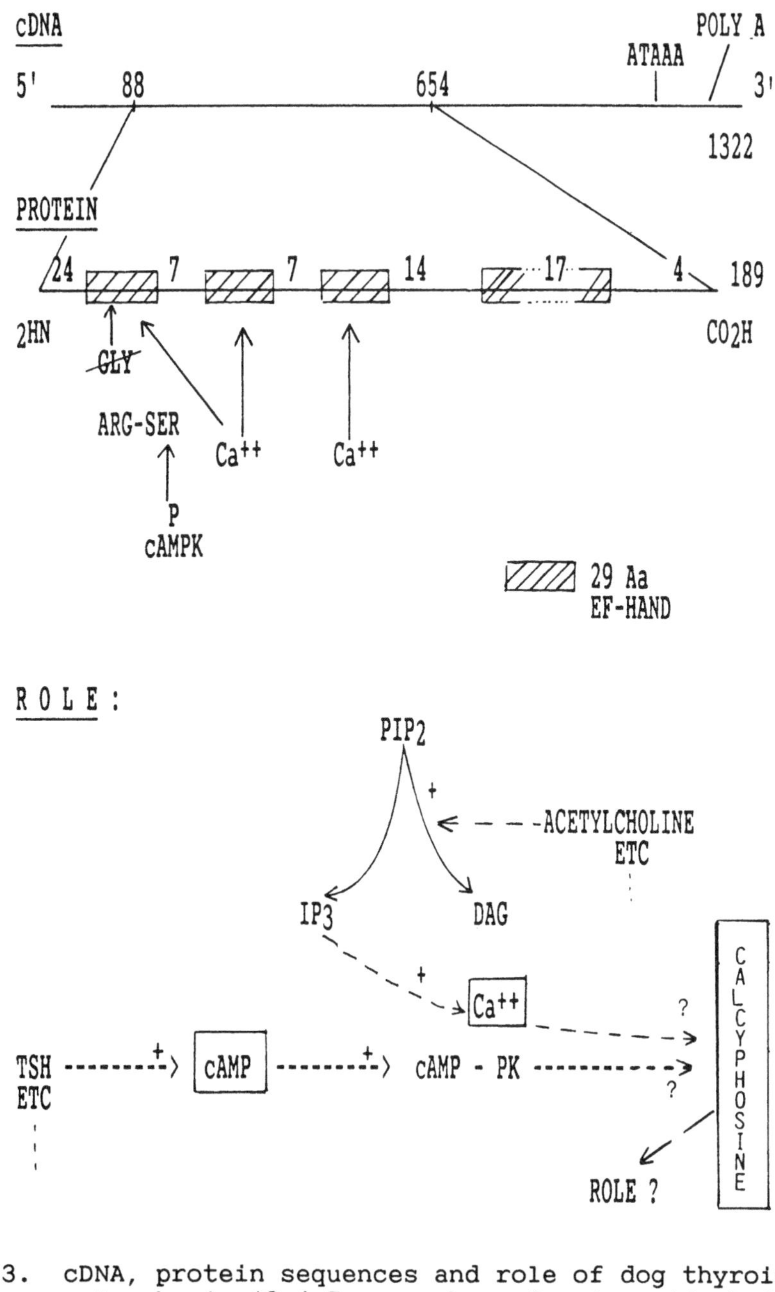

Fig. 3. cDNA, protein sequences and role of dog thyroid calcyphosin (2,4,7,.. number of aminoacids before, between, in and after the EF-hands).

peptide which binds Ca^{++}. Phosphorylase kinase presents similar properties but on different subunits. It is hypothesized that calcyphosin, as a common target of the cyclic AMP and the Ca^{++} phosphatidylinositol cascade, might be involved in the control of a general cell biology process (such as ion transport or motility) or in the reciprocal control of these two cascades. Whatever its role, this protein presents a great interest as such !! (Lefort et al, 1989).

Thus quite apart from the insights they give on the control of thyroid cell function, proliferation and differentiation and on the role of these controls in disease, our results have led, by serependity, to interesting new general findings which may have a larger medical interest : the discovery of severe selenium deficiency in Central Africa (Goyens et al., 1987), the cloning of 4 new receptors, the cloning of calcyphosin, a common protein target of the cyclic AMP and the Ca^{++} phosphatidylinositol cascades.

REFERENCES

Baserga, R. 1985 The Biology of Cell Reproduction, Harvard University Press.

Björkman, U, and Ekholm R. Accelerated exocytosis and H_2O_2 generation in isolated thyroid follicles enhance protein iodination, Endocrinology 122:488 (1988)

Contor, L., Lamy, F., Lecocq, R., Roger, P.P. and Dumont, J.E. Differential protein phosphorylation in induction of thyroid cell proliferation by thyrotropin, epidermal growth factor, or phorbol ester. Mol. Cell. Endocrinol 8:2494 (1988)

Corvilain, B., Van Sande, J. and Dumont, J.E. Inhibition by iodide of iodide binding to proteins: the "Wolff-Chaikoff-effect" is caused by inhibition of H_2O_2 generation. Biochem. Biophys. Res. Commun. 154:1287 (1988)

Decoster, C., Mockel, J., Van Sande, J., Unger, J., and Dumont, J.E. The role of calcium and guanosine 3':5'-monophosphate in the action of acetylcholine on thyroid metabolism. Eur. J. Biochem. 104:199 (1980)

Dere, W.H., and Rapoport, B. Control of growth in cultured rat thyroid cells. Mol. Cell. Endocrinol. 44:195 (1986)

Dumont, J.E., Jauniaux, J.C., Roger, P.P. The cyclic AMP-mediated stimulation of cell proliferation. Trends Biochem. Sci. 14:67 (1989)

Dumont, J.E., Takeuchi, A., Lamy, F., Gervy-Decoster, C., Cochaux, P., Roger, P., Van Sande, J., Lecocq, R., and Mockel, J. Thyroid control: an example of a complex cell regulation network. Adv. Cyclic Nucl. Res., 14:479 (1981).

Goyens, P., Golstein, J., Nsombola, B., Vis, H., and Dumont, J.E. Selenium deficiency as a possible factor in the pathogenesis of myxoedematous endemic cretinism. Acta Endocrinol. 114:497 (1987)

Graff, I., Mockel, J., Laurent, E., Erneux, C., and Dumont, J.E. Carbachol and sodium fluoride, but not TSH, stimulate the generation of inositol phosphates in the dog thyroid. FEBS Letters 210:204 (1987).

Lamy, F., Roger, P., Contor, L., Reuse, S., Raspé, E., Van Sande, J., and Dumont, Control of thyroid cell proliferation: the example of the dog thyrocyte, Horm. Cell Regul.153(11):169 (1987)

Lamy, F., Roger, P.P., Lecocq, R., and Dumont, J.E. Differential protein synthesis in the induction of thyroid cell proliferation by thyrotropin, epidermal growth factor or serum. Eur. J. Biochem. 155:265 (1986)

Laurent, E., Mockel, J., Van Sande, J., Graff, I., and Dumont, J.E. Dual activation by thyrotropin of the phospholipase C and cyclic AMP cascades in human thyroid. Mol. Cell. Endocr. 52:273 (1987)

Lefort, A., Lecocq, R., Libert, F., Lamy, F., Swillens, S., Vassart, G., and Dumont, J.E. Cloning and sequencing of a calcium-binding protein regulated by cyclic AMP in the thyroid. EMBO J. 8:111 (1989)

Libert, F., Parmentier, M;, Lefort, A., Dinsart, C., Van Sande, J., Maenhaut, C., Simons, M.J., Dumont, J.E., and Vassart, G. Selective amplification and cloning of four new members of the G protein-coupled receptor family. Science (in press) (1989)

Mockel, J., Van Sande, J., Decoster, C., and Dumont, J.E. Tumor promoters as probes of protein kinase C in dog thyroid cell : inhibition of the primary effects of carbamylcholine and reproduction of some distal effects. Metabolism 36:137 (1987).

Raspé, E., Roger, P., and Dumont, J.E. Carbamylcholine, TRH, PGF_2 and fluoride enhance free intracellular Ca^{++} and Ca^{++} translocation in dog thyroid cells. Biochem. Biophys. Res. Commun. 141:569 (1986)

Roger, P.P., Servais, P., and Dumont, J.E. Induction of DNA synthesis in dog thyrocytes in primary culture : synergistic effects of thyrotropin and cyclic AMP with epidermal growth factor and insulin. J. Cell. Physiol. 130:587 (1987)

Roger, P.P., Taton, M., Van Sande, J., and Dumont J.E. Mitogenic effects of thyrotropin and cyclic AMP in differentiated human thyroid cells in vitro. J. Clin. Endocrinol. Metab. 66:1158 (1988)

Sheela Rani C.S., Boyd A.E., and Field, J.B. Effects of acetylcholine, TSH and other stimulators on intracellular calcium concentration in dog thyroid cells. Biochem. Biophys. Res. Commun. 131:1041 (1985)

Unger, J., Ketelbant, P., Erneux, C., Mockel, J., and Dumont, J.E. Mechanism of cholinergic inhibition of dog thyroid secretion in vitro. Endocrinology 114:1266 (1984)

Valente, W.A., Vitti, P., Kohn, L.D., Brandi, M.L., Rotella, C.M., Toccafondi, R., Tramontano, D., Aloj, S.M., Ambesi-Impiombato, F.S. The relationship of growth and adenylate cyclase activity in cultured thyroid cells. Separate bioeffects of thyrotropin. Endocrinology 112:71 (1983)

Van Sande, J., Lamy, F., Lecocq, R., Mirkine, N., Rocmans, P., Cochaux, P., Mockel, J. and Dumont J.E. Pathogenesis of autonomous thyroid nodules : in vitro study of iodine and adenosine 3',5'-monophosphate metabolism. J. Clin. Endocrinol. metab. 66:570 (1988)

Van Sande, J., Mockel, Boeynaems, J.M., Dor, P., Andry, G., and Dumont, J.E. Regulation of cyclic nucleotide andprostaglandin formation in normal human thyroid tissue and in autonomous nodules. J. Clin. Endocrinol. Metab. 50:776 (1980)

DISCUSSION

Corbin: It did not look like that there was a very big consensus sequence for cAMP dependent protein kinase, I mean are you sure that that is the site, that gets phosphorylated? There was an acidic residue.

Dumont: Well we are not sure, in fact what we are doing now is taking out the protein, cutting it with trypsine and we are trying to see where the phosphate is, but we are not sure yet. I mean looking at the sequence we have the question is there something that looks like a consensus sequence, it looks like it but we have no evidence that I can show you.

Cooke: Is your protein secreted by the cell or is it purely intracellular?

Dumont: That is a good question, we do not find it when we concentrate the incubation medium, so we have no evidence for it and if you look at the -these are preliminary data- localization by electron microscopy you see it is in the cytosol. It is also a very acidic protein and it has no putative transmembrane domain, we have not got the gene yet so we do not know if it has any signal peptide.

Corbin: I have one more question, it is about the TSH receptor, I suppose it is possible that the membrane binding domain is totally different from the β-adrenergic receptor.

Dumont: You see the power of the method is astonishing because we could pick up receptors that are probably very very scarce in our preparations, but on the other hand probably the choice of the probe that you make will select just a few molecules. There might be very few of them but it will select these, and to make our probes we started mostly with the consensus sequence of β1,β2, α2, muscarinic and serotonine receptors but it was a little bit biased towards the β-adrenergic receptor. So we came out with that type of receptor. It is very possible that if we change these probes a little bit to look like say substance K or any other type of peptide hormone receptor, that we come with other receptors. It is on the one hand tremendously powerful on the other hand it selects and obviously it did not select right for the TSH receptor.

Clark: Over what range of calcium concentrations does this protein bind calcium?

Dumont: We do not know. We have not produced it yet. We are trying to express it, to make that kind of studies.

Cooke: Can I go back to the initial actions of TSH, in the dog you have shown that it is only working through the cAMP pathway, whereas in the human you have got the IP3 pathway as well. Is there any difference in the pattern of cAMP production, in other words is there more desensitization in the human than there is in dog, or do you get desensitization.

Dumont: There is desensitization, but it is very low compared to other systems. If we want to get a good desensitization we need to incubate our cells at least two hours with the hormone and then the desensitization is at best a decrease of 50 %.

Cooke: And there is no difference between the two species?

Dumont: Not striking but I would not say that we have studied that carefully. But I should remind you that in the dog thyroid everything seemed to be mediated by cAMP but we have a discrepancy which was already noticed by Bob Mitchell in his '75 review, where he was studying the PI turnover, and he said 'something is wrong with the tyroid, it does not fit the pattern'! In the dog tyroid we have an other effect, which is not cAMP mediated and which is increased 32P labeling of phosphatidylinositol. So there is something else going on that we are now studying and which is different from the classical PIP2 response; it may be synthesis of PI or something like that but obviously the two types of measurements give the same answers. And it is astonishing because the last time we wanted to put all the stuff together, one of the referees said 32P turnover and phosphatidylinositol incorporation is old stuff we do not take that any more!!

It is an old methodology, but still it is a methodology that tells us something else than the IP3 method. So there is something going on there but we do not know all of it. And the funny thing is that there has been an article recently about the dog thyroid by an other worker and he found increased IP3 generation in dog thyroid cells after TSH stimulation. And you can measure that but if you look at the ratio between IP3 and PIP2 which is the precursor there is no effect. In fact as you label more your PI and your PIP2 if you stimulate the cells just the normal rate of IP generation will give you a higher activity. It does not mean there is activation of the cascade it means that we have labeled more the precursor.

Clark: Given the biochemical properties of calmodulin, it seems you could go back and do some obvious biochemical approaches; have you thought of doing that, like calmodulin is resistant to boiling, and TCA and, you could almost, if you were lucky, go back and now do the biochemistry and pull out lots of protein.

Dumont: Yes but the main point in that is to have enough of the protein, and that is why we are trying to express it in bacteria. I mean you do not have that much in thyroid, I mean you can not collect say 5 kilo's of thyroid to start a purification so really we have to produce it and we are just trying to do that.

ACTIVATION OF THE REGULATED SECRETORY PATHWAY IN NEURONAL CELLS DURING IN VITRO DIFFERENTIATION

E. Sher, S. Denis-Donini, A. Zanini, C. Bisiani, E. Biancardi, F. Clementi
CNR Center of Cytopharmacology
Department of medical Pharmacology
University of Milano
via Vanvitelli 32
20129 - Milano
Italy

Secretion is a fundamental process that characterizes several cells. It is present early in evolution and it has been maintained in both animal and plant cells. The secretion process can be schematically classified in two types: constitutive and regulated (Burgess & Kelly, 1987). In constitutive secretion the secretory material is synthesized and continuously secreted; in regulated secretion the secretory material is first accumulated in discrete organelles and then released in the extracellular space after an appropriate stimulus. The latter pathway is typical of neurons which release neurotransmitters and neuropeptides upon receptor activation.

The regulated secretory pathway is a complex process involving several subsequent steps, namely: synthesis of secretory material, storage, receptor activation, signal transduction, and finally secretion by exocytosis.

It is interesting to understand whether the activation of this secretory pathway occurs by the activation of a single program or is the result of a parallel and coordinate activation of different events, the sum of which can induce the final functional pathway.

To approach this point, we searched for a suitable cell model and we screened several neuron-like cell lines growing in vitro. Finally, we selected the IMR32 human neuroblastoma cell line because these cells can be "educated" to acquire the regulated secretory pathway by means of a drug-induced differentiation.

NATO ASI Series, Vol. H 44
Activation and Desensitization of Transducing Pathways
Edited by T. M. Konijn, M. D. Houslay, P. J. M. Van Haastert

IMR32 cell line

IMR32 cells derive from a human neuroblastoma adapted to grow in vitro (Tumilowicz, et al., 1970). These cells maintain several characteristics of neurons, such as ability to extend neurites, voltage-operated Ca^{2+} channels (VOCCs), neurotransmitter receptors, uptake and synthesis of neurotransmitters.
The IMR32 cell line (obtained from the American Type Culture Collection) was grown as previously described (Clementi et al., 1986) in Eagle's minimum essential medium containing Earle's salts, 10% heat-inactivated fetal calf serum, 100 IU/ml penicillin and 100 μg/ml streptomycin. The cells were never grown more than three months consecutively.

Characteristics of undifferentiated IMR32 cells

a) Membrane receptors and ion channels
We have investigated cholinergic muscarinic and nicotinic receptors with binding techniques, using ^{125}I-α–Bungarotoxin (α–Bgtx) and ^{3}H-scopolamine, as radioligands, and by electrophysiology, measuring membrane currents with the patch clamp technique in whole-cell and cell attached configurations (Gotti et al., 1987).
We have characterized in IMR32 cells three types of cholinergic receptors. The first is a muscarinic receptor with low affinity for pirenzepine (≈2,000 receptors/cell); the second is an αBgtx binding site, with a nicotinic pharmacological profile, which is not coupled to an Na^{+} channel and is present in relatively high amounts (≈3,000 receptors/cell) (Table 1). The third is an acetylcholine-operated Na^{+} channel which is present in very low amounts, is blocked by curare but is not blocked by αBgtx (Gotti et al., 1986).
Muscarinic receptor activation triggers both Ca^{2+} influx into the cells and Ca^{2+} redistribution from the intracellular stores, as revealed by experiments using the fluorimetric Ca^{2+} probe Fura2 (Fig.1).
We have also observed that IMR32 cells have receptors for α-Latrotoxin (αLtx) (Fig.1), the major toxin of the venom of the black widow spider, which is a potent neurotransmitter secretagogue, and which is able to open cation channels in nerve and neurosecretory cells (Rosenthal and Meldolesi, 1989).
As far as VOCCs are concerned, we found that IMR32 cells express different kinds of Na^{+} and K^{+} channels (Gotti et al., 1987).

TABLE 1: CHARACTERISTICS OF ^{125}I-αBgtx AND ^{3}H-SCOPOLAMINE BINDING IN CONTROL AND DIFFERENTIATED CELLS

Treatment	^{125}I-αBgtx		^{3}H-Scopolamine	
	B_{max}	Kd	B_{max}	Kd
None	189±24	7.70±1.7	473± 73	3.32±0.77
BrdU (2.5 μM)	370±29*	8.11±1.4	1116±15O*	4.80±1.5
Bt_2cAMP (1 mM)	319±36*	7.35±4.6	423± 80[a]	3.85±1.7

B_{max} values are expressed as femtomoles per mg of membrane protein; Kd values are expressed in nanomoles. All values are the mean ± SE of five to eight experiments performed on control and differentiated cells after 13 days of culture with or without the drugs. [a]Not significant (vs.controls). * P<0.01 (vs. controls)

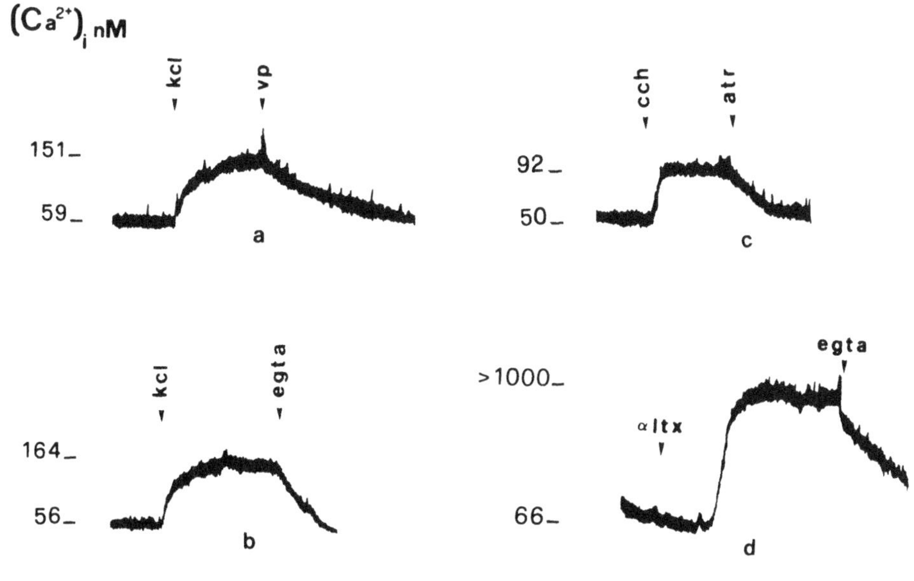

Figure 1 - Quin2 measurements of $[Ca^{2+}]$ changes in IMR32 cells exposed to a) KCl (50 mM) followed by verapamil (2μM), b) KCl followed by EGTA (1 mM), c) carbachol (500μM) followed by atropine (1μM), d) αLTx (2mM) followed by EGTA (1mM).

The presence of VOCCs and their properties have been studied in details with binding techniques, Fura2, and electrophysiology (Sher et al., 1988b).
IMR 32 cells, like other neurons, express different classes of VOCCs. In particular, they express both low voltage-activated ("T") and high voltage-activated ("N" + "L") VOCCs. The peptide neurotoxin ω-conotoxin binds with high affinity to IMR32 membranes (Sher et al., 1988a) and blocks specifically a subclass ("N") of high voltage activated calcium currents. These last channels are thought to play a crucial role in the control of neurotransmitter secretion (Miller, 1987).
The results of these experiments indicate that IMR32 cells express a complement of functioning membrane receptors and ion channels, that could trigger and modulate the regulated neurotransmitter secretion.

b) Neurotransmitter synthesis and secretion

In previous studies (Gotti et al., 1987) we have found that noradrenaline, dopamine (DA), and serotonin are present in IMR32 cells (Table 2). We have also demonstrated that non differentiated IMR32 cells are able to take up ^{3}H-DA in a specific, high-affinity and saturable fashion (Sher et al., 1989).

TABLE 2: NEUROTRANSMITTER CONTENT IN IMR32 CELLS

Treatment	Noradrenaline	Dopamine	Serotonin
	(pmol/mg protein)		
None	195± 24	8083±2167	3030± 298
BrdU (2.5 μM)	5614± 681*	28724±5936*	31290±8497*
Bt_2cAMP (1 mM)	17349±2783*	20161±1769*[a]	45000±7000*[a]

Neurotransmitter content was determined by HPLC. All values represent the mean ± SE of four to nine determinations. Cells were cultured for 13 days with the indicated concentrations of the drug.
[a]Not significant (vs.BrdU-treated cells). * P<0.01 (vs. controls)

However, when these cells were tested with several secretagogues, they were not able to release any of their catecholamine content, although the secretagogues

used activate different pathways: a) Ca^{2+}-dependent release (ionomycin); b) Ca^{2+}-independent release possibly related to protein kinase C activation, plus synergy with Ca^{2+}-dependent release (phorbol 12-Myristate 13-Acetate (TPA) c) Ca^{2+}-dependent plus Ca^{2+}-independent release mediated by unknown mechanisms (αLTx) (Meldolesi et al., 1987).
We conclude that these cells have functioning membrane receptors and ion channels, are able to take up and synthesize neurotransmitters, but are **not** able to release them upon stimulation.

The secretory response is acquired after cell differentiation.

IMR32 cells have the interesting property of differentiating in vitro, in response to pharmacological agents, towards a "mature" neuronal phenotype, which can be evaluated not only morphologically but also biochemically or functionally (Gotti et al., 1987; Gupta et al., 1985; Prasad, 1975).
We have investigated whether the ability to release ^{3}H-DA could be acquired by IMR32 cells after exposure to differentiating agents, such as N6-O2-dibutyrryl cyclic adenosine 3' 5' monophosphate (Bt_2cAMP) or 5'-bromo-2' deoxyuridine (BrdU), that are known to induce a characteristic neuron-like differentiation of these cells (Gotti et al., 1987)

a) Neurotransmitter release
As we mentioned above, non differentiated IMR32 cells do not release ^{3}H-DA. On the other hand, we found that BrdU-differentiated cells are able to release ^{3}H-DA in response to both ionomycin and αLTx (Table 3). TPA alone induced only a slight ^{3}H-DA release which was additive with the ionomycin effect. In contrast, Bt_2cAMP differentiated IMR32 cells were not able to release ^{3}H-DA in response to both the calcium ionophore ionomycin, and TPA (Table 3). Only αLTx was able to induce a significant release of ^{3}H-DA from such cells, although at a reduced level when compared to BrdU differentiated cells. Since the level of spontaneous release was equivalent in control and differentiated cells, we can exclude the possibility that the different ability to release ^{3}H-DA in response to secretagogues could be due to a modification of the basal rate of release in the different phenotypes. αLTx in Bt_2cAMP and BrdU treated IMR32 cells, as in other systems (Rosenthal and Meldolesi, 1989), was able to induce neurotransmitter release even in the absence of extracellular calcium and at levels only slightly lower than those observed in the presence of extracellular calcium (Table 3).

TABLE 3: RELEASE OF ^{3}H-DA FROM NON DIFFERENTIATED (A) AND DIFFERENTIATED (B,C) CELLS

Drugs	None	Bt_2cAMP	BrdU
	increase of release (%)		
	A	B	C
α-Latrotoxin ($2x10^{-9}$M)	2.0 ± 1	16 ± 2	42 ± 2
IONOMYCIN ($1x10^{-6}$M)	4.1 ± 1.5	2 ± 1	27 ± 3
TPA ($0.1x10^{-6}$M)	3.0 ± 1	1 ± 0.5	6 ± 2
IONOMYCIN + TPA	6.0 ± 2	3 ± 2	33 ± 6
α-Latrotoxin + EGTA	1.5 ± 0.7	10 ± 2	31 ± 5

The cells were loaded with ^{3}H-DA (0.8μM) and the amount of release determined after 8 minutes. Values are expressed as % increase of release over basal ±S.E., and represent the average of 5 to 8 different experiments.

b) Secretory granules

We then investigated whether the described differences in stimulus-secretion coupling between control and differentiated cells could rely on the ability of the cells to assemble secretory organelles and store in them ^{3}H-DA taken up from the medium. To test this hypothesis, the storage of DA was evaluated autoradiographically, biochemically and by immunofluorescence. Control and differentiated cells were incubated for 1 h with ^{3}H-DA (0.5μM), in the absence of pargyline. Under such conditions only ^{3}H-DA stored in secretory granules could be detected since, unlike cytoplasmic ^{3}H-DA, it is protected from degradation by monoamine oxidases. We have found that very few control cells were labeled, probably only the spontaneously differentiating cells (Fig. 2a). On the contrary, a greater proportion of cells was labeled in both Bt_2cAMP and BrdU differentiated cultures (Fig. 2, b,c). The intracellular labeling was not uniform, suggesting that

^{3}H-DA was concentrated in discrete subcellular structures. In Bt_2cAMP treated cells labeling was concentrated at the periphery of the cell body, in particular in the short neurites (Fig.2b). In BrdU treated cells, which exhibit a more differentiated phenotype, the labeling was concentrated along neurites and in varicosities (Fig.2c). The same type of compartmentalization could also be observed in BrdU differentiated cells by immunofluorescence using anti-DA antiserum (Fig.2d), suggesting that both endogenous and exogenous DA was stored in similar structures.

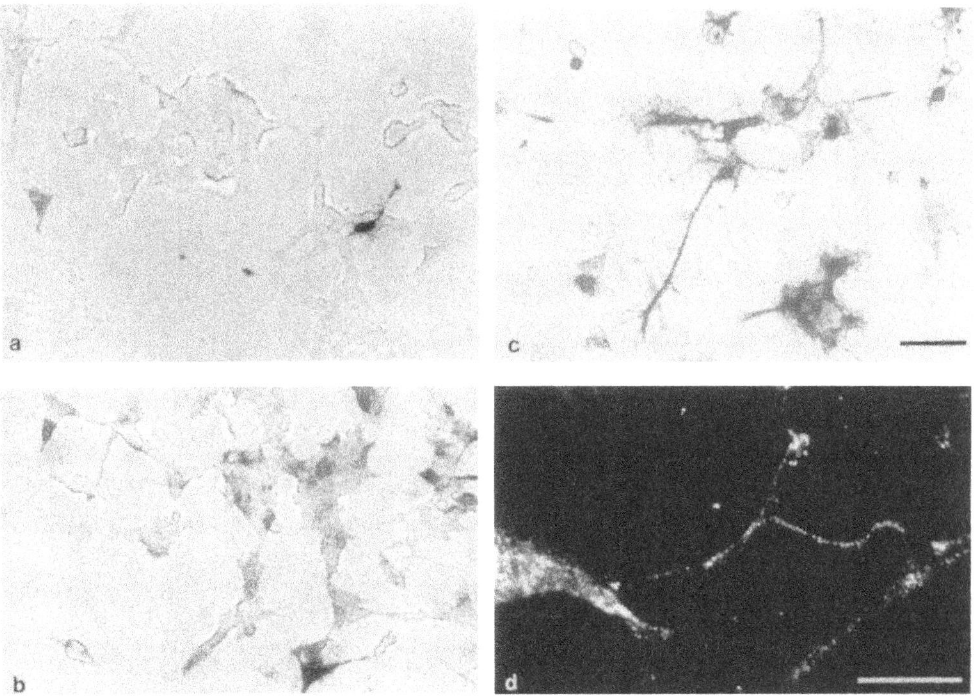

Figure 2 - Autoradiography of ^{3}H-DA taken up from the medium and immunofluorescence with anti-DA antiserum. Control (a) Bt_2cAMP (b) and BrdU (c) treated IMR32 cells were loaded with ^{3}H-DA (0.5μM, 60 min at 37°C) and processed for autoradiography. Very few cells are labeled in control cultures (a), whereas both types of differentiated cultures accumulate a great amount of neurotransmitter (b,c). Note the particular compartmentalization of labeled structures, in the short neurites in Bt_2cAMP treated cells (b) and along the long processes of BrdU treated cells (c). In (d) the immunofluorescence pattern obtained with the anti-DA antiserum in BrdU treated cells is shown. A punctate labeling in the cytoplasm and along neurites is clearly noticeable. Bars: 50μm.

c) Ultrastructure of secretory granules

A striking difference was found between control and differentiated IMR32 cells at the ultrastructural level. Control cells had none or very few secretory vesicles in the cytoplasm or in the short neurites, whereas in both Bt_2cAMP and 5BrdU treated cells many secretory organelles were present (Fig.3).

The great majority of these secretory organelles had a "dense core".; few "clear" small vesicles, possibly related to synaptic vesicles, were found scattered throughout the cytoplasm and in the neurites.

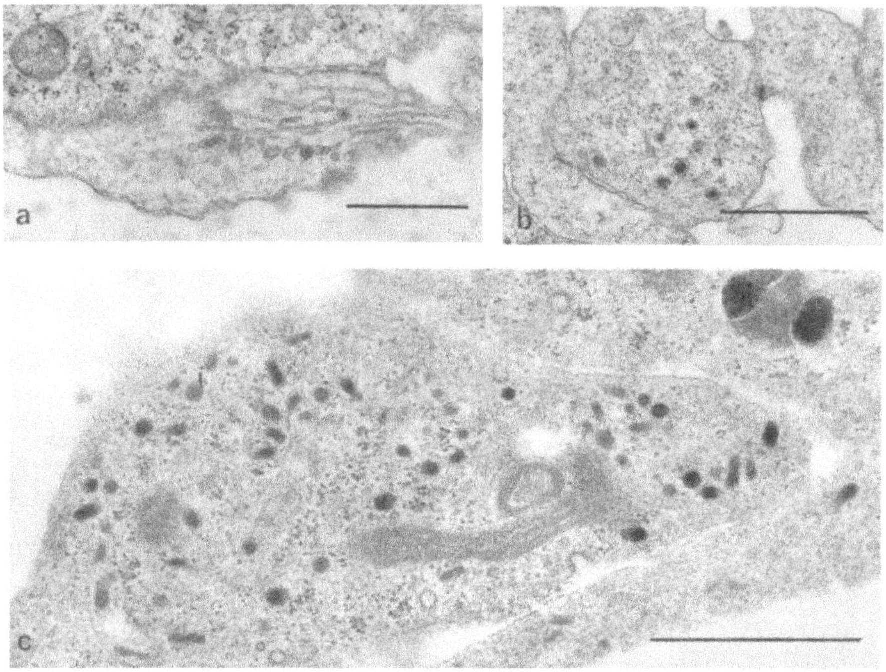

Figure 3 - Transmission electron microscopy of control and differentiated IMR32 cells.

Control cells (a) are almost devoid of secretory granules. In contrast Bt_2cAMP (b) and BrdU (c) treated cells contain many secretory organelles with a typical dense core, which are particularly concentrated in the neurites. Bar = 1 μm.

d) Detection of secretory organelle proteins

Immunofluorescence experiments with an anti-chromaffin granule antiserum, which recognizes both chromogranin A and B (Pelagi et al., 1989) revealed that in control IMR32 cells only low levels of granule proteins were detectable in a few punctate structures mainly at the periphery of the cell (Fig.4a).

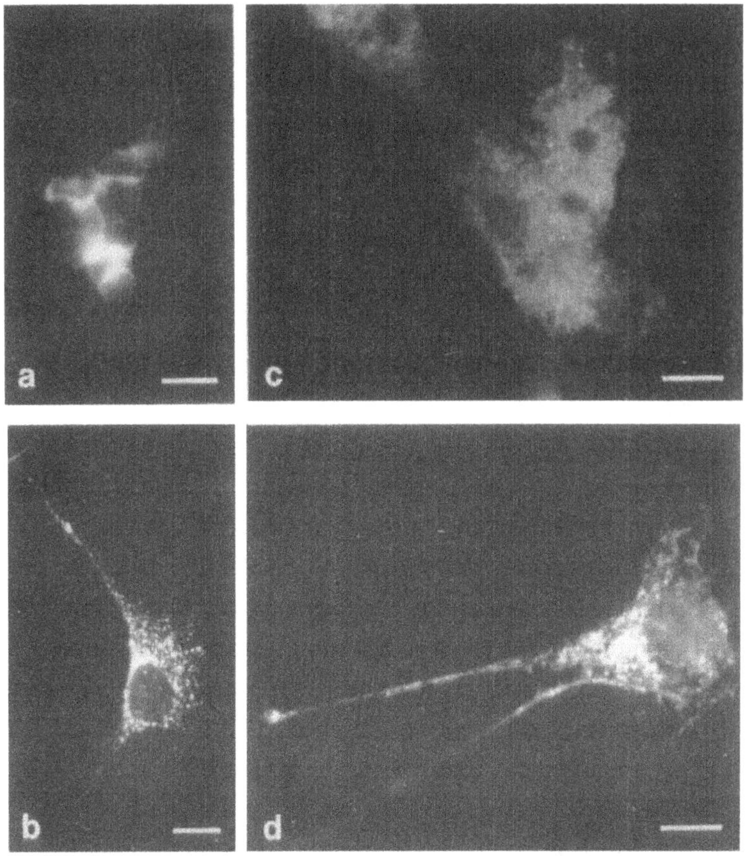

Figure 4 - Immunolocalization of chromogranins and synaptophysin in control and BrdU differentiated cells.
In control cells (a and c) a pale immunolabeling for both chromogranins (a) and synaptophysin (c) is present. In BrdU differentiated cells (b,d) a typical punctate immunolabeling, with similar distribution for both chromogranins (b) and synaptophysin (d), is clearly evident in Golgi area and in neurites. Bars = 25μm.

On the contrary, a very strong dotted signal was found in BrdU differentiated cells at the level of the Golgi apparatus, near the plasma membrane, along neurites and inside varicosities and "terminals" (Fig.4b). Similar results were obtained in Bt_2cAMP treated cells.
Moreover we followed the regulation of expression of a marker of secretory vesicles membranes, by using an affinity purified antibody directed against synaptophysin, a Ca^{2+}-binding, transmembrane glycoprotein present on different types of secretory organelles (Navone et al., 1986; Obendorf et al., 1988). Differentiated IMR32 cells expressed more synaptophysin than control cells

(Fig.4c,d); furthermore this protein was localized in discrete organelles, mainly in the Golgi region, in neurites and in nerve ending-like structures (Fig.4d).
In conclusion, our morphological and biochemical data demonstrate that IMR32 cells are able, after differentiation, to express coordinately secretory granules components, to assembly them in mature secretory organelles (that can store neurotransmitters) and to release neurotransmitters through exocytosis upon stimulation.

Is the secretory pathway similar in IMR32 cells differentiated by means of different agents?

A very interesting finding is the peculiar behaviour of Bt_2cAMP differentiated cells, which reflects important molecular differences from both non differentiated and BrdU differentiated cells. Bt_2cAMP differentiated cells (just as BrdU differentiated cells) contain more catecholamines than control cells (Table 2), and express a higher ability to store ^3H-DA in secretory organelles. However, the Bt_2cAMP differentiated cells do not release ^3H-DA in response to the calcium ionophore ionomycin, in the presence or absence of TPA, but they do release it in response to αLTx, even in a calcium free medium (Table 3).
This suggests that some step, distal to calcium influx and protein kinase C activation, is not functioning in this particular phenotype and that, therefore, Ca^{2+} is not efficient as stimulus-secretion coupling agent. Much evidence is accumulating that exocytosis in several cell systems can proceed in the virtual absence of Ca^{2+} in the medium and without activation of protein kinases (Bittner et al., 1986; Gomperts, 1986; Vallar et al., 1987; Neher, 1988). In the same systems GTP analogues can trigger exocytosis by activating a yet unknown GTP-binding protein. Since in these cell models the experiments were performed by means of considerable manipulation of the cell ionic homeostasis their physiological relevance is still to be clarified. We have here demonstrated that in intact cells an alternative pathway of secretion, independent from Ca^{2+} and protein kinase C, exists.

Conclusion

We have shown that in IMR32 cells, differentiation-inducing agents induce substantial modifications both in morphological and in biochemical and physiological features, and that these modifications may differ depending on the drug used to induce differentiation.

The mechanisms by which the two tested drugs induce differentiation are not completely known although they probably act differently. Bt_2cAMP mimics a dramatic increase in cAMP content, and this is known to induce differentiation in a large number of cells (De Laat et al., 1982; Prasad, 1975). The effect of BrdU may be dependent on the block of DNA synthesis and on cell division (Shubert and Jacob, 1970), and on cAMP accumulation (Prasad et al., 1973), but may also involve other modifications, such as the adhesion of cells to the surface of culture dishes, which may indirectly trigger the differentiation of neuroblasts (Shubert and Jacob, 1970).

All these findings suggest that the process of differentiation in IMR32 neuroblastoma cells, in relation to the secretory properties, is probably a multistep phenomenon that proceeds not in cascade, but along parallel pathways, the coordination of which may produce the complex features of functional differentiation.

IMR32 cells are, therefore, an interesting cellular model for studying the different steps of the secretory pathways, and for unraveling the process of neuronal differentiation.

REFERENCES

Bittner MA, Holz RW, Neubig RR (1986) Guanine nucleotide effects on catecholamine secretion from digitonin-permeabilized adrenal chromaffin cells. J Biol Chem 261:10182-10188

Burgess TL, Kelly RB (1987) Constitutive and regulated secretion of proteins, Ann Rev Cell Biol 3:243-293

Clementi F, Cabrini D, Gotti C, Sher E (1986) Pharmacological characterization of cholinergic receptors in a human neuroblastoma cell line. J Neurochem 47: 291-297

De Laat SW, van der Saag PT (1982) The plasma membrane as a regulatory site in growth and differentiation of neuroblastoma cells. Int Rev Cytol 74:1-54

Gomperts B (1986) Calcium shares the limelight in stimulus-secretion coupling. Trends Biol Sci 11:290-292

Gotti C, Wanke E, Sher E, Fornasari D, Cabrini D, Clementi F (1986) Acetylcholine operated ion channel and α-Bungarotoxin binding site in a human neuroblastoma cell line reside on different molecules. Biochem Biophys Res Comm 137: 1141-1147

Gotti C, Sher E, Cabrini D, Bondiolotti G, Wanke E, Mancinelli E, and Clementi F (1987) Cholinergic receptors, ion channels, neurotransmitter synthesis, and neurite outgrowth are independently regulated during the in vitro differentiation of a human neuroblastoma cell line. Differentiation 34:144-155

Gupta M, Notter MD, Felten S, Gash DM (1985) Differentiation characteristics of human neuroblastoma cells in the presence of growth modulators and antimitotic drugs. Dev Brain Res 19: 21-29

Meldolesi J, Pozzan T, Ceccarelli B (1987) Exo-endocytosis: mechanism of drug and toxin action. Hand Exp Pharmacol 83: 339-359

Miller R (1987) Multiple calcium channels and neuronal junction. Science 235: 46-52

Navone F, Jahn R, Di Gioia G, Stukenbrok H, Greengard P, De Camilli P (1986) Protein P38: an integral membrane protein specific for small vesicles of neurons and neuroendocrine cells. J Cell Biol 103: 2511-2527

Neher E (1988) The influence of intracellular calcium concentration on degranulation of dyalized mast cells from rat peritoneum. J Physiol 395:193-214

Obendorf D, Schwarzenbrunner V, Fisher-Colbrie R, Loslop A, Winkler H (1988) In adrenal medulla synaptophysin (Protein p38) is present in chromaffin granules and in a special vesicle population. J Neurochem 51: 1573-1580

Pelagi M, Bisiani C, Gini A, Bonardi MA, Rosa P, Mare P, Viale G, Cozzi MG, Salvadore M, Zanini A, Siccardi AG, Buffa R (1989) Preparation and characterization of anti-human chromogranin A and chromogranin B (secretogranin I) monoclonal antibodies. Mol Cell Probes, in press

Prasad KN, Mandel B, Kumar S (1973) Human neuroblastoma cell culture: effect of 5-bromodeoxyuridine on morphological differentiation and levels of neural enzymes. Proc Soc Exp Biol Med 44:38-42

Prasad KN (1975) Differentiation of neuroblastoma cells in culture. Biol Rev 50:129-165

Rosenthal L, Meldolesi J (1989) α-latrotoxin and related toxins. Pharmacol Ther 42:115-134

Sher E, Pandiella A, Clementi F (1988a) Ω-conotoxin binding and effects on calcium channel function in human neuroblastoma and rat pheochromocytoma cell lines. FEBS Lett 235:178-182

Sher E, Gotti C, Pandiella A, Madeddu L, Clementi F (1988b) Intracellular calcium homeostasis in a human neuroblastoma cell line: modulation by depolarization, cholinergic receptors and α-latrotoxin. J Neurochem 50:1708-1713

Sher E, Denis-Donini S, Zanini A, Bisiani C, Clementi F (1989) Human neuroblastoma cell acquire regulated secretory properties and different sensitivity to Ca^{2+} and α-latrotoxin after exposure to differentiating agents. J Cell Biol, in press

Shubert D, Jacob F (1970) 5-Bromodeoxyuridine-induced differentiation of a neuroblastoma. Proc Natl Acad Sci USA 67: 247-254

Tumilowicz JJ, Nichols WW, Cholan JJ, Greene AE (1970) Definition of a continuous human cell line derived from neuroblastoma. Cancer Res 30: 2110-2118

Vallar L, Biden TJ, Wollheim CB (1987) Guanine nucleotides induce Ca^{2+} independent insulin secretion from permeabilized RINm5F cells. J Biol Chem 262:5049-5056

DISCUSSION

Dumont: What is the mechanism of action of α-latrotoxin?

Clementi: It induces a massive stimulation of asynchronous quantal release of neurotransmitter in several synapses of vertebrates. The initial step is binding to a specific receptor that activates a cation channel with a cell depolarization and an increase of cytoplasmic Ca^{2+}. However toxin can stimulate release of neurotransmitter in the absence of extracellular Ca^{2+}. The mechanism of this Ca^{2+} independent stimulation of exocytosis is not known and probably involves a G protein.

Chabre: Could you tell us what the toxin is chemically speaking?

Clementi: The toxin is the major component of the black widow spider venom.

Chabre: I want to know chemically what it is. Is it a peptide?

Clementi: Yes it is a peptide of high molecular weight, 130 K daltons, and it is a monomer.

Chabre: The sequence is known?

Clementi: No. However the toxin has been purified and its biochemical and functional activity well studied, and have been recently reviewed.

Chabre: Is it a hydrophobic peptide or hydrophylic? Does it go easily through the membrane?

Clementi: There are hydrophobic parts in the toxin and it probably can insert itself in the membrane. In fact it can make channels in liposomes. However these channels are different from the channels which are possible to measure in living cells, containing toxin receptors, or in liposomes reconstituted with the purified receptor.

Chabre: You said it makes channels; you mean it makes leaks on liposomes or specific channels.

Clementi: In large amount toxin alone induces leaks in liposomes or at least channels with large conductances. But, if you make a reconstitution in liposomes with the purified receptor and the proper amount of the peptide, you obtain discrete channels with a characteristic conductance of about 15 pS. And this toxin does not work in cells when the receptor is not present.

Mulle: I want to ask a question about the nicotinic receptor and the muscarinic receptor and the differentiation in your cell line. Do you see any increase in the nicotinic binding?

Clementi: We were originally interested in these cells because they express nicotinic receptors. Bij molecular and cell biology techniques, we have shown that they express at least 3 types of nicotinic receptors. The most abundant is the α-bungarotoxin binding site (about 3000 toxin binding sites per cell), probably not correlated with an ion channel. They have also nicotinic channels (1 per cell) with the pharmacological characteristics of the nervous nicotinic receptor. During differentiation, we have an increase of both types of nicotinic molecules. We have not studied yet the other nicotinic receptor that we know is expressed in these cells (the α5 type).

Mulle: The question was whether the dibutyryl-cAMP can increase all the receptors present in these cells.

Clementi: There is an increased expression of α-bungarotoxin binding sites but not of muscarinic receptors. Other receptors, i.e. the adenosine receptors, are decreased.

Corbin: What is the mechanism of bromodeoxyuridine? How does it stimulate differentiation?

Clementi: We really don't know. It decreases dramatically the growth and this could be one of the stimuli; in fact, many neuronal cells in vitro, when serum deprived, stop to grow and differentiate. We think that this is only one stimulus, but other effects of this drug can be relevant, e.g. those on plasma membrane. Moreover, this drug does not differentiate all neuron-like cells, and it is not a very common drug to use for differentation, probably because it has a very peculiar mechanism. Other neuroblastomas can be differentiated with more physiological drugs, such as retinoic acid or increase of cAMP.

Corbin: Simply it can raise the cAMP level. Say it works through the dibutyryl cAMP pathway. Is that possible?

Clementi: There is an increase of adenylate cyclase after bromodeoxyuridine treatment, but it is not so large as when you use the dibutyryl-cAMP. Furthermore, the bromodeoxyuridine pattern of differentiation is different from that obtained with cAMP.

Premont: If you would take both agents together, both the bromodeoxyuridine and dibutyryl-cAMP, if they are operating through different mechanisms, does a combination give you a higher percentage of cells differentiating in a culture?

Clementi: Yes, that is true. We did not perform a very careful study at this point, but we have studied the Ω -conotoxin binding sites, a marker for neuronal type of calcium channels, and we found that a treatment with both agents induces the expression of these proteins much earlier and in much larger quantities than the two agents alone and furthermore they induce a morphological differentiation much more clearly.

NATO ASI Series H

Vol. 1: **Biology and Molecular Biology of Plant-Pathogen Interactions.** Edited by J.A. Bailey. 415 pages. 1986.

Vol. 2: **Glial-Neuronal Communication in Development and Regeneration.** Edited by H.H. Althaus and W. Seifert. 865 pages. 1987.

Vol. 3: **Nicotinic Acetylcholine Receptor: Structure and Function.** Edited by A. Maelicke. 489 pages. 1986.

Vol. 4: **Recognition in Microbe-Plant Symbiotic and Pathogenic Interactions.** Edited by B. Lugtenberg. 449 pages. 1986.

Vol. 5: **Mesenchymal-Epithelial Interactions in Neural Development.** Edited by J.R. Wolff, J. Sievers, and M. Berry. 428 pages. 1987.

Vol. 6: **Molecular Mechanisms of Desensitization to Signal Molecules.** Edited by T.M. Konijn, P.J.M. Van Haastert, H. Van der Starre, H. Van der Wel, and M.D. Houslay. 336 pages. 1987.

Vol. 7: **Gangliosides and Modulation of Neuronal Functions.** Edited by H. Rahmann. 647 pages. 1987.

Vol. 8: **Molecular and Cellular Aspects of Erythropoietin and Erythropoiesis.** Edited by I.N. Rich. 460 pages. 1987.

Vol. 9: **Modification of Cell to Cell Signals During Normal and Pathological Aging.** Edited by S. Govoni and F. Battaini. 297 pages. 1987.

Vol. 10: **Plant Hormone Receptors.** Edited by D. Klämbt. 319 pages. 1987.

Vol. 11: **Host-Parasite Cellular and Molecular Interactions in Protozoal Infections.** Edited by K.-P. Chang and D. Snary. 425 pages. 1987.

Vol. 12: **The Cell Surface in Signal Transduction.** Edited by E. Wagner, H. Greppin, and B. Millet. 243 pages. 1987.

Vol. 13: **Toxicology of Pesticides: Experimental, Clinical and Regulatory Perspectives.** Edited by L.G. Costa, C.L. Galli, and S.D. Murphy. 320 pages. 1987.

Vol. 14: **Genetics of Translation. New Approaches.** Edited by M.F. Tuite, M. Picard, and M. Bolotin-Fukuhara. 524 pages. 1988.

Vol. 15: **Photosensitisation. Molecular, Cellular and Medical Aspects.** Edited by G. Moreno, R.H. Pottier, and T.G. Truscott. 521 pages. 1988.

Vol. 16: **Membrane Biogenesis.** Edited by J.A.F. Op den Kamp. 477 pages. 1988.

Vol. 17: **Cell to Cell Signals in Plant, Animal and Microbial Symbiosis.** Edited by S. Scannerini, D. Smith, P. Bonfante-Fasolo, and V. Gianinazzi-Pearson. 414 pages. 1988.

Vol. 18: **Plant Cell Biotechnology.** Edited by M.S.S. Pais, F. Mavituna, and J.M. Novais. 500 pages. 1988.

Vol. 19: **Modulation of Synaptic Transmission and Plasticity in Nervous Systems.** Edited by G. Hertting and H.-C. Spatz. 457 pages. 1988.

Vol. 20: **Amino Acid Availability and Brain Function in Health and Disease.** Edited by G. Huether. 487 pages. 1988.

NATO ASI Series H

Vol. 21: **Cellular and Molecular Basis of Synaptic Transmission.**
Edited by H. Zimmermann. 547 pages. 1988.

Vol. 22: **Neural Development and Regeneration. Cellular and Molecular Aspects.**
Edited by A. Gorio, J.R. Perez-Polo, J. de Vellis, and B. Haber. 711 pages. 1988.

Vol. 23: **The Semiotics of Cellular Communication in the Immune System.**
Edited by E.E. Sercarz, F. Celada, N.A. Mitchison, and T. Tada. 326 pages. 1988.

Vol. 24: **Bacteria, Complement and the Phagocytic Cell.**
Edited by F.C. Cabello und C. Pruzzo. 372 pages. 1988.

Vol. 25: **Nicotinic Acetylcholine Receptors in the Nervous System.**
Edited by F. Clementi, C. Gotti, and E. Sher. 424 pages. 1988.

Vol. 26: **Cell to Cell Signals in Mammalian Development.**
Edited by S.W. de Laat, J.G. Bluemink, and C.L. Mummery. 322 pages. 1989.

Vol. 27: **Phytotoxins and Plant Pathogenesis.**
Edited by A. Graniti, R.D. Durbin, and A. Ballio. 508 pages. 1989.

Vol. 28: **Vascular Wilt Diseases of Plants. Basic Studies and Control.**
Edited by E.C. Tjamos and C.H. Beckman. 590 pages. 1989.

Vol. 29: **Receptors, Membrane Transport and Signal Transduction.**
Edited by A.E. Evangelopoulos, J.P. Changeux, L. Packer, T.G. Sotiroudis, and K.W.A. Wirtz. 387 pages. 1989.

Vol. 30: **Effects of Mineral Dusts on Cells.**
Edited by B.T. Mossman and R.O. Bégin. 470 pages. 1989.

Vol. 31: **Neurobiology of the Inner Retina.**
Edited by R. Weiler and N.N. Osborne. 529 pages. 1989.

Vol. 32: **Molecular Biology of Neuroreceptors and Ion Channels.**
Edited by A. Maelicke. 675 pages. 1989.

Vol. 33: **Regulatory Mechanisms of Neuron to Vessel Communication in Brain.**
Edited by F. Battaini, S. Govoni, M.S. Magnoni, and M. Trabucchi. 416 pages. 1989.

Vol. 34: **Vectors as Tools for the Study of Normal and Abnormal Growth and Differentiation.**
Edited by H. Lother, R. Dernick, and W. Ostertag. 477 pages. 1989.

Vol. 35: **Cell Separation in Plants: Physiology, Biochemistry and Molecular Biology.**
Edited by D.J. Osborne and M.B. Jackson. 449 pages. 1989.

Vol. 36: **Signal Molecules in Plants and Plant-Microbe Interactions.**
Edited by B.J.J. Lugtenberg. 425 pages. 1989.

Vol. 37: **Tin-Based Antitumour Drugs.** Edited by M. Gielen. 226 pages. 1990.

Vol. 38: **The Molecular Biology of Autoimmune Disease.**
Edited by A.G. Demaine, J-P. Banga, and A.M. McGregor. 404 pages. 1990.

Vol. 39: **Chemosensory Information Processing.** Edited by D. Schild. 403 pages. 1990.

Vol. 40: **Dynamics and Biogenesis of Membranes.**
Edited by J.A.F. Op den Kamp. 367 pages. 1990.

Vol. 41: **Recognition and Response in Plant-Virus Interactions.**
Edited by R.S.S. Fraser. 467 pages. 1990.

NATO ASI Series H